Springer Series in Computational Neuroscience

Volume 14

Series Editors
Alain Destexhe
CNRS, UPR -2191, UNIC, Gif-sur-Yvette, France

Romain Brette
Institut de la Vision, Paris, France

Computational Neuroscience gathers monographs and edited volumes on all aspects of computational neuroscience, including theoretical and mathematical neuroscience, biophysics of the brain, models of neurons and neural networks, and methods of data analysis (e.g. information theory). The editors welcome suggestions and projects for inclusion in the series.

About the Editors:

Alain Destexhe is Research Director at the Centre National de la Recherche Scientifique (CNRS), France, and Romain Brette is Research Director at the Institut National de la Santé et de la Recherche Médicale (INSERM), France.

More information about this series at http://www.springer.com/series/8164

Basabdatta Sen Bhattacharya
Fahmida N. Chowdhury
Editors

Validating Neuro-Computational Models of Neurological and Psychiatric Disorders

Editors
Basabdatta Sen Bhattacharya
School of Engineering
University of Lincoln
Lincoln
United Kingdom

Fahmida N. Chowdhury
Office of Multidisciplinary Activities
National Science Foundation
Arlington
Virginia
USA

ISSN 2197-1900   ISSN 2197-1919 (electronic)
Springer Series in Computational Neuroscience
ISBN 978-3-319-20036-1   ISBN 978-3-319-20037-8 (eBook)
DOI 10.1007/978-3-319-20037-8

Library of Congress Control Number: 2015951468

Springer Cham Heidelberg New York Dordrecht London
© Springer International Publishing Switzerland 2015
This work is subject to copyright. All rights are reserved by the Publisher, whether the whole or part of the material is concerned, specifically the rights of translation, reprinting, reuse of illustrations, recitation, broadcasting, reproduction on microfilms or in any other physical way, and transmission or information storage and retrieval, electronic adaptation, computer software, or by similar or dissimilar methodology now known or hereafter developed.
The use of general descriptive names, registered names, trademarks, service marks, etc. in this publication does not imply, even in the absence of a specific statement, that such names are exempt from the relevant protective laws and regulations and therefore free for general use.
The publisher, the authors and the editors are safe to assume that the advice and information in this book are believed to be true and accurate at the date of publication. Neither the publisher nor the authors or the editors give a warranty, express or implied, with respect to the material contained herein or for any errors or omissions that may have been made.

Printed on acid-free paper

Springer International Publishing AG Switzerland is part of Springer Science+Business Media (www.springer.com)

Foreword

There is a long and successful tradition of using mathematical modelling to advance understanding in the neurosciences. The seminal example being the work of Alan Hodgkin and Andrew Huxley on the action potential dating back to the 1950s. As well as being seen as a tool for complementing our understanding of fundamental biological processes, computational modelling is now a widely accepted practice for gaining knowledge when direct experimental probing is either infeasible, impractical or ethically unsound. A case in point is the understanding of human diseases of the brain and psychiatric disorders. Hand-in-hand with the use of computational modelling and model validation comes the use of other tools from the arsenal of theoretical neuroscience, including dynamical systems theory, time-series analysis and multi-scale modelling. In much the same way that these essentially mathematical tools have helped shed light on understanding single cell and small network behaviour, they are now beginning to be wielded effectively at the whole brain level. This compilation exemplifies such an approach, and one sees much use of so-called neural mass models to develop frameworks that can encompass not only forward modelling of neuroimaging signals but the utilisation of the growing wealth of data from imaging modalities, such as EEG, MEG, and fMRI to construct biologically meaningful models for real brain states. It is a delight to see this framework being pushed to tackle issues in epilepsy, deep brain stimulation, and Alzheimer's disease, with a concomitant emphasis on the development of next generation whole brain models and their integration with statistical methods to develop powerful new approaches relevant to clinical applications. Indeed this book is an important step along a path that also poses plenty of challenges for the wider modelling community, including the incorporation of extracellular space, models of non-excitable glial cells, neurovascular coupling, and metabolic regulation, to name but a few processes to incorporate within the next generation of whole brain models.

Centre for Mathematical Medicine and Biology,   Stephen Coombes
School of Mathematical Sciences,
University of Nottingham

Preface

This book originated from a curious and serendipitous mix of personal and professional factors several years ago. We (Basabdatta Sen Bhattacharya and Fahmida N Chowdhury) first met in 2010, during a luncheon at the WCCI (World Congress in Computational Intelligence) in Barcelona, Spain. At that conference, Fahmida had co-organized and presented an NSF-NIH (US National Science Foundation and US National Institutes of Health) grant-writing workshop with NIH scientist Patricia Mabry. Some of the potential funding areas from both of those US agencies could involve neuro-computational modeling of various disorders. Basabdatta was already working on using computational models in understanding biomarkers in brain-signals corresponding to neurological disorders. Against this backdrop, our conversation turned to health-related applications of mathematical and computational modeling; and it is during these discussions that we questioned whether there were any uniform standards for validating models with real data obtained from experimental psychology (or from other sources). We continued this discussion via email, which eventually led to the development of a workshop at the 2013 OCNS (Organisation for Computational Neurosciences) Annual Meeting in Paris, France; the workshop was titled "Validating neuro-computational models of neurological and psychiatric disorders". The workshop was run jointly by Basabdatta Sen Bhattacharya and Rosalyn Moran, and was an interesting mix of talks—speakers included Piotr Suffczyński, Dimitris Pinotsis, Udo Ernst, Andre Marreiros, Ingo Bojak and Rosalyn Moran with an opening overall introductory talk by Basabdatta Sen Bhattacharya. Post-workshop, the idea was presented to Springer, who has a 'tie-up' with the OCNS annual meetings and offers interested parties to forward proposals for edited compilations on the workshops organised for its series on Computational Neurosciences, eds. A. Destexhe, R. Brette. We do not claim that this book represents a comprehensive overview or analysis of existing model validation procedures used by research groups and laboratories across the field of computational neuroscience. Rather, the book is a snapshot of some of the major work in modeling and validation, coming from a diverse group of researchers in the field.

We are convinced that this effort is very timely, because the field of computational neuroscience has now matured to the point where proper validation of the various models proposed by different research groups would be required in order for the field to grow in a more cohesive and meaningful fashion. Otherwise, we run the risk of getting stagnated by the incompatibility of data formats, problems with reproducibility, and, most importantly, lack of any real impact in terms of usability in the fields of health and medicine.

The issue of model validation has received attention from many traditional fields of science and engineering; indeed, there is general agreement that validation is the process of determining to what extent a model represents the real world from the perspective of the intended uses of the model. A look at a 1978 report entitled "Methods and Examples of Model Validation—An Annotated Bibliography", by J. Gruhl and N. Gruhl from MIT Energy Laboratory, reveals just how extensive the topic is: even 36 years ago, the bibliography in this report was 74 pages long. It is worth quoting a short paragraph from this report, where the authors provide a list for validation approaches:

- Comparison with other empirical models
- Comparison with theoretical or analytical models
- Comparison with hand calculations or reprogrammed versions of model components
- Data splitting on observed data, by time or region
- Obtain new estimation/prediction data with time, new experiments, or in-simulated environments
- Examination of reasonableness and accuracy, that is, comparison with understanding
- Examination of appropriateness and detail.

This list was generated by and for engineers, but the concept is quite applicable to other fields, including computational neuroscience. Indeed, many researchers are already using these ideas, although we did not find in the published literature any collective presentation and explicit discussion of the issues involved. Perhaps this book will fill that gap. To that effect, the Introductory Chapter of this book (by Rosalyn Moran) captures the essence and the connections between the various modeling and validation approaches presented in the book. Thus, it is strongly recommended that the Introductory Chapter be read first, to set the context for the rest of the book. The sequence of the chapters follows the exposition presented in the introduction; however, each chapter is self-contained, and can be read independently. It is worth mentioning—at the risk of stating the obvious—that ultimately, validation of these models will have to come from the user communities, and therefore, validation techniques must be developed in close cooperation with the experimental scientists, with clear understanding of end-user goals. Our sincere and humble hope is that this book will bring model validation issues to the forefront and serve as a significant step toward generating a more unified and coordinated effort toward this crucial issue in our research enterprise.

We could not have done this work without support and encouragement from our employers, the US National Science Foundation and the University of Lincoln, UK. We are grateful to Rosalyn Moran, who not only played a leading part in the CNS workshop but provided immense help, support and guidance during planning and editing of the book; suffice it to say that the book would not be what it is without her input. Basabdatta Sen Bhattacharya would like to thank Péter Érdi for his review of the first draft of the workshop proposal; the final title as well as the 'body' of the workshop theme was a modified version based on his feedback. We are grateful to our families, friends and colleagues for their support and understanding.

Basabdatta Sen Bhattacharya
Fahmida N. Chowdhury

Acknowledgements

This book was produced while one of the Editors (Chowdhury) was employed by the US National Science Foundation. Any opinions, findings, and conclusions or recommendations expressed in this material are those of the authors and do not necessarily reflect the views of the National Science Foundation.

Contents

1 **Introduction** 1
   Rosalyn Moran

2 **Discovery and Validation of Biomarkers Based on Computational Models of Normal and Pathological Hippocampal Rhythms** 15
   Péter Érdi, Tibin John, Tamás Kiss and Colin Lever

3 **DCM, Conductance Based Models and Clinical Applications** 43
   A. C. Marreiros, D. A Pinotsis, P. Brown and K. J. Friston

4 **Computational Models of Closed–Loop Deep Brain Stimulation** 71
   Yixin Guo and Kelly Toppin

5 **A Multiscale "Working Brain" Model** 107
   P. A. Robinson, S. Postnova, R. G. Abeysuriya, J. W. Kim,
   J. A. Roberts, L. McKenzie-Sell, A. Karanjai, C. C. Kerr,
   F. Fung, R. Anderson, M. J. Breakspear, P. M. Drysdale,
   B. D. Fulcher, A. J. K. Phillips, C. J. Rennie and G. Yin

6 **How to Render Neural Fields More Realistic** 141
   Axel Hutt, Meysam Hashemi and Peter beim Graben

7 **Multilevel Computational Modelling in Epilepsy: Classical Studies and Recent Advances** 161
   Wessel Woldman and John R. Terry

8 **Computational Modeling of Neuronal Dysfunction at Molecular Level Validates the Role of Single Neurons in Circuit Functions in Cerebellum Granular Layer** 189
   Shyam Diwakar

9 **Modelling Cortical and Thalamocortical Synaptic Loss and Compensation Mechanisms in Alzheimer's Disease** 221
   Damien Coyle, Kamal Abuhassan and Liam Maguire

10 **Toward Networks from Spikes** 277
   Mark Hereld, Jyothsna Suresh, Mihailo Radojicic, Lorenzo L. Pesce, Janice Wang, Jeremy Marks and Wim van Drongelen

11 **Epileptogenic Networks: Applying Network Analysis Techniques to Human Seizure Activity** 293
   Sofija V. Canavan, Tahra L. Eissa, Catherine Schevon, Guy M. McKhan, Robert R Goodman, Ronald G. Emerson and Wim van Drongelen

Index 313

Contributors

R. G. Abeysuriya School of Physics, University of Sydney, Sydney, NSW, Australia

Center for Integrated Research and Understanding of Sleep, University of Sydney, Glebe, NSW, Australia

Cooperative Research Center for Alertness, Safety, and Productivity, University of Sydney, NSW, Australia

NeuroSleep: The Center for Translational Sleep and Circadian Neurobiology, University of Sydney, Glebe, NSW, Australia

Brain Dynamics Center, Westmead Millennium Institute, Westmead, NSW, Australia

Kamal Abuhassan School of Computing and Intelligent Systems, Faculty of Computing and Engineering, Magee Campus, Ulster University, Derry, Northland Road, UK

R. Anderson School of Physics, University of Sydney, Sydney, NSW, Australia

Peter beim Graben Department of German Studies and Linguistics, Humboldt-Universität zu Berlin and Bernstein Center for Computational Neuroscience, Berlin, Germany

M. J. Breakspear QIMR Berghofer Medical Research Institute, Herston, QLD, Australia

P. Brown Nuffield Department of Clinical Neurosciences, John Radcliffe Hospital, University of Oxford, Oxford, UK

The Wellcome Trust Centre for Neuroimaging, University College London, Queen Square, London, UK

Sofija V. Canavan Department of Computational Neuroscience, University of Chicago, Chicago, IL, USA

Damien Coyle School of Computing and Intelligent Systems, Faculty of Computing and Engineering, Magee Campus, Ulster University, Derry, Northland Road, UK

Shyam Diwakar Amrita School of Biotechnology, Amrita Vishwa Vidyapeetham (Amrita University), Clappana P.O., Kerala, India

Wim van Drongelen Department of Pediatrics, University of Chicago, Chicago, IL, USA

Departments of Neurology, University of Chicago, Chicago, IL, USA

The University of Chicago, Chicago, IL, USA

P. M. Drysdale School of Physics, University of Sydney, Sydney, NSW, Australia

Tahra L. Eissa Department of Neurobiology, University of Chicago, Chicago, USA

Ronald G. Emerson Department of Neurology, Hospital for Special Surgery, Weill Cornell Medical Center, New York, USA

Péter Érdi Center for Complex Systems Studies, Kalamazoo College, Kalamazoo, Michigan, USA

Institute for Particle and Nuclear Physics, Wigner Research Centre for Physics, Hungarian Academy of Sciences, Budapest, Hungary

K. J. Friston The Wellcome Trust Centre for Neuroimaging, University College London, Queen Square, London, UK

B. D. Fulcher School of Physics, University of Sydney, Sydney, NSW, Australia

F. Fung School of Physics, University of Sydney, Sydney, NSW, Australia

Robert R Goodman Department of Neurosurgery, St. Luke's-Roosevelt and Beth Israel Hospitals, New York, NY, USA

Yixin Guo Department of Mathematics, Drexel University, Philadelphia, PA, USA

Meysam Hashemi INRIA Grand Est—Nancy, Team NEUROSYS, Villers-lès-Nancy, France

CNRS, Loria, Vandoeuvre-lès-Nancy, France

Universitè de Lorraine, Loria, Vandoeuvre-lès-Nancy, France

Mark Hereld Argonne National Laboratory, The University of Chicago, Argonne, IL, USA

Axel Hutt INRIA Grand Est—Nancy, Team NEUROSYS, Villers-lès-Nancy, France

CNRS, Loria, Vandoeuvre-lès-Nancy, France

Universitè de Lorraine, Loria, Vandoeuvre-lès-Nancy, France

Tibin John Center for Complex Systems Studies, Kalamazoo College, Kalamazoo, Michigan, USA

A. Karanjai School of Physics, University of Sydney, Sydney, NSW, Australia

Contributors xvii

C. C. Kerr School of Physics, University of Sydney, Sydney, NSW, Australia

Brain Dynamics Center, Westmead Millennium Institute, Westmead, NSW, Australia

Department of Physiology and Pharmacology, Downstate Medical Center, State University of New York, Brooklyn, NY, USA

J. W. Kim School of Physics, University of Sydney, Sydney, NSW, Australia

Center for Integrated Research and Understanding of Sleep, University of Sydney, Glebe, NSW, Australia

Cooperative Research Center for Alertness, Safety, and Productivity, University of Sydney, Sydney, NSW, Australia

NeuroSleep: The Center for Translational Sleep and Circadian Neurobiology, University of Sydney, Glebe, NSW, Australia

Brain Dynamics Center, Westmead Millennium Institute, Westmead, NSW, Australia

Tamás Kiss Institute for Particle and Nuclear Physics, Wigner Research Centre for Physics, Hungarian Academy of Sciences, Budapest, Hungary

Neuroscience Research Unit, Pfizer Global Research and Development, Cambridge, MA, USA

Colin Lever Department of Psychology, Durham University, Durham, UK

Liam Maguire School of Computing and Intelligent Systems, Faculty of Computing and Engineering, Magee Campus, Ulster University, Derry, Northland Road, UK

Jeremy Marks The University of Chicago, Chicago, IL, USA

A. C. Marreiros Max Planck Institute for Biological Cybernetics, Tübingen, Germany

Nuffield Department of Clinical Neurosciences, John Radcliffe Hospital, University of Oxford, Oxford, UK

L. McKenzie-Sell School of Physics, University of Sydney, Sydney, NSW, Australia

Guy M. McKhan Department of Neurological Surgery, Columbia University Medical Center, New York, NY, USA

Rosalyn Moran Virginia Tech Carilion Research Institute & Bradley Department of Electrical and Computer Engineering, Virginia Tech, Roanoke, VA, USA

Department of Psychiatry & Behavioral Medicine, Virginia Tech Carilion School of Medicine, Roanoke, VA, USA

Lorenzo L. Pesce The University of Chicago, Chicago, IL, USA

A. J. K. Phillips Division of Sleep Medicine, Brigham and Women's Hospital, Boston, MA, USA

D. A Pinotsis The Wellcome Trust Centre for Neuroimaging, University College London, Queen Square, London, UK

S. Postnova School of Physics, University of Sydney, Sydney, NSW, Australia

Center for Integrated Research and Understanding of Sleep, University of Sydney, Glebe, NSW, Australia

Cooperative Research Center for Alertness, Safety, and Productivity, University of Sydney, Sydney, NSW, Australia

NeuroSleep: The Center for Translational Sleep and Circadian Neurobiology, University of Sydney, Glebe, NSW, Australia

Mihailo Radojicic The University of Chicago, Chicago, IL, USA

C. J. Rennie School of Physics, University of Sydney, Sydney, NSW, Australia

Brain Dynamics Center, Westmead Millennium Institute, Westmead, NSW, Australia

J. A. Roberts QIMR Berghofer Medical Research Institute, Herston, QLD, Australia

P. A. Robinson School of Physics, University of Sydney, Sydney, NSW, Australia

Center for Integrated Research and Understanding of Sleep, University of Sydney, Glebe, NSW, Australia

Cooperative Research Center for Alertness, Safety, and Productivity, University of Sydney, Sydney, NSW, Australia

Center for Integrative Brain Function, University of Sydney, Sydney, NSW, Australia

NeuroSleep: The Center for Translational Sleep and Circadian Neurobiology, University of Sydney, Glebe, NSW, Australia

Brain Dynamics Center, Westmead Millennium Institute, Westmead, NSW, Australia

Catherine Schevon Department of Neurology, Hospital for Special Surgery, Weill Cornell Medical Center, New York, USA

Jyothsna Suresh The University of Chicago, Chicago, IL, USA

John R. Terry College of Engineering, Mathematics and Physical Sciences, University of Exeter, Exeter, UK

Kelly Toppin Department of Mathematics, Drexel University, Philadelphia, PA, USA

Janice Wang The University of Chicago, Chicago, IL, USA

Wessel Woldman College of Engineering, Mathematics and Physical Sciences, University of Exeter, Exeter, UK

G. Yin School of Physics, University of Sydney, Sydney, NSW, Australia

Chapter 1
Introduction

Rosalyn Moran

Abstract Computational approaches to neuroscience have provided new breakthroughs in understanding brain function through a diverse array of applications—from the development of fine grained white matter fibre tracking (Behrens et al., Neuroimage, 34(1):144–55, 2007), to the analysis of neural codes responsible for higher cognitive function (Canolty et al., Science, 313(5793):1626n1628, 2006). Models in computational neuroscience are motivated by the need to succinctly but sufficiently summarize the complex processes underlying neural function and subsequent behavior. Capturing a low-dimensional space—in a model—in order to adequately mimic the high-dimensional reality is, of course, context dependent and could appear overwhelming when considering the number of elements that comprise the central nervous system and our behavioral repertoires. One area where these models could be most applicable, and indeed where focused representations are aided by a symptomatology, is in the understanding, diagnosis, prognosis and control of neurological and psychiatric disease. This book has been written to address the current progress in computational models of neural pathology and to ascertain whether these models are fit for purpose. Specifically the authors address the validity of their models' outputs and the varied approaches that have been taken in order to corroborate model-based scientific discoveries in this field. In this introduction I present the commonalities and distinctions of the model validation approaches presented in the coming chapters where applications range from sleep disturbance to epilepsy.

Keywords Neuroscience · Neurophysiology · Models · Disease · Disorders · Validation

---

R. Moran (✉)
Virginia Tech Carilion Research Institute & Bradley Department of Electrical and Computer Engineering, Virginia Tech, 2 Riverside Circle, Roanoke, VA 24016, USA

Department of Psychiatry & Behavioral Medicine, Virginia Tech Carilion School of Medicine, 2 Riverside Circle, Roanoke, VA 24016, USA
e-mail: rosalynj@vtc.vt.edu

Computational models have been extensively employed in the study of neurological and psychiatric disorders. For example, in Alzheimer's disease (AD), molecular level pathology has been explored using probabilistic models of statistical energy interactions among amino acid bond pairs to identify candidates contributing to the protein misfolding that can lead to plaque aggregation [6]. Modeling AD at the cellular and behavioral level, Hebbian learning rules have been employed to simulate the effects of runaway synaptic modifications in associative memory networks, mimicking AD deficits in mnemonic processing and offering a framework for understanding specific degenerative patterns [7]. In psychiatric disorders, models of decision making and probabilistic learning have provided novel quantitative assessment of complex pathologies [10] including borderline personality disorder [9] and depression [8]. Even perhaps traditionally considered derisive disorders like 'hysteria' are becoming newly appreciated through better models that conjoin behavioral and perceptual learning accounts [5] with potential failures in neurobiological signaling [3].

This book cannot cover the wealth of modeling approaches that are being usefully applied in clinically motivated neuroscience today. Rather, the following chapters deal mainly with biophysical models at the cellular and cell population levels. Some of these treatments manifest behavioral indices in addition to neuronal dynamics. For example in Chap. 5, Robinson and colleagues harvest measures of sleep times from their multiscale dynamic model of brain arousal. Overall, the book features models with an electrophysiological and neuroimaging perspective, where clinical data are currently readily available to proffer characterizations of the dynamics of brain activity *in vivo* in the diseased state and provide a direct challenge to the modeler to explain, diagnose and potentially control aberrancies in brain function.

Systems Neurophysiology

In Chap. 2, Érdi and colleagues present a motivation for the study of cellular population activity, adopted in chapters that follow. From Systems Biology—a field that aims to "combine system-level description with microscopic detail" has emerged Systems Neuroscience: "a field devoted to understanding whole systems of the brain, such as those involved in sensation, movement, learning and memory" by integrating pertinent aspects of brain structure, dynamic function and information processing. They extend this concept to Quantitative Systems Neuropharmacology—a discipline they have developed that uses models to query the potential effects of a novel pharmacological agent on system brain dynamics *in silico*. Given that these types of agents are the first and often only treatment option in neurological and psychiatric disorders, their chapter highlights the potential to use biophysical models to test whether currently available therapeutics could be readapted and applied to other 'off-label' diseases. This theme is continued in Chap. 5 where Robinson and colleagues challenge their models with caffeine to examine effects of adenosine and other receptor activity on sleep profiles and in Chap. 6 where Hutt and colleagues explore how anaesthetic agents induce the particular EEG profile observed in clinical settings. Érdi and colleagues also introduce the importance of scale and how no particular level of detail (molecular, cellular or system) should take precedence when considering the processes of pathology. Similarly in Chap. 7 Woldman and Terry describe epilepsy cell dynamics at the micro, meso and macroscales. Models of epileptic networks are later explored statistically in Chap. 11, where network motifs are constructed using coherence, Granger causal and graph theoretic approaches. This analysis by Canavan et al. of multi-electrode arrays implanted in human cortex, highlights the validity of these approaches given the commonalities observed in their analyses—where connections out of seizure zones are confirmed by visual assessment and proffer mechanistic adjuvants, in terms of understanding the pathways seizure progression.

Models in the book do consider all scales including that of of neuronal populations—e.g. through summary of the interactions of multiple single cell models (hundreds to thousands), such as that provided in Chap. 4 by Guo and Toppin to simulate closed-loop deep brain simulations and in Chap. 8 by Diwakar to simulate cerebellar outputs that emerge as a result of dysfunction in granular layer signaling. In these treatments, the dynamics of the population itself are considered, for example, through the use of statistical mechanical simplifications in Chap. 3 by Marreiros and colleagues; where large cell populations (hundreds of thousands) that underlie EEG and LFP time series are summarized by a proposed probability density, with dynamics prescribed by the Fokker-Planck evolution of connected states in Dynamic Causal Models for electrophysiology. Hutt and colleagues perhaps most explicitly support a scale of brain investigation at the level of population dynamics through analogy with the famous Navier Stokes equations in fluid mechanics—which, they argue, supports the idea that single molecular substrates need not be incorporated into models when studying mass action in the central nervous system. In Chap. 9 Coyle and colleagues present two comprehensive modeling accounts of Alzheimer's disease, involving different implementations of a large-scale neural network model. With dynamics based on Izhikevich's single cell equations and developed with detailed biological plausibility.

Complementary to EEG and LFP simulations, this book develops approaches for *in vivo* physiological modeling of cell cultures and proposes methodologies to assess single-cell causality. In Chap. 10, Hereld and colleagues present a strategy for assessing the quality of information about connectivity that can be extracted from a sparse multi-channel sampling of active neuronal networks. This scale will likely bridge molecular and human physiological models in the future.

Biophysical Models

The book is also streamlined in terms of its technical details with all chapters utilizing some form of ordinary or partial differential equations to describe ongoing fluctuations in neuronal membrane potential and output firing regimes—summarized in

Woldman & Terry's Chapter. Two particular flavors of these models arise again and again and are rehearsed briefly here.

Conductance Based Models

Conductance based models use time derivatives to describe the flow of current into cells and cell populations; these straightforward equations appeal to Ohm's Law and the Hodgkin-Huxley parallel resistor-capacitor (RC) circuit model to describe ion flow into and out of cells. The models can accommodate active, neurotransmitter dependent ion flow through receptors as well as leaky ion channels and applied current. The kinetics encoding postsynaptic responses are formulated as an equivalent RC circuit with the time derivative of the membrane potential given as:

$$C\frac{dV}{dt} = \sum_m g_m(t)\,(V_m - V(t)) + I(t) \qquad (1.1)$$

Where V represents the membrane potential, C represents the membrane capacitance, g_m—the conductance of a particular ion-channel, V_m is this ion's reversal potential and $I(t)$ represents the external input to the cell. Then conductance states—describing the state of an active ion channel can also be given appropriate dynamic constraints based on receptor time constants and the number of open channels at that cell:

$$\frac{dg_m}{dt} = \kappa_m(S(V) - g_m) \qquad (1.2)$$

Where S can represent a sigmoidal activation function describing population firing averages or can be replaced with a Heaviside function to mimic the all-or-nothing properties of single cell firing—both dependent on the afferent's cell membrane depolarization V. Different ionic currents can be incorporated when modelling different brain regions—for example in Chap. 4 these type of currents are used to describe connected networks of the thalamus and basal ganglia where currents through the Globus Pallidus differ from those exciting thalamus, in accordance with known neurophysiology.

Neural Field Models

These types of formulations do not include a description of the spatial characteristics of cell activity. For this reason, some authors employ partial differential equations in what are called in this book—neural field models. Hutt and colleagues in Chap. 6 introduce a simple example of this—the so-called Amari model—whereby synaptic temporal dynamics are augmented with a spatial kernel representing a d-dimensional cortical manifold;

1 Introduction

$$\tau \frac{dV(x,t)}{dt} = -V(x,t) + \int K(x,y)S(V(y,t))dy + I(x,t) \qquad (1.3)$$

Here, the membrane potential is described in both space x, and time t—and can accommodate the same type of applied inputs $I(x, t)$ as above, while also allowing for spatially distributed inputs through kernel K. Simplifications, such as spatial homogeneity and isotropic input can be applied with a simple kernel function: $K(x, y) = K(\|x-y\|)$, while more elaborate constraints, for example specific intralaminar profiles of connectivity can be incorporated into these neural field models as explicated in Marreiros and colleagues in the dynamic causal models for neural fields.

Validation Approaches

So what makes a good model? Famous lines of thought on this issue include George Box's epigram—"*Essentially, all models are wrong, but some are useful*" [2], or Rosenblueth and Wiener's—"*The best material model of a cat is another, or preferably the same cat*" [11]. This—of Rosenblueth and Wiener—is in fact the truncated (and more famous) version of their full writing which was preceded "*That is, in a specific example, the best material model…*". This oft omitted qualifier highlights the delicate balance that must be achieved by models of neuronal processes that have adequate detail to describe the key features under study while also being able to generalize to other data sets and potentially to other disorders. This is a clear achievement of the authors of this book—who highlight the generalizability of their modeling approaches by illustrating different applications of the underlying biophysical fundaments to different brain regions and cell types, different control regimes and to multiple diseases or brain states.

In addition, the authors develop their own internal validation criteria. In Chap. 5, Robinson and colleagues outline particular criteria that should be met when developing and applying biophysical models of neuronal dynamics. These include (1) *that the model be based on anatomy and physiology and incorporate different spatial and temporal scales, (2) that they provide quantitative predictions that can be experimentally corroborated, (3) that parameters of the model can be constrained through independent manipulations of the brain, (4) that they generalize to multiple brain states and (5) that they be invertible*. While the models applied in the chapters that precede and follow Robinson et al. all conform to criteria 1–4 only certain authors discuss and implement model invertibility; criteria 5. However model invertibility may serve as a useful criterion for measuring model goodness and grandfather other criteria for a number of reasons. Firstly an invertible model is a model where parameter values can be recovered from simulated or empirical data—hence it directly assesses model complexity and generalizability. If the model is too simple then empirical data from a variety of sources should reveal features of the data that are not adequately captured by the model—and in the other direction—if the model is too complex then simulated changes in parameter values will not be recovered in a multi-start inversion scheme since parameter redundancy should be revealed by changes in "the wrong" parameter.

This fifth criterion is explicitly addressed by Dynamic Causal Models (Chap. 3). This model framework uses a variational Bayesian inversion scheme to recover the underlying parameter space from real and simulated data. In this context, we learn in Chap. 3 that DCMs are subject to three forms of validation, namely tests of *(1) face validity, (2) construct validity and (3) predictive validity.* The first here, face validity is a test of model invertibility. DCMs specify the dynamics of layered neuronal ensembles comprising interconnected with distinct cell types and ion channels with parameters that encode, for example, synaptic connectivity and receptor time constants. By simulating across regions of parameter space and applying the inversion routine, the repertoire of these models are revealed—e.g. different spectra as well as the goodness of parameterization. Formally the inversion procedure produces a metric of model-goodness known as the negative Free Energy, which is an approximation to the model evidence and allows competing models or hypotheses to be tested. This metric is similar to the Akaike Information Criterion or the Bayesian Information criterion but can incorporate *a priori* parameter values and co-dependencies among parameters. Construct validity is the test of whether estimates of parameter values reflect the underlying biological reality. This is linked to Robinson's criterion 4– where alterations in parameters of the model that lead to changes in model output should be testable using some independent manipulation of the brain component that that parameter is thought to represent.

Throughout the book we see examples of this construct validity using pharmacological manipulations as independent verification. For example in Chap. 2, Érdi and colleagues report that zolpidem—a benzodiazepine that agonizes inhibitory $GABA_A$ receptors has been independently shown to alter (decrease) the frequency of theta frequency in hippocampal circuits—a finding they uncovered using their model in a comparison of AMPA vs. GABA receptor parameter function. Similarly in Chap. 6, Hutt and colleagues describe how neural field models recapitulate empirical EEG spectra under general anaesthesia by increasing their models $GABA_A$ receptor time constant. In Chap. 9 Coyle and colleagues present a similar validation; recapitulating key features of EEG spectra in Alzheimer's disease.

Face validity is a vital first step in proposing a de novo model and is explored deeply in Chap. 10 by Herald and colleagues. They take on a very high-dimensional challenge—to uncover networks from spikes. Their ambitious and remarkable work is keenly motivated by the sampling problem—how can a systematic causality be ascribed to spiking neurons in samples where only "a handful" of spikes can be distinguished from a cultures containing as many as ten thousand members? Their results are highly promising—using a "shift set" metric they build measures of pairwise causality and choose careful surrogate techniques to illustrate the ability to sample above stochastic process noise. Useful for researchers in this wide field is their provision of a temporal limit on the length of time a recording will render useful pairwise statistics—it's about 850 msec.

1 Introduction

Table 1.1 A summary of model validation criteria addressed

Model features (Robinson and colleagues)	Test of validity (Marreiros and colleagues)	Application
Anatomical & physiological plausibility	Face validity	Chaps. 2,3,4,5,6,7,8,9, 10, 11
Quantitative predictions, experimentally corroborated	Face validity	Chaps. 2,3,5,6,7,8,9,10,11
Independent corroboration of parameter estimates through independent manipulation	Construct validity	Chaps. 2,3,4,5,6
Generalizable to multiple brain states	Construct validity	Chaps. 2,3,5,6,10,11
Invertible	Face validity	Chaps. 3 & 5
–	Predictive validity	Chap. 3 & outlined in Chap. 4

Predictive validity is potentially the most useful but the least tested aspect of model validation in this book—indicative of a still maturing field. In this book the authors describe particular features of predictive validity both in terms of prior and future work. Predictive validity refers explicitly to the application of a model—what have you designed it to do? In essence, it is the final test of a model, where that model generates a quantitative measurement that is unknown after model inversion or simulation, but will be revealed as correct or incorrect after some additional empirical data have been acquired. (Table 1.1)

One excellent example of the promise of a predictive validation is given in Chap. 4, where a model of closed-loop deep-brain stimulation suggests that a particular online stimulation protocol with multiple electrode contacts should outperform open-loop and coordinated reset protocols in terms of ameliorating thalamic cell firing pathology in Parkinson's disease leading directly to motor improvements over and above the current protocols. This ambitious work and final test is not yet complete, but demonstrates the application and timeliness of the field of biophysical modelling in clinical neuroscience.

In general when piecing together empirical findings in clinical neuroscience, it could be noted that every approach harbors some form of model. Differential equations—for example—represent a dynamic process model which requires some elementary mathematical background—while other approaches to experimental testing require less background—for example a two sampled t-test applied to some metric from a group of patients compared to a group of controls. But even this simple analysis countenances an abstraction that is a model—the underlying hypothesis being that these two groups differ along the measured dimension in a homogenous way. In this regard, the chapters in this book serve to provide a template for adjudicative scientists to self-reflect on their own underlying assumptions and offers different perspectives on how to validate the veracity of those assumptions.

Disordered States of the Brain

While the authors all provide multiple instances of their models application to neurological and psychiatric disorders—here we briefly outline some of the highlighted applications.

Alzheimer's Disease

In Chap. 2, Érdi and colleagues use a network of spiking neurons to model the septo-hippocampal system and its oscillatory properties. The validity of their model is supported by the application of various cell types in hippocampal region CA1 (pyramidal cells, basket cells and oriens lacunosum-moleculare—OLM—interneurons); that are known to exist in this structure and by verified anatomical pathways that they embed in the connection structure between CA1 and its afferents. Here the primary afferent projection cells are GABAergic projections from the medial septum. Given this circuit—the dynamics engendered by the modelled post-synaptic membrane ion channels can exhibit a variety of rhythmic profiles. However the authors tune parameters of the model in order to generate the theta oscillation that has been shown to be generated empirically by this circuit and which is the primary focus of their modeling aims. *In particular the authors aim to establish whether theta frequency can be used as a biomarker of Alzheimer's pathology.*

There are several steps to establishing this aim that are built upon the septo-hippocampal model circuit. Overall, the role of the model is to provide a theoretical foundation/reason for why theta frequency disruption could reflect underlying pathological processes; in other words to link theta frequency markers to specific pathogenic mechanisms. First, we notice that the model is tuned to theta frequencies—is this valid? The authors crucially test their parameterization by building a series of simulation studies designed to probe alterations around this tuned theta frequency. The chapter thus provides evidence that this model has construct validity by generating perturbations in simulated oscillatory responses that have been independently verified by pharmacological experimentation. Namely the authors show that their model's theta frequency is sensitive to GABAergic time constants but not to AMPA time constants and that the slowing of hyperpolarizing potassium currents, leads to an altered intercept of theta frequency vs external stimulation plot without changing the slope. Both of these simulation results are then shown to be supported by effects of specific anxiolytics.

With relation to AD biomarkers, the authors then explain future extensions to the original model that will include other medial temporal lobe and hippocampal substructures to mimic pathological effects of amyloid-beta plaque aggregation on theta oscillations. Specifically, they describe a roadmap where simulations of early amyloid beta-induced neuronal hyperexcitability and synaptic loss in the performant pathway (from the entorhinal cortex to dentate gyrus and from CA3 to CA1) in their augmented model will be tested against animal models of early disease processes. Given the veracity of the simulations to this point—the authors have thus provide a corroborating set of studies that lay credence to studying these amyloid beta-induced effects in their modelling framework.

In Chap. 9, Coyle and colleagues present a remarkably detailed and motivated model of circuit disruption in Alzheimer's disease. Their goal is ambitious—not only to model the dynamics that emerge from local and long-range (including thalamic) connection pathologies—but also to simulate potential compensatory mechanisms of synaptic loss. Their chapter in many ways exemplifies the direction of the field—to offer at times counter-intuitive and novel predictions on how healthy brain function could be maintained.

Parkinson's Disease

In Chap. 3, Marreiros and colleagues provide an overview of Dynamic Causal Models for electrophysiology and its application to Parkinson's disease (PD). Similar to the modelling approach of Érdi and colleagues, the authors here provide a review of supportive studies that validate parameter estimates and model inference in DCM, from applications outside of the PD disease process. In particular they mention pharmacological and microdialysis measurements used to provide construct validity and highlight the predictive validity of DCM in another study where behavioural assessments of working memory could be predicted on the basis of ion channel functioning assessed via model inversion.

The authors explain the fundaments of the modeling approach—that of a mean-field reduction rather than multiple single cell currents, with application to animal and human LFP recordings and non-invasive EEG and MEG. They also highlight the anatomical connectivity structure that has been embedded in the model dynamics; namely the laminar structure (particular afferent and efferent cells) of intra and inter area connections. These connectivity rules, as deduced by tracing studies and post mortem analysis are most fully treated in the neural field models; that forms one of the DCM framework's biophysical model options. Given the background validation the authors then highlight DCM's recent application to Parkinson's disease. The goal of the study here was to *"provide a way of identifying pathology and potential therapeutic targets at the synaptic level"* in PD.

In particular the authors attempt to deconstruct a key hallmark of the Parkinsonian state—the exaggerated beta oscillation that propagates throughout cortex, basal ganglia and the thalamus, with unknown etiology. Using convolution based models that capture the key physiological properties of the cortico-basal gangli thalamo-cortical looping circuits, with partially observed network dynamics from human patient recordings, the authors demonstrate that certain pathways in the loop may underlie hypersynchronization in the beta band. In particular they find the "hyperdirect" pathway from cortex to the subthalamic nucleus (STN) to be overactive in patients off their dopaminergic medications—so too the indirect pathways between external globus pallidus and STN. They then show how the face validity of these predictions was acquired by creating a "virtual lesion" in the model's STN. This simulated ablation did decrease the beta band response in accordance with inactivation effects observed in empirical studies. Importantly their findings thus bring the beta band synchrony in line with current treatments of PD -including deep brain stimulation.

In Chap. 4, Guo and Toppin continue the exploration of pathological signal transmission in Parkinson's disease and develop a conductance based, single compartment model of basal ganglia and thalamic neurons. The work developed in this chapter begins by producing a model that mimics the pathological neuronal activity in these nuclei and *then testing in simulation whether these pathological properties of PD affected circuits can be controlled via intervention with feedback-adjusted stimulation.* The goal is to develop more effective and low-power deep brain stimulation (DBS) protocols for use in human patients.

In this *in silico* setting, model validation takes a different tone. Specifically the authors address whether a composite metric of firing in the thalamic source is a good measure of pathological control and potential symptom improvement. To do this the authors cleverly compare this metric across "baseline" pathological settings, under high frequency stimulation (HFS), coordinated reset stimulation (CRS) and their multi-site delayed feedback stimulation (MDFS). By illustrating an ameliorative effect of HFS and CRS the authors critically establish that the thalamic error index is sensitive to perturbations known, in the real world, to ameliorate symptoms. Moreover they demonstrate the superiority of MDFS in lowering this index.

Their model builds on previous work by their group to include larger populations of neurons with greater biological plausibility in the connectivity between modelled neurons. In constructing the baseline pathology setting, the model parameters are tuned so that thalamic cells produce 12 Hz oscillation—their quiescent natural state. Then the authors add a simulated periodic input at a beta frequency to further entrain the thalamic cells. By virtue of its connectivity with the globus pallidus, the authors demonstrate that this model setting will also produce the burst firing patterns observed in empirical studies of STN dynamics in PD. Further validity of the model-based findings is provided by readjusting baseline parameter settings by altering thalamic ion channel conductances. These newly tuned models still retain physiologically plausible values but aid in addressing the reliability of the model output. They show that these perturbations do not affect the ability of the stimulation protocols to disrupt pathological firing patterns. The authors conclude by examining other in vivo tests of closed loop DBS and put forward strategies to further validate their model before following suit.

Sleep Disturbances

In Chap. 5, Robinson and colleagues present a fascinating set of insights into human arousal using one large-scale biophysical model. Theirs is a neural field model designed for application to human EEG. The authors build from first principles, in accordance with their set of criteria for a 'good model', the physiological and anatomical components of their system. This population based dynamic description includes cortex as well as thalamic relay nuclei and the reticular formation and implicitly employs a mean field approximation, using a sigmoidal function to mimic and output the average rate of firing of the manifold. Appropriate delays are also incorporated—including the thalamo-cortical delay which is optimized with other parameters to produce the alpha wave in its quiescent state.

Arousal state modeling is achieved by considering the ascending arousal system (AAS) in conjunction with thalamo-cortical dynamics. A 'sleep drive' parameter that reflects the build-up of metabolites such as Adenosine is included as a modulator of the AAS interactions. Then the cortico-thalamic model is augmented by an AAS model which includes sleep drive parameter-modulated dynamics on a separate wake promoting dynamic population (representing monoaminergic nuclei) and a sleep promoting population (representing the ventrolateral preoptic nucleus). Using a neural mass (point process) approximation to the ensuing full dynamic system, the authors perform a stability analysis across levels of the sleep parameter and uncover two stable regimes separated by an unstable branch—with the stable branches reported to reflect waking and sleeping.

Having established an intuitive and biologically grounded validation of their AAS-thalamo-cortical model, they then employ the model to simulate the effects of sleep deprivation. To do this they drive the sleep-parameter to a high magnitude to reflect the build-up of sleep associated metabolites while holding the monoaminergic and cortical average membrane potentials at high values. They then release these voltages and show how recovery sleep lasts longer than their normal sleep rhythm, with recovery sleep of about 13 h on the first "night". Sleep latencies only return to normal around night four. The authors go on to predict jetlag recovery times after specific time-zone shifts and report a 1.5 day per hour in circadian change to renormalize sleep cycles—a value that has predictive validity from empirical experiments and which of course one can try oneself!

Anaesthesia

The model of Robinson and colleagues is built upon in Chap. 6 by Hutt and colleagues who use the thalamo-cortical neural field to interrogate clinically observed EEG spectra under general anaesthesia. In particular, the authors are motivated by recent empirical work that suggests that anaesthetics can have multiple cites of action—in both the cortex and thalamus. *Their model is designed to test which regions of the brain are most affected by this ubiquitous and yet still mysterious class of drug.* The authors begin by providing construct validity to the approach by demonstrating how a simple version of the neural field can generate different EEG spectral responses depending upon the spatial correlation of incoming spatiotemporal noise. They show that under widely correlated 'global noise' the system exhibits one predominant frequency model but that under less extended 'local noise', two separate modes emerge which contribute to a bimodal response in the EEG power spectrum. The types of effects they highlight are typical of EEG under external stimulation when global noise loses predominance.

Employing this field model then to general anaesthesia, the authors augment the firing properties of pyramidal cell populations and alter parameters associated with GABAergic decay. These physiological effects of GA on $GABA_A$ receptors are known, however their site of action remains a matter of dispute. To resolve this, the authors deploy separate parameters of decay on the GABAergic receptors of cortical excitatory neurons (p1), cortical inhibitory neurons (p2) and at thalamic relay neurons (p3). A stability analysis then reveals that three separate resonances emerge, which depend on the relative magnitudes of the separate decay parameters. In particular by increasing p2– the time constant of the inhibitory interneurons IPSP—relative to the others, a pair of zero-frequency resonances collide, and gradually move leading to magnitude increases in delta and theta power in addition to an enhanced alpha resonance.

Epilepsy

Woldman and Terry in Chap. 7 provide a comprehensive overview of the biophysical models applied in the study of epilepsy. This disease may serve as a prototype for the application of circuit-based modeling approaches. Importantly the authors introduce the disorder as a collection of disorders where distinct pathological mechanisms can manifest in the pathological synchronized neuronal firing that characterizes the epileptic state.

The authors return to the question of scale and review the micro, meso and macro scale approaches that have been put forward to understand this complex disease of neural dynamics. In their consideration of properties of dynamical systems generally, they highlight how structural failures could potentially result abnormal dynamics. Linking power laws, with systems near criticality the authors point out an exceedingly interesting and perhaps counterintuitive suggestion about brain connectivity in epilepsy. Specifically they review graph-theoretic studies of synchronization and show how the removal of edges from a large-scale dynamic model could lead to hyper rather than hypo synchronization.

The authors specifically outline three potential causes of the epileptic state with the aim of developing models *to advance our understanding of this significant neurological disorder"*, to test the contributive importance of physiology vs. anatomy and to ultimately *classify people with epilepsy according to their likely response to anti-epileptic drugs*. Validating the three distinct mechanisms proposed will obviously be crucial to achieving these aims—the tools for such validation are developed in their chapter. These mechanisms are currently described in dynamic terms—comprising possible (1) noise induced transients, (2) bifurcations or (3) instability-induced transients that persist under pathological structural connectivity.

1 Introduction

In Chap. 11, Canavan and colleagues model directly in data space using seizures recorded on a multi-electrode array (MEA) implanted in human cortex. Here the authors employ multiple techniques to build networks from data features—with the goal of allowing a direct comparison of their outputs (from measures including graph theoretic and Granger causal analysis) with standard clinical assessments c.f to validate them against visual evaluation. These two chapters approach the dynamics of epilepsy from bottom-up and top-down perspectives respectively and together provide complementary tools for the understanding and characterization of this most common of neurological disorders.

Vestibulo-Cerebellar Learning Problems

In Chap. 8, Diwakar moves us from our comfortable cortical view to that oft-neglected structure—the cerebellum. Using a biophysical multi-compartmental model of granular layer micro-circuitry, the chapter develops a detailed description of centre surround inhibition via mossy fiber transmission to granule cells. These preliminary details are used to demonstrate the mechanisms of long-term potentiation and long-term depression via spike-time dependent plasticity.

Ataxia was investigated in this model by mimicking genetic mutant effects on voltage-gated sodium recovery. Animal models of mutations to fibroblast growth factor homologous factors (FHF) produce a failure of granule cells to fire along with specific behavioral traits of ataxia—specifically an inability to retain gait and balance on narrow edges. In the model—altering the particular dynamics of voltage-gated sodium channels also led to depresses firing patterns. In the aggregate, cell population's LFPs similarly reflected an abnormality in wave characteristics as observed in the in vivo case. In the same model, the disability of NMDA receptors—also implicated in ataxias (the NR2A subunits)—produced a different abnormality, causing a disruption in the temporal sequence of inputs, rather that the "quantal loss" induced by the sodium mutations. Together these analyses demonstrate the subtle and distinct mechanisms that can contribute to a so-called spectrum of disorders that require delicate links back to behavioral disorder.

Together this compilation of work stands at the cutting edge of current systems understanding of complex brain disorders. Of vital importance is that our models continue to undergo independent validation, revision, pruning and augmentation as our knowledge develops—and a rigorous roadmap for this future work is central to each contributed chapter. What is certain is that the methodologies supported by years of computational and physiological research and embodied in the chapters that follow by a variety of different models, are reaching their full potential. With models of single synaptic compartments now bridging all the way to cognitive decline in AD it is clear that the approach is now ready for primetime.

References

1. Behrens T, Berg HJ, Jbabdi S, Rushworth M, Woolrich M. Probabilistic diffusion tractography with multiple fibre orientations: what can we gain? Neuroimage. 2007;34(1):144–55.
2. Box GE, Draper NR. Empirical model-building and response surfaces. New York: John Wiley & Sons; 1987.
3. Brown H, Adams RA, Parees I, Edwards M, Friston K. Active inference, sensory attenuation and illusions. Cogn Process. 2013;14(4):411–27.
4. Canolty RT, Edwards E, Dalal SS, Soltani M, Nagarajan SS, Kirsch HE, Berger MS, Barbaro NM, Knight RT. High gamma power is phase-locked to theta oscillations in human neocortex. Science. 2006;313(5793):1626–8.
5. Edwards MJ, Adams RA, Brown H, Pareés I, Friston KJ. A bayesian account of 'hysteria'. Brain. 2012;135(11):3495–512.
6. Fitzpatrick AW, Knowles TP, Waudby CA, Vendruscolo M, Dobson CM. Inversion of the balance between hydrophobic and hydrogen bonding interactions in protein folding and aggregation. PLoS Comput Biol. 2011;7(10):e1002169.
7. Hasselmo ME. Runaway synaptic modification in models of cortex: implications for Alzheimer's disease. Neural Netw. 1994;7(1):13–40.
8. Huys QJ, Vogelstein J, Dayan P (2008) Psychiatry: Insights into depression through normative decision-making models. In: Advances in neural information processing systems, pp 729–736.
9. King-Casas B, Sharp C, Lomax-Bream L, Lohrenz T, Fonagy P, Montague PR. The rupture and repair of cooperation in borderline personality disorder. Science. 2008;321(5890):806–10.
10. Kishida KT, King-Casas B, Montague PR. Neuroeconomic approaches to mental disorders. Neuron. 2010;67(4):543–54.
11. Rosenblueth A, Wiener N. The role of models in science. Philos Sci. 1945;12(4):316–21.

Chapter 2
Discovery and Validation of Biomarkers Based on Computational Models of Normal and Pathological Hippocampal Rhythms

Péter Érdi, Tibin John, Tamás Kiss and Colin Lever

Abstract Quantitative systems pharmacology is an emerging field with the goal of offering new methodologies for drug discovery based on concepts that grew out of systems theory. Oscillation is a central topic of dynamical systems theory, and neural oscillations are related to both normal and pathological behavior. The role of abnormal neural oscillation in several dynamical diseases is briefly reviewed. Two special cases were investigated. The possible mechanisms of anxiolytic drugs on hippocampal electric patterns were analyzed by combined physiological and computational methods. A network of neuron populations that generates septo-hippocampal theta rhythm was modeled using a compartmental modeling technique. The effects of cellular and synaptic parameters were studied. Pyramidal hyperpolarization-activated (I_h) conductance and decay time constant of inhibitory post-synaptic current have significant effects on frequency. A biophysically realistic model of the electrical activity of the hippocampus, an early target of Alzheimer's disease is also manipulated on a synaptic and cellular level to simulate biochemical effects of amyloid-β accumulation. This can help elucidate a mechanism of age-dependent theta oscillation changes in AD mouse models, reflecting changes in synchronous synaptic activity that could mediate oscillation-dependent memory deficiencies and serve as a biomarker for amyloid-β accumulation.

P. Érdi (✉)
Center for Complex Systems Studies, Kalamazoo College, Kalamazoo, Michigan, USA

Institute for Particle and Nuclear Physics, Wigner Research Centre for Physics, Hungarian Academy of Sciences, Budapest, Hungary
e-mail: Peter.Erdi@kzoo.edu

T. John
Center for Complex Systems Studies, Kalamazoo College, Kalamazoo, Michigan, USA

T. Kiss
Neuroscience Research Unit, Pfizer Global Research and Development,
Pfizer Inc, 700 Main Street, Cambridge, MA 02139, USA

Institute for Particle and Nuclear Physics, Wigner Research Centre for Physics, Hungarian Academy of Sciences, Budapest, Hungary

C. Lever
Department of Psychology, Durham University, Durham, UK

© Springer International Publishing Switzerland 2015
B. S. Bhattacharya, F. N. Chowdhury (eds.), *Validating Neuro-Computational Models of Neurological and Psychiatric Disorders,* Springer Series in Computational Neuroscience 14, DOI 10.1007/978-3-319-20037-8_2

Key words: Quantitative systems pharmacology · Neural oscillation · Dynamical diseases · Theta rhythm · Anxiety · Alzheimer's disease

Dynamic and Computational Neuropharmacology (Quantitative and Systems Pharmacology)

The Partially Admitted Renaissance of Systems Theory and of Cybernetics

Systems Theory and Cybernetics

Systems theory was proposed by Ludwig von Bertalanffy (1901–1972), a biologist who worked on the basic principles of life and searched for universal laws of organization [80]. The basic concepts of the systems approach are: (i) a system is a whole that functions by virtue of the interaction of its parts, and (ii) is defined by its elements and the relationship among these elements; (iii) the systems approach integrates the analytic and synthetic methods by taking into account the interaction of the system with its environment; (iv) living structures are open systems, which interact with other systems outside of themselves. Bertalanffy's conceptual model of the living organism as an open system has had revolutionary implications for life and behavioral sciences.

Cybernetics, as a scientific discipline has been named by Norbert Wiener (1894–1964). It was the title of his book with the subtitle Control and Communication in the Animal and the Machine [86]. Cybernetics was a pluralistic theory and an interdisciplinary movement of a number of leading intellectuals. The term cybernetics goes back to Plato, when he explained the principles of political self-governance. goal-directed behavior

Wiener himself emphasized the role of feedback mechanisms in the goal-oriented systems. While the physiologists already knew that the involuntary (autonomous) nervous systems control Bernard's internal milieu, he extended the concept suggesting that the voluntary nervous system may control the environment by some feedback mechanisms and searched for a theory of goal-oriented behavior. This theory supplemented with the concept of **circular causality** promised a new framework to understand the behavior of animals, humans, and computers just under design and construction that time.

The other supporting pillar of cybernetics is the brain-computer analogy suggested by the spirit of the McCulloch-Pitts Neuron (MCP neuron). An MCP [55] neuron is a formal model, and it can be identified as a binary threshold unit. A neuron initiates an impulse if the weighted sum of their inputs exceeds a threshold, otherwise it remains in silence.The MCP model framework wanted to capture the logical basis of neural computation, and intentionally contains neurobiological simplifications. The state is binary, the time is discrete, the threshold and the wiring are fixed. Chemical and electrical interactions are neglected, glial cells are also not taken into consideration. McCulloch and Pitts showed that a large enough number of synchronously updated neurons connected by appropriate weights could perform many possible computations. From retrospective we see that bottom up models of brain regions can be built based on networks of interconnected single cell models.

McCulloch (1898–1969) served as the chairman of a series of conferences (Macy conferences held between 1946–1953 sponsored by and named after the Macy Foundation, where at the beginning Wiener also played an important role. The main topics of the conferences were: (i) Applicability of a Logic Machine Model to both Brain and Computer, (ii) Analogies between Organisms and Machines; (iii) Information Theory; (iv) Neuroses and Pathology of Mental Life and (v) Human and Social Communication.

Systems theory and cybernetics emphasized the importance of **organization principles** and the have anticipated the use of abstract **computational models** in biology to study *normal* and *pathological* phenomena.

Genetic Determinism, Biological Complexity and Systems Biology

It is a mere coincidence that the last Macy conference was held in the same year (1953) when the Watson–Crick [11] paper was published. The research program of the new "molecular biology" suggested that the replication, transcription and translation of the genetic material should and could be explained by chemical mechanisms. Crick's central dogma of molecular biology stated that there was a unidirectional information flow from DNA via RNA (ribonucleic acid) to proteins.

While the central dogma was enormously successful in discovering many detailed chemical processes of life phenomena, philosophically it suggested, as Crick himself wrote [10], that "the ultimate aim of the modern movement in biology is to explain all biology in terms of physics and chemistry".[1]

The central dogma led to **genetic determinism**. While certain phenotypes can be mapped to a single gene, the extreme form of genetic determinism, which probably nobody believes, would state that all phenotypes are purely genetically determined. Genetic determinism has lost its attraction as a unique explanation for the appearance of specific phenotypic traits. After 60 years of extensive research in molecular biology, there is a very good understanding of the intricate mechanisms that allow genes to be translated into proteins. However, this knowledge has given us very little insight about the **causal chains** that link genes to the morphological and other phenotypic traits of organisms [48]. Also, human diseases due to genetic disorders are the results of the interaction of many gene products. One generally used method

---

[1] As Jean-Pierre Dupuy [14] analyzes, it is one of the most striking ironies in the history of science, that the big attempt of molecular biology to reduce biology to physics happened by using the vocabulary of the cybernaticians. "Cybernetics, it seems, has been condemned to enjoy only posthumous revenge" [14], p. 78.

to understand the performance of a complex genetic networks is the *transgenic* techniques.

In the spirit of systems theory and of cybernetics Robert Rosen (1934–1998) [58, 59] gave a formalism, which connected *phenotype* (i.e. what we can observe directly about an organism) and *genotype* (the genetic makeup). In particular, phenotype is interpreted as being "caused" by genotype. He also argued that to understand biological phenotype, in addition to the Newtonian paradigm, the organizational principles should be uncovered. He realized that a crucial property of living systems, that while they are thermodynamically open systems, organizationally they should be closed. To put it in another way, all components, which are subject of degradation due to ordinary wear and tear, should be repaired or resynthesized within the cell. Rosen gave a mathematical framework to show how it is possible to do.

Systems biology is an emergent movement to combine system-level description with microscopic details. It might be interpreted as the renaissance of cybernetics and of system theory, materialized in the works of Robert Rosen. In an excellent review Olaf Wolkenhauer [87] explained how the concepts of systems theory, and of cybernetics were applied by Rosen to biology, and how his ideas returned now under the name of systems biology. For a very good new introductory textbook on systems biology, see [30].

Genetic reductionism, in particular, has been abandoned as a useful explanatory scheme for understanding the phenotypic traits of complex biological systems. Genes are increasingly studied today because they are involved in the genetic program that unfolds during development and embryogenesis rather than as agents responsible for the inheritance of traits from parents to offspring [73].

As a reaction to something that some people might have seen as the "tyranny of molecular biology", the systems thinking has been revitalized in the last several years. Systems thinking correctly states that while reductionist research strategy was very successful, it underestimates the complexity of life. It is clear, that decomposing, dissecting and analyzing constituents of a complex system is indispensable and extremely important. Molecular biology achieved a lot to uncover the structures of many chemical molecules and chemical reactions among the molecules behind life processes. The typical molecular biologist's approach suggests that there is an "upward causation" from molecular states to behavior. The complex systems perspective [15] does not deny the fundamental results of molecular biology, but emphasizes other principles of biological organization.

One of the pioneers of systems biology, Denis Noble offers ten principles of systems biology [52]:

1. Biological functionality is multi-level
2. Transmission of information is not one way
3. DNA is not the sole transmitter of inheritance
4. The theory of biological relativity: there is no privileged level of causality
5. Gene ontology will fail without higher-level insight
6. There is no genetic program
7. There are no programs at any other level

8. There are no programs in the brain
9. The self is not an object
10. There are many more to be discovered; a genuine "theory of biology" does not yet exist

Systems biology emphasizes (i) the interactions among cell constituents and (ii) the dynamic character of these interactions. Biological systems are paradigmatic of hierarchical dynamical systems. For such systems, levels are often connected by some feedback control mechanism. Famously, protein channels carry current that changes the membrane potential of a cell, while the membrane potential changes the protein channels. This mechanism implements circular causality.

Generally, what we see is that systems biology, partially unwittingly, returned to its predecessors, systems theory and cybernetics.

Systems Neuroscience and Systems Neuropharmacology

Systems Neuroscience is a field devoted to understanding whole systems of the brain, such as those involved in sensation, movement, learning and memory, attention, reward, decision-making, reasoning, executive functions, and emotions. The structural, functional and dynamic aspects are integrated into a coherent framework [3]. It also deals with the principles of information processing, storage and retrieval at the systems level.

The study of brain systems includes the analysis of individual regions, as well as hierarchical levels of information processing. In terms of methodologies, it benefits from diverse techniques from single-cell recording to high-resolution imaging of brain activity. Systems neuroscience also uses computational studies to organize data into a coherent picture.

While the standard structure-based design of drugs for psychiatric disorders is based on drug-receptor interactions, the systems physiology perspective [7] emphasizes the effects of drugs on spatiotemporal brain activities. The theoretical framework for understanding normal and pathological spatiotemporal activities should be dynamical system theory to identify targets, and then, in the next stage, chemists should design molecules to modify the desired target.

Dynamical system theory and computational neuroscience combined with traditional molecular and electrophysiological methods would open new avenues in drug discovery that may lead to genuinely new neurological and psychiatric therapies [1]. More specifically, these model-based highly valuable techniques are able to integrate multiple disciplines at different spatiotemporal scales [16].

Recently the concepts and methods of systems biology have been used in pharmacology [67] and psychiatry [70]. Specifically Noble [52] states that there is no privileged level of causality, and there is no reason to assume that the molecular/genetic level uniquely determines mental activities, so the levels should be integrated.

Our own belief is [72]: "…Multi-scale modeling has multiple meanings and goals including different time and spatial scales, levels of organization, even multi-stage processing. While the significance and importance of describing neural phenomena at different levels simultaneously is clear in many cases, we certainly don't have a single general mathematical framework. Mostly we have specific examples for coupling two or more levels. The understanding and control of normal and pathological behavior, the transfer of knowledge about the brain function and dynamics to establish new computational and technological devices needs the integration of molecular, cellular, network, regional and system levels, and now the focus ins on elaborating mathematically well-founded and biologically significant multi-scale models."

Quantitative Systems Pharmacology

The main goal of **computational neuroscience** is to build biologically realistic mathematical models, and test how they match and predict experimental research by using computer simulations. Its main mathematical tool is **dynamical system theory** which helps to understand the neural mechanisms of temporal and spatio-temporal neural activities.The discipline of **computational neuropharmacology** emerged [1] as pharmaco-therapeutic strategies were suggested by constructing biophysically realistic mathematical models of neurological and psychiatric disorders.

Nowadays the term "Quantitative Systems Pharmacology" QSP is used and defined as "…an approach to translational medicine that combines computational and experimental methods to elucidate, validate and apply new pharmacological concepts to the development and use of small molecule and biologic drugs. QSP will provide an integrated 'systems- level' approach to determining mechanisms of action of new and existing drugs in preclinical and animal models and in patients. QSP will create the knowledge needed to change complex cellular networks in a specified way with mono or combination therapy, alter the pathophysiology of disease so as to maximize therapeutic benefit and minimize toxicity and implement a 'precision medicine' approach to improving the health of individual patient.…". As an NIH White Paper by the QSP Workshop Group writes [67]. "…The concept of QSP in CNS is tightly linked to the term of 'computational neuropharmacology', an area inspired by seminal work of the late Leif Finkel [19] and further developed as 'computational neuropharmacology'" [1, 21]. QSP promises to build network and circuit level models based on detailed models of single neurons and synaptic transmission bu using morphological, biophysical and cellular electrophysiological data. The effects of drugs acting on intrinsic neuron and synaptic parameters are directly incorporated into the model.

In this paper we adopt the approach of quantitative systems pharmacology to get a better understanding the effects and neural mechanisms of potential anxiolytic drugs on hippocampal theta oscillations and study the relationship between altered network connectivity and hippocampal pattern, which may be related to the eventual early diagnosis of Alzheimer's disease.

Abnormal Neural Oscillation

Neural rhythmicity is known to be involved in cognition and motor behavior [7,83] and pathological symptoms are related to altered temporal patterns. The understanding of the underlying cellular, synaptic and network level mechanisms of these altered patterns contribute to obtain more knowledge about the neural mechanisms of certain neurological and psychiatric diseases, and may help to the elaboration of more advanced therapeutic strategies. The iterative combination of experiments and computational studies was recently offered [9, 18, 60] to achieve this goal. Some relationship between disorders and altered temporal pattern will be very briefly mentioned here.

Epilepsy

A variety of temporal pattern is associated to different types of epileptic behaviors [43,56] explained by the *dynamical diseases* paradigm. Both macroscopic, population level models championed by Lopes Da Silva [43] and biophysically detailed epilepsy models exist (famously reflected in the works of Roger Traub, say [69]). The prediction and control of epileptic seizures became a hot, and controversial topic: [49] has 273 citations (*April*19*th*, 2014). Dynamical models may predict seizure development and the administration of drugs could be designed accordingly providing novel therapeutic strategies for epileptic patients.

Abnormal Cortical Rhythms in Parkinson Disease

Abnormal synchronised oscillatory activities may contribute to bradykinesia in patients with Parkinson's disease [4,17,57]. Neuronal correlates of PD include a shift to lower frequencies and enhanced synchronized oscillations at low theta (3–7 Hz) and beta 10–30 Hz regions in the basal ganglia, thalamus, and cortex.

Phase-amplitude coupling (PAC), a phenomenon that may emerge in spatially distributed oscillating network [33]. It describes the coupling between the phase of a slow oscillation and the amplitude of a fast oscillation, with the highest amplitude occurring at the so-called preferred coupling phase. It was suggested [12] that exaggerated PAC in the primary motor cortex is associated to PD. Pathological beta oscillation may have striatal origin, and combined physiological and computational models (e.g. [44]) contributed the uncover some details of the pathophysiological mechanism.

Abnormal neural oscillations and synchrony in schizophrenia

There are accumulating evidences that alterations in the synchronized neural oscillatory activity may have an important role in the pathophysiology of schizophrenia. Abnormal beta and gamma oscillations emerge due to the impaired synaptic transmission in the network GABAergic [71]. The "GABAergic origin hypothesis" [51] suggests that hypofunction of N-methyl-d-aspartic acid-type glutamate receptors (NMDARs) at GABAergic interneurons is sufficient for generating schizophrenia-like effects. Mostly parvalbumin containing interneurons are involved, as computational studies [79] also suggest. Not only high frequency oscillations, but low frequency rhythms including delta patterns may be related to the pathophysiology of schizophrenia [39, 47].

Just as oscillations can provide potentially useful temporal organization of processing within a region, so they can also support efficient inter-regional processing. It has been hypothesised that inter-region coherence may be a general mechanism for task-led inter-regional communication [20, 81]. Sigurdsson et al. (2010) investigated hippocampal-prefrontal coupling in a mouse model of schizophrenia. They contrasted the choice (test) runs vs. the sample runs in a discrete-trial, T-maze, working memory task. In wild-type mice, replicating the findings of Jones and Wilson (2005) in rats, they showed that both theta-band coherence of the two regions LFPs, and phase-locking of medial prefrontal neuronal firing to CA1 theta, was higher during choice runs. In the schizophrenia-model mice, however, this theta-band coupling was impaired.

Abnormal Brain Rhythms in ADHD

There are newer developments in the understanding of the of the normal and pathological mechanisms of attention related to the entrainment of internal neural oscillations to external rhythms [8,42]. Phase-amplitude coupling plays also role as a mechanism of entrainment. Abnormal neural oscillations and selective attention deficits occur in Attention Deficit Hyperactivity Disorder (ADHD) and in other psychiatric disorders. Findings of abnormal entrainment may lead to new to therapeutic interventions.

Anxiolytic Drugs and Altered Hippocampal Theta Rhythms

Background

In the behaving mammal, hippocampal theta, a sinusoidal 4–12 Hz fluctuation in the hippocampal local field potential, is very prominent during behaviors which involve spatial translation of the head [53, 74] or arousal and anxiety [31, 62, 65]. Studies focusing on theta amplitude/power have long noted that hippocampal theta has an atropine-sensitive component and atropine-resistant component [41]. The general observation is that systemic injection of non-specific muscarinic antagonists such as atropine and scopolamine eliminate the theta (type II component) that is observed during alert immobility (aroused/anxious states,) and certain anaesthetised states, but fairly minimally affect the theta (type I component) observed during locomotion [6, 41]. Lesions to the septum eliminate both components of hippocampal theta. Hippocampal theta plays a key role in two functions associated with the hippocampus: (1) spatial cognition and memory; (2) anxiety and anxiolytic drug action. We focus on the latter here.

Figure 2.1 shows the empirical finding. Figure 2.1a shows that increasing the strength of the reticular stimulation increases the frequency of the theta oscillation, and that the relationship is broadly linear. Figure 2.1b shows the effects of various drugs upon this relationship. Only the anxiolytic drugs (chlordiazepoxide, diazepam, alprazolam, and amylobarbitone) reduce theta frequency, while the antipsychotic drugs (haloperidol, chlorpromazine) do not. Figure 2.1c shows the reduction of theta frequency by buspirone, a later-discovered anxiolytic acting at 5HT-1A receptors. As a group, anxiolytics reduce the slope and/or intercept of the relationship between stimulation strength and theta frequency.

To enhance the validity of the theta-frequency reduction assay and to provide a test bed for theta-behaviour relationships, it is important to extend the assay to the freely behaving animal. Accordingly, we recently tested two well-established anxiolytic drugs (chlordiazepoxide, a benzodiazepine, and buspirone, a 5HT-1A agonist) and a putative anxiolytic drug (O-2545, a CB1 agonist) in the awake and locomoting rat in a foraging task. Figure 2.1d, e, f shows the results. Notably, there is a broadly linear relationship between theta frequency and running speed. Interestingly, all three anxiolytic drugs elicited a specific form of theta frequency reduction; they reduced the intercept of this relationship, as measured by the intercept of the frequency-speed regression at 0 cm/s. The specificity of this anxiolytic-elicited effect is suggested by the contrast to the reduction of the slope of the frequency-speed regression (Fig. 2.1g) elicited by introducing the rats to a new spatial context (a dissociation and effect predicted by a theta oscillatory interference model of grid cells and place cells (see [5] and [85] for more details)).

The approach of quantitative systems pharmacology can be used to study the events underlying the altered theta rhythms resulting from anxiolytics involving multi-level effects including molecular and neural network phenomena. In the behaving animal, hippocampal theta frequency reliably increases with running speed, in a broadly linear fashion, likely tied to spatial coding. Interestingly, anxiolytic drugs reliably reduce hippocampal theta frequency in a consistent way, reducing the offset of the frequency-running speed relationship [85] (see Fig. 2.1) while only minimally affecting the slope of this relationship. Theta rhythm can be studied in the absence of interaction with environment using electrical stimulation to the brainstem in anesthetized rats to activate ascending pathways that control the type II theta component. In this model, increasing strength of stimulation results in a broadly linear increase in theta frequency (See Fig. 2.1b), and all anxiolytic

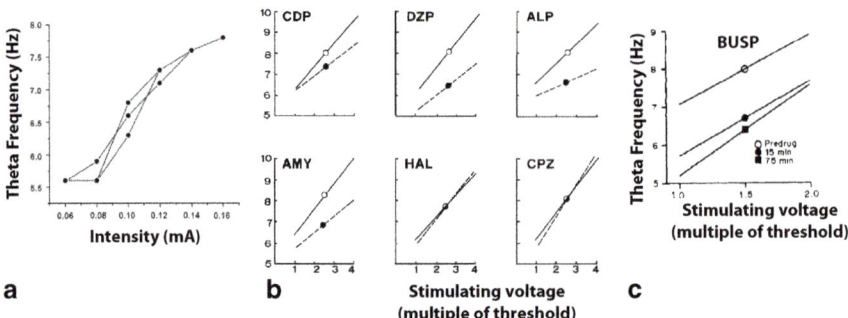
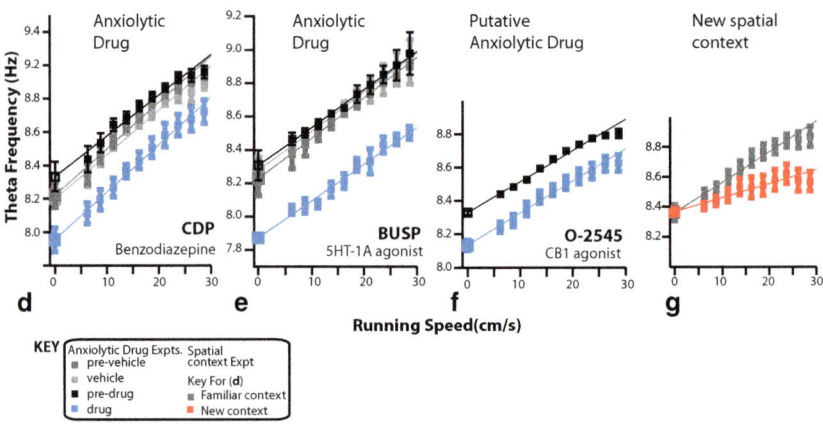

Fig. 2.1 Anxiolytic drugs reduce hippocampal theta frequency. In the anaesthetised rat, anxiolytic drugs reduce frequency of reticular-stimulated hippocampal theta. **a** Broadly linear relationship between reticular formation stimulation intensity and theta frequency in the anaesthetised rat. Adapted from [66]. **b** Anxiolytic drugs reduce theta frequency. Open circles indicate baseline, closed circles effect of drug. Adapted from [22]. **c** Buspirone, later discovered anxiolytic drug acting on 5HT-1A receptors. Adapted from [45]. In the freely moving rat, anxiolytic drugs reduce the intercept of the theta frequency to running speed relationship (**d–f**), while spatial novelty reduces its slope (**d**). Three neurochemically-different anxiolytic drugs (all i.p. injections: D, CDP, benzodiazepine agonist, 5mg/kg; E, Buspirone, 5HT-1A agonist, 1mg/kg; F, O-2545, putative anxiolytic, CB1 agonist, 100 μg/ml, 0.5ml/kg) have the common effect of reducing the zero cm/s intercept of the theta frequency to running speed relationship. In contrast, exploration of a novel spatial context reduces the slope of this relationship (**g**), which then recovers as the novel spatial context itself becomes familiar (not shown). Parts **f** and **g** present data from the same rats (i.e. a within-subjects double dissociation of intercept and slope effects is observed). Open squares indicate y-intercept of regression lines. All recording sites are from CA1. Adapted from [85]

drugs tested in this model [22, 46] reduce theta frequency, whether by reducing the slope or offset of the theta-frequency-to-stimulation relationship. This connection between theta frequency reduction and anxiolytic efficacy suggested to us the use of a spiking neural network model based on experimental data and conductance-based modeling formalisms to help explain and predict biophysical manifestations of anxiolysis. This is investigated in the anxiolytic drug efficacy-predicting context of modifications to the linear relationship between theta frequency and brainstem stimulation level.

Methods

A network of spiking neuron models based on time and voltage dependent membrane conductances in the neural simulator software GENESIS is used to investigate the dynamics of theta rhythm frequency control by potential anxiolytic drugs.

Cell Models and Network Structure

Three inhibitory neuron populations were modeled as consisting of single compartment units corresponding to the assumption that spatial integration of inputs to these populations has less contribution to theta oscillation than does space-dependent input to pyramidal cells, which were represented using a multi-compartment model. Time and voltage dependent changes in the conductance of both the spatially uniform inhibitory cell membranes and the spatially distributed pyramidal cell membranes were modeled as time and voltage resistors in resistor-capacitor circuits. The resistance, capacitance, and electromotive force parameters of these circuits are equivalent to experimentally determined membrane properties for each neuron and spatial compartment [24].

Cells modeled that make up the septo-hippocampal system included hippocampal CA1 basket interneurons, CA1 oriens-lacunosum-moleculare neurons, GABAergic neurons projecting from the medial septum to the hippocampus, and CA1 pyramidal cells. Each of these individual neuron models displays many of the firing properties of the neuron they represent, and the connection probability amongst instances of these neuron models come from experimental studies along with a parameter search for a region synaptic strength space demonstrating robust generation of theta rhythm [24].

Original model specifications and parameter settings are given at geza.kzoo.edu/theta/theta. This model was modified to incorporate the recently verified rhythmic glutamatergic input from the medial septum to the hippocampus contributing to theta activity [28] using a spiking input with rhythmic firing rate to pyramidal somata and medial septum GABAergic cells via AMPA receptors at theta range frequencies directly related to ascending pathway stimulation level.

Simulating Electrical Stimulation of Ascending Pathways

To simulate the assay that reliably predicts the clinical efficacy of anxiolytics, electrical stimulation to the nucleus Pontis Oralis (nPO) region of the brainstem was modeled as depolarizing current to pyramidal, basket, and medial septal GABAergic (MS-GABA) neurons, as well as by the frequency of the rate modulation of the medial septal glutamatergic (MS-Glu) spiking object. Increasing stimulus to the nPO was modeled by increasing depolarizing current to pyramidal, basket and MS-GABA populations while maintaining the ratio between the currents over equal increments. Increasing nPO stimulus also involved increasing the spike rate modulation frequency of the MS-Glu object within the physiologically relevant theta range of 3–10 Hz [28], roughly corresponding to pyramidal cell firing rate.

Recording Synthetic Local Field Potential

Septo-hippocampal theta rhythm is recorded as a population activity measured as the extracellular field potential around pyramidal cells, which by their arrangement form an open field so that voltage differences between extracellular points along each neuron approximately sum to give a measure of the time-course of total synaptic activity at all pyramidal cells. This was calculated by arranging the spatial coordinates of the pyramidal cells along a 400 μm radius circle as they are arranged in hippocampal CA1. The transmembrane current at each compartment of the cells is recorded and divided by the compartment's distance from the recording electrodes, multiplied by factor accounting for extracellular conductivity. The resulting local field potential (LFP) was low-pass filtered using a 10th-order butterworth filter, and Fourier transform was performed on this time series to visualize LFP oscillations as a power spectrum in the frequency domain. The predominant theta-range frequency of the LFP was determined as the frequency corresponding to the peak of the power spectrum in the range 2–12 Hz.

Simulating Synaptic Parameter Changes

Post-synaptic membrane ion channels are modeled as parallel resistor branches on single cell model circuits. The time-dependence of the conductance of these channels follows first-order kinetics when a pre-synaptic spike occurs according to the equation

$$\frac{ds}{dt} = \alpha F\left(V_{\text{pre}}\right)(1-s) - \beta s$$

with s representing the fraction of open synaptic channels at the post-synaptic membrane, weighted by a maximal conductance g_{syn} representative of the post-synaptic receptor density and potential difference driving force, yielding the magnitude of the post-synaptic current. The multiplicative inverse of the β term in this synaptic

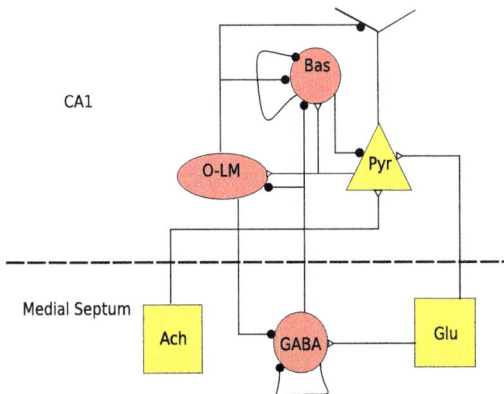

Fig. 2.2 a Structure of septo-hippocampal network model. Three hippocampal neuron populations and one medial septal population are modeled explicitly. Two medial septal populations are modeled indirectly by their rhythmic or tonic effects on the system (*squares*). *Red symbols* and *black circles* indicate inhibitory populations and synapses, respectively; *yellow symbols* and *open triangles* indicate excitatory populations and synapses

model is referred to here as the decay time constant of the post-synaptic conductance, τ_{syn} [84]. This time constant of post-synaptic conductance was increased for GABA synapses and decreased for AMPA synapses to modulate the dynamics of different loops within the septo-hippocampal system and analyze the resulting network frequencies. The maximal synaptic conductance g_{syn} was increased at GABA synapses to model effects of positive allosteric modulators of $GABA_A$ receptors.

Results

Modeling Stimulation of Ascending Pathways to Hippocampus

Ascending pathway stimulation, experimentally achieved by injecting depolarizing current to the nucleus Pontis Oralis (nPO) of the brainstem, was modeled by increasing depolarizing current to the neuron populations with afferents from these pathways, representing increased cholinergic input from the medial septum (Fig. 2.2) and the indirect effects of ascending excitatory afferents to hippocampal CA1 [77]. Both firing synchrony for pyramidal neurons and recorded LFP served as reference points for observing the resulting septo-hippocampal theta activity. A representative LFP exhibited increasing frequency and amplitude of oscillation in the 3-10 Hz theta range (Fig. 2.3a). The increasing frequency and power of the main frequency component of an LFP can be observed more clearly in the peaks of power spectra in a representative trial of increasing modeled brainstem stimulation (Fig. 2.3b). The frequency of synchronous firing amongst pyramidal cells increased with increased nPO stimulation (Fig. 2.3c). An approximately linear increase in mean theta frequency with nPO stimulus level was observed (Fig. 2.4; black, $R^2 = 0.69$, $p = 0.010$).

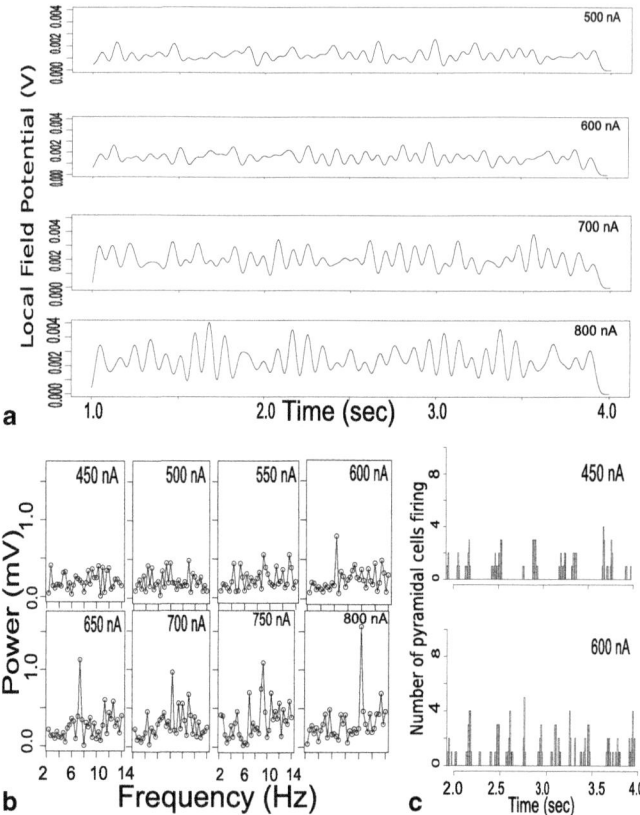

Fig. 2.3 a Synthetic LFP recordings from pyramidal cells for increasing nPO stimulation in model. **b** Representative power spectra of LFP in the theta band for increasing nPO stimulation. **c** Representative firing histograms showing increasing frequency of synchronous firing in pyramidal somata

Increasing Decay Time Constant of GABA-receptor Conductance Affects nPO Stimulus-theta Frequency Relationship

Simulating a longer lasting conductance change at inhibitory synapses but not a shorter excitatory change affected the oscillation frequency of the network (data not shown). Upon doubling the synaptic time constant of inhibition, the stimulus-frequency relationship remained significantly linear ($R^2 = 0.97$, $p < 0.0001$, Fig. 2.4; blue). However, the intercept of the regression line was significantly reduced when the decay time constant was doubled while statistically controlling for the effects of varying stimulation level on theta frequency (ANCOVA, $F(1,236) = 6.44$, $p = 0.012$, Fig. 2.4). Doubling the time-constant of GABA-receptor mediated synapses reduced theta frequency at low levels of nPO stimulation.

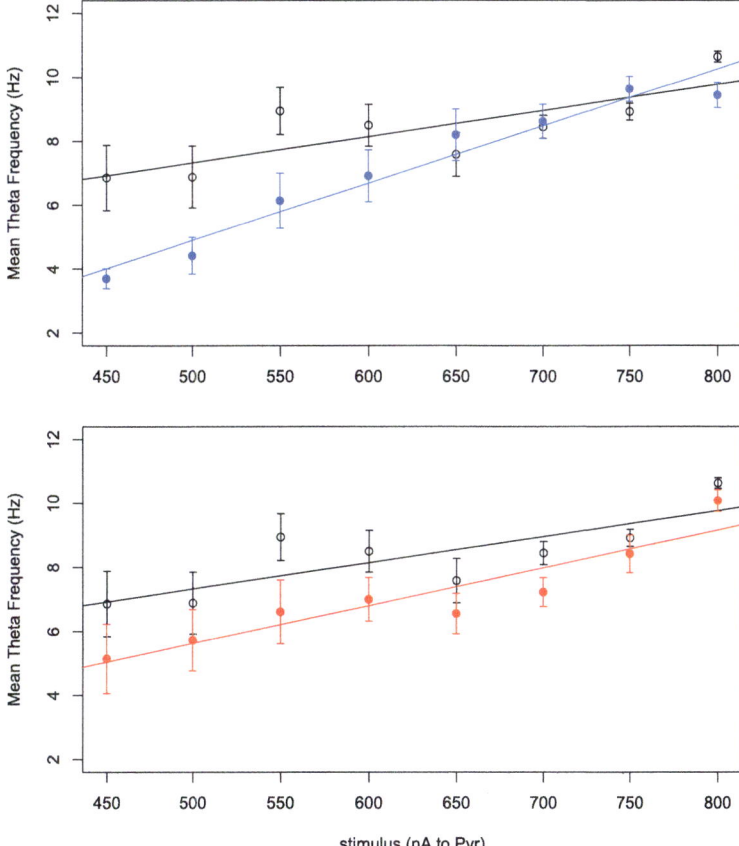

Fig. 2.4 *Upper* panel, Modulation of synaptic dynamics affects frequency of nPO stimulation elicited theta rhythm. Mean frequency and standard error are shown with default parameter settings (*black*) and with GABA decay time constant doubled (*blue*) for $n=16$ runs of the simulation per point. *Lower* panel, Slowing down pyramidal hyperpolarization-activated (I_h) current dynamics slightly lowers intercept of nPO elicited theta frequency relationship. Mean theta frequency and standard error are shown for a range of nPO stimulation levels with default parameter settings (*black*) and with I_h conductance rise and fall rates cut in half (*red*) for 16 runs of the simulation per point. Models effect of potentially selective anxiolytic drug

Slowing Hyperpolarization-activated (I_h) Current Dynamics Lowers Intercept of nPO Stimulus-theta Frequency Relationship

One cellular parameter intrinsic to pyramidal neurons that was implicated in frequency modulation was the influence of the hyperpolarization-activated current I_h [90]. To simulate an decreased influence of I_h on the electrical dynamics of the pyramidal cell membrane, the rates of I_h conductance rises and falls in the somata and dendritic regions of pyramidal cells in the model were each halved.

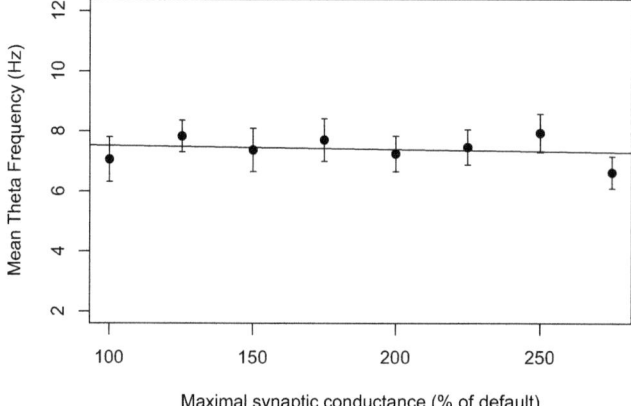

Fig. 2.5 Modulation of maximal synaptic conductance associated with GABA receptors within septo-hippcampal system has negligible effect on theta frequency. Error bars indicate standard error of the mean of $N = 16$ runs of the simulation Depolarizing currents of 600 nA, 1.4 μA, and 2.2 μA to pyramidal somata, basket cells, and MS-GABA cells was used, respectively, with a MS-Glu population firing rate modulation of 6 Hz

When the time course of I_h was thus extended within a range assumed to be physiologically relevant, the stimulus-frequency relationship remained significantly linear ($R^2 = 0.92$, $p < 0.001$, Fig. 2.4), but with a reduced mean frequency at each level of stimulation and a significantly reduced intercept (ANCOVA, F(1,252) = 12.66, $p < 0.001$, Fig. 2.4, red). A marginally significant difference in slope is consistent with the interpretation that the magnitude of post-manipulation theta frequency reduction was somewhat greater at lower stimulation levels than at higher stimulation levels.

Modulation of Maximal Synaptic Conductance Associated with GABA Receptors within Septo-hippcampal System Had No Significant Effect on Theta Frequency

While known anxiolytics such as benzodiazepines are characterized by their interaction with GABA receptors as positive allosteric modulators, the effects of these interactions when incorporated into the septo-hippocampal system do not necessarily reflect these low level effects on the network level. The action of positive allosteric modulation on $GABA_A$ receptors was investigated in the septo-hippocampal model by increasing maximal synaptic conductance at GABA-receptor mediated synapses relative to default settings (see Methods). No significant difference in mean network frequency was detected between any level of positive allosteric modulator action (ANOVA, F(7,112) = 1.302, $p = 0.256$, Fig. 2.5).

Discussion

The frequency of hippocampal theta rhythm measured in the synthetic local field potential of CA1 pyramidal cells was reduced by modulating synaptic and intrinsic cellular parameters in a biophysically realistic model of the septo-hippocampal system. Further, the nPO stimulation assay was modeled with increasing depolarizing currents to evaluate potential effects of anxiolytic drugs, the septo-hippocampal model was expanded to begin to account for the role of medial septal glutamatergic population in hippocampal theta production, and potential biophysical effects of selective anxiolytics were identified.

It was suggested that lengthening the time course of inhibition at individual synapses lowers network theta frequency at low levels of depolarizing stimulus from ascending pathways. Slowing the conductance decay at GABA synapses might be combined with other mechanisms of frequency reduction to have effects resembling a more complete anxiolytic profile. This result is comparable to the effects of zolpidem, which increases the synaptic time constant of $GABA_A$ receptors [23] and is known to have weak anxiolytic effects.

Decreasing the influence of the I_h channel on the time evolution of the membrane potential was suggested to result in significantly lower theta frequency across the range of stimulation intensities, consistent with an anxiolytic profile. I_h contributes to depolarization of the neural membrane in response to hyperpolarization and its slowing could change the resonance properties of pyramidal cells and contribute to changing the frequency of the network theta activity. The systems level effect on theta rhythm frequency of this ion channel level change is an untested prediction of our model. Interestingly, a recent study [90] examined the effects of blocking I_h current using the drug ZD7288 in the nPO-stimulated hippocampal theta paradigm (Fig. 2.6). ZD7288 reliably lowered the intercept of the stimulation-frequency relationship, and showed some sign of an anxiolytic effect in that it increased open arm entries in the elevated plus maze model of anxiety, supporting our result.

While positive allosteric modulators of GABA receptors may be predicted to reduce hippocampal theta frequency based on the drug-receptor interaction level, our results suggest that cellular and network properties of the septo-hippocampal can preclude the interaction's effect on theta rhythm frequency. Our model suggests that the frequency-modulating effects of positive allosteric modulators of GABA receptors on hippocampal theta is mediated by synapses outside of the septo-hippocampal system because of the lack of an influence of maximal synaptic conductance of inhibitory synapses on the network theta frequency. This prediction is consistent with evidence suggesting that the reduction of reticular-elicited hippocampal theta by systemic benzodiazepine injection is largely or wholly mediated by medial hypothalamic sites. For instance, chlordiazepoxide infusions to the medial supramammillary nucleus mimic those of systemic injections of chlordiazepoxide [88].

Upstream regulators of theta rhythm frequency should also be incorporated to expand on the systems level approach to discovering mechanisms and biomarkers of selective anxiolytics. The biophysical synaptic and cellular parameters of the

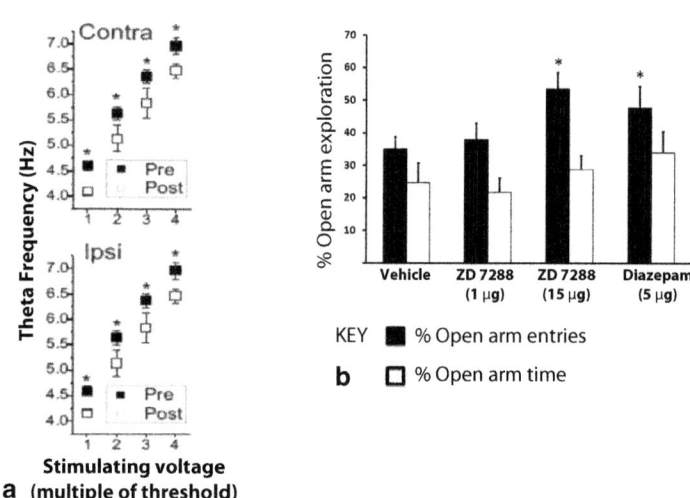

Fig. 2.6 Ih blocking drug (ZD7288) reduces theta frequency and may have anxiolytic effects **a** ZD7288 reduces offset of theta frequency-stimulation strength regression. **b** Higher dose of ZD7288 (15 μg) produces effect in anxiolytic direction (higher % open arm entries) which is equivalent to that of Diazepam (5 μg). Lower dose of ZD7288 (1 μg) was without effect. All drugs directly infused into the hippocampus. (Adapted from Yeung et al. (2013))

candidate frequency controlling input structure to the medial septum, the supramammillary nucleus (SUM), could have further consistent influence on hippocampal rhythms. Experimental results suggest this direction as lesioning the SUM has anxiolytic effects in different animal models of anxiety [2, 89], and that benzodiazepine administration into the SUM lowers theta frequency and behavioral inhibition [88], a model of anxiety-like behavior. Therefore, modeling the role of the SUM in theta frequency modulation, which is predicted to affect information processing in the septo-hippocampal system by changing oscillatory phase relationships controlling the direction of synaptic plasticity amongst its elements (see [37] for review), could further our understanding of the biophysical requirements for anxiolysis. Systems level pharmacological simulations in realistic computational models such as this can account for the interaction of effects across levels in disease and thus hold promise for effective drug discovery.

Validation

Certain validation techniques were used in the development of this model of nPO-stimulation evoked theta rhythm frequency control to ensure that the model was suitable for its intended purpose [61]. Some techniques employed include evaluating face validity, considering internal validity, and using operational graphics [63].

Face validity was established iteratively along with model development—the behavior of the model was considered for its realism by the authors experienced with empirically determined and modeled properties of hippocampal theta rhythm to either remove or keep an element of the model as it was being developed, keeping theta frequency control as the goal of each additional model component. Internal validity was also subjectively confirmed with variance of frequency data points being small compared to the range of frequencies observed in a given assay, which resembled the range observed in empirical studies [46]. The variance in frequency resulting from different initial membrane potential distributions and connectivities could thus resemble that resulting from the different connectivities of the corresponding mouse model brain regions, supporting the use of this model in analyzing conserved properties of theta rhythm frequency.

Finally, operational graphics were used to visualize the effects of the manipulations corresponding to anxiolytic drug action or model additions at multiple levels of analysis. In addition to the overall result of the reduction of the frequency of the local field potential of the network in the theta frequency range, visualizations that helped substantiate the modeled frequency changes included the effect on individual current dynamics, individual cell voltage traces (data not shown), and firing histograms of individual populations (Fig. 2.3c). The results of these tests support the applicability of this model to understanding changes in theta frequency caused by changes in biophysical dynamics of membrane conductance in the septo-hippocampal microcircuitry.

Comparisons to previous models of theta rhythm control also provide insight into the present model's accuracy in predicting potential pharmacological mechanisms of theta rhythm frequency reduction. Previous experimental and computational studies have examined the control of hippocampal theta rhythm frequency at different neural pathways and levels of abstraction with various conclusions [13,38,82]. Kirk and McNaughton [38] used procaine injections to suggest that the frequency of hippocampal theta is primarily encoded in the supramammillary nucleus rather than in the medial septum. Reinforcing this result, Denham and Borisyuk [13]3 showed that the frequency of hippocampal theta in a Wilson-Cowan type differential equation model of the population activities of the septo-hippocampal system was relatively robust to physiologically relevant changes in the time constants and connection strengths associated with each population. However, Wang [82] found that increasing the synaptic conductance of cross-population inhibition or decreasing the synaptic conductance of recurrent medial septal inhibition lowers the frequency of septo-hippocampal theta rhythm in a biophysically realistic model of two inhibitory populations. Our study, incorporating the biophysical realism of the latter computational study with the systems-level detail of the study by Denham and Borisyuk [13], suggests that changes in parameters within the septo-hippocampal system can lower the frequency of hippocampal theta rhythm. Thus, incorporating the more detailed dynamics of individual neurons as exemplified in the latter study by Wang [82] seems to reveal a conductance-based control of theta rhythm frequency, further supporting the validity of our biophysically detailed model to be able to describe the events resulting in theta rhythm frequency changes through mechanisms of anxiolytic drugs.

Alzheimer's Disease and Altered Hippocampal Theta Rhythms

The correlation between characteristics of elicited theta rhythm and anxiolytic drug efficacy points toward the rhythm's sensitivity to changes in neurotransmission at the synapses that help mediate elicited theta. Extending this assumption to other dynamical disorders involving hippocampal theta suggests that the reduced hippocampal function and correlated decrease of the power of hippocampal theta rhythm observed in amyloid-β overproducing mice [64] or following local hippocampal amyloid-β injection in rats [78] might be understood in terms of synaptic and cellular changes relevant to early AD implemented in a similar computational model of the septo-hippocampal system.

We note that functionality associated with theta frequency variation is also likely to be disrupted by Alzheimer-type pathology. Jeewajee et al. [32] found that environmental novelty reduces theta frequency. Interestingly, this was examined in an AD model; Villette et al. [78] showed that while object novelty elicited theta frequency reduction in the wild-type mouse, no such reduction was seen in the AD rats.

Background

In the early 1990s Alzheimer's Disease (AD) was postulated to be casually linked to the accumulation of the peptide fragment amyloid-β ($A\beta$) that constitutes the disease's characteristic amyloid plaques [26,50]. Such amyloid deposition has been suggested to represent an early hallmark of AD prior to the clinical observation of declining cognitive abilities [68]. Changes in brain electrical activity is believed to reflect or probably even precede declining cognition. Therefore, multiple clinical research efforts have been undertaken during the past several decades to use electroencephalographic and magnetoencephalographic signals to diagnose and classify patients [29, 75]. These studies have shown for example that resting state EEG rhythms can be used to differentiate between the normally aging population, patients with mild cognitive impairment, and AD patients. In general, the hallmark of EEG abnormalities in AD patients is an overall slowing of electrical oscillations resulting in a lower mean frequency, decreased complexity of the signals, and decreased coherence among cortical regions [34], which are thought to be the result of functional disconnection between cortical regions.

Interestingly enough, McNaughton and colleagues suggested that while the frequency of reticular-elicited hippocampal theta oscillation could be used to screen antianxiety drugs, an increase in the power of this oscillation might be a reliable indicator of procognitive drug action [46]. Indeed, while probably less theoretically founded than the behavioral inhibition theory-based argument for anxyolitic action, it was shown that drugs, which result in improved cognition, like the acetylcholinestherase inhibitor donepezil [36], significantly increase the power of the reticular-elicited hippocampal theta oscillation. Conversely, drugs that are known to cause cognitive disruption, like the non-specific NMDA blocker dizocilpine [36], or cholinergic antagonists, like scopolamine significantly decrease the power of theta [40]. One striking example that leads back to Alzheimer's disease is conveyed by the results obtained following administration of semagacestat, a γ-secretase inhibitor to mice. Following a Phase III study that failed because subjects on semagacestat showed worse cognitive performance than subjects on placebo, Hajós and colleagues found that semagacestat significantly decreased power of reticular-elicited theta oscillations [25]. Interestingly, reticular-elicited theta rhythm in anesthetized mice have recently been suggested as an early biomarker of the overproduction $A\beta$, while the relevant synaptic and cellular mechanisms mediating these changes are not known [64].

Methods

The above mentioned biophysically realistic model of the electrical activity of the hippocampus, an early AD target, is manipulated on a synaptic and cellular level to simulate biochemical effects of amyloid-β accumulation. This can help elucidate a mechanism of age-dependent theta oscillation changes, which reflects changes in synchronous synaptic activity.

Connecting Biochemical Events to Oscillatory Electrical Signals

For the computational approach to the discovery of AD biomarkers, we plan to build on our biophysically realistic neural network model of the CA1 hippocampus and medial septum to account for separate mechanisms of $A\beta$ toxcitiy most likely related to theta rhythm in the hippocampus proper, medial septum/diagonal band of Broca (MS/DBB), dentate gyrus (DG), and entorhinal cortex (EC). Observations on effects of $A\beta$ on the relevant network include a reduced rhythmicity of the "pace-making" medial septum GABAergic neurons [78], decreased septo-hippocampal innervation from MS-GABAergic neurons, increased inhibition in the dentate gyrus following $A\beta$-induced neuronal hyperexcitability [54], and synaptic loss in the termination of the perforant path from the EC primarily to the DG, as well as in Schaffer collateral terminals from hippocampal CA3 to CA1 regions [27].

After incorporating these structures shown to be affected in very early AD and important for theta oscillation in addition to hippocampal CA1 and the MS/DBB, we can simulate early $A\beta$ induced network dysfunction by manipulating the specific parameters corresponding to each of these $A\beta$-induced changes. We would like to answer the following question in order to predict the most salient biophysical manifestations of early-AD related theta-dependent memory deficits: What disease progression-ordered subset of these mechanisms best replicates the disease progression course of hippocampal theta rhythm typical of mice exhibiting $A\beta$ accumulation [64]?

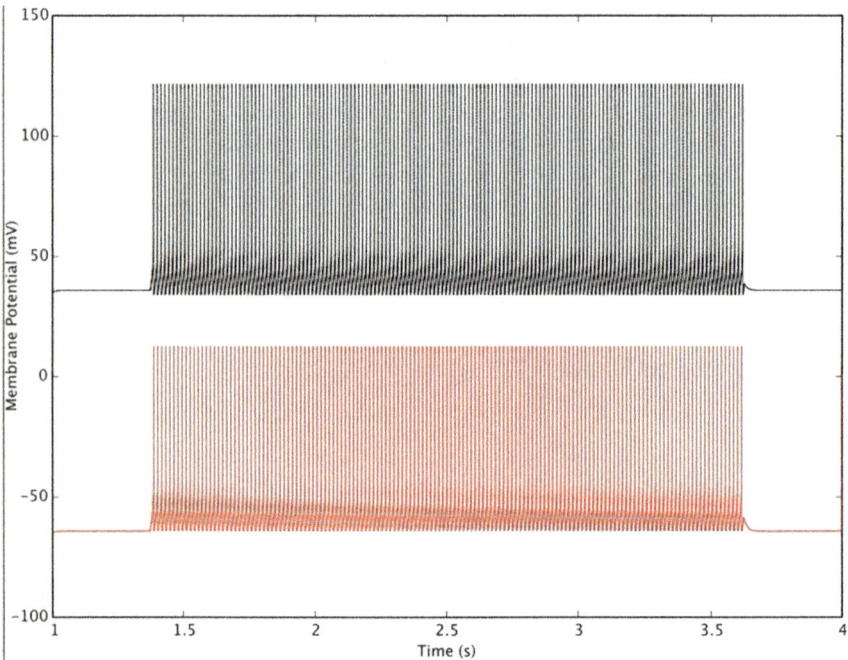

Fig. 2.7 Primarily basket cells but also O-LM cells are interneurons expressing PV in the present hippocampal CA1 model. The above voltage traces show the response of basket cells to a step current of 10 pA. These cells modeled as single compartments with a fast sodium channel and delayed rectifier potassium current both in this simulation and when in the network. Upper trace (shifted up 100 mV) shows baseline cell response and lower trace shows impaired cell with sodium channel density reduced by 15 %, exhibiting approximately 10 % decrease in action potential amplitude (baseline, 86.7 mV; impaired, 77.7 mV), resembling experimental results

Preliminary Results

Recent investigation of the neuronal mechanism behind hippocampal gamma rhythm changes associated with $A\beta$ accumulation suggests an impairment in parvalbumin (PV) expressing interneurons in the hippocampus. In particular, a reduced expression of the pore-forming subunit of voltage-gated sodium channels in these interneurons has been suggested to mediate both reduced gamma activity as well as memory loss in transgenic mice [76]. This effect is implemented in the current septo-hippocampal model to evaluate this mechanism in the context of theta rhythm control.

Reducing the density of fast sodium current channels in basket cells and oriens-lacunosum-moleculare interneurons resulted in an increase in theta rhythm amplitude in the 8–10 Hz range (Fig. 2.8). Distinct effects on single cell firing patterns included a reduced action potential amplitude in basket cells included a reduced action potential amplitude (Fig. 2.7) as reported experimentally [76]. Further biochemical mechanisms are currently being investigated that can reduce theta rhythm amplitude or affect frequency control as experimentally observed.

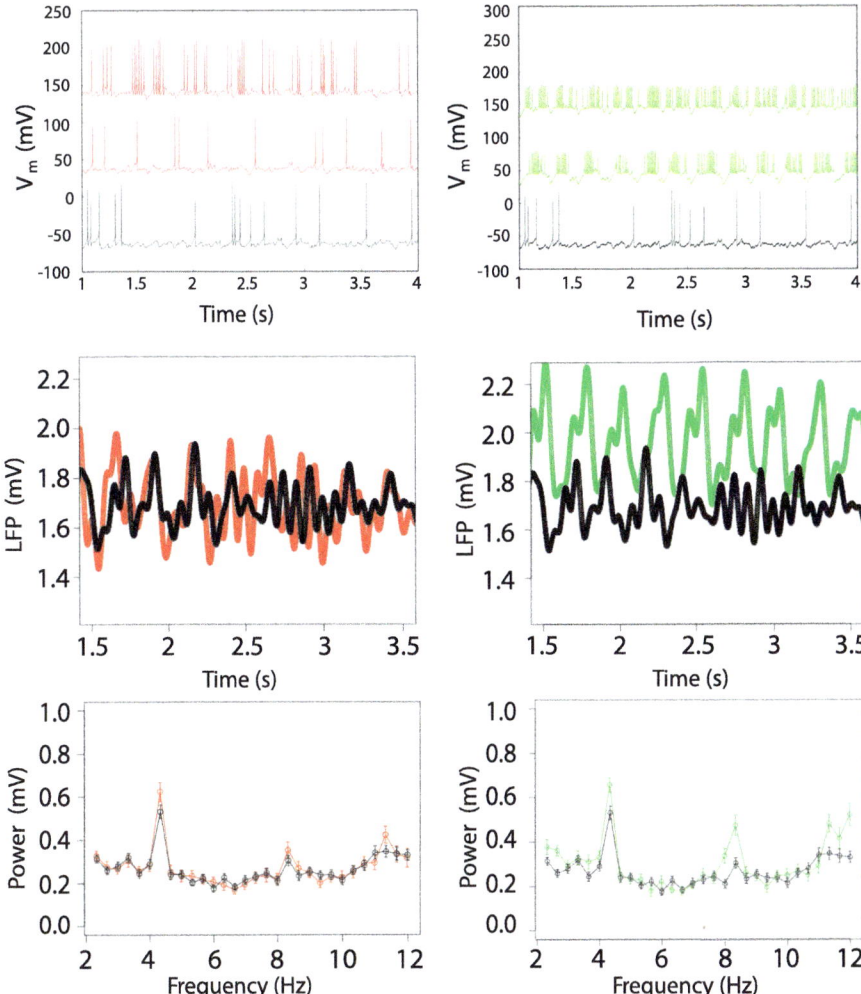

Fig. 2.8 Effects of lower voltage-gated sodium channel densities of PV-expressing interneurons within septo-hippocampal network. *left*, Electrical patterns resulting from 15 % reduction in Na channel density in basket cells only. *right*, 50 % reduction in Na channel density in both basket and O-LM cells, representing a stronger effect. *Top*, Voltage traces of basket cell soma activity within network. Two effected cells are shifted up 10 and 20 mV, demonstrating differential (*left*) or synchronous firing (*right*), and black trace is baseline. *Middle*, Average local field potential (LFP) measured along pyramidal cells with biochemical effects implemented as specified ($n = 24, 48$). Bottom, Average power spectra of local field potentials quantifying the amplitude of oscillations in the LFP as a function of frequency in the theta rhythm band (4–12 Hz, error bars indicate SEM)

Validation

Specific validation criteria and corresponding tests will also need to be determined and performed to ensure its accuracy in being applied to theta rhythm changes involved in amyloid-β overproduction. These should include the types of tests performed on the model for the context of theta frequency modulation mentioned above but applied to theta power control relevant to amyloid-β effects. The validation process should also include analysis of the sensitivity of model results to parameter variability [63] as new susceptible components relevant to theta rhythm generation are incorporated, particularly synaptic conductances when connecting these populations, to ensure that the rhythm's generation is still robust. This would be necessary to validate the physiological relevance of the resulting theta rhythm to amyloid-β overproducing mice models of the hippocampal formation in AD.

Conclusions

Our intention in this paper is to illustrate the potential power of the concepts and methods of the emerging field of quantitative systems pharmacology. The discipline can be considered as the indirect intellectual descendant of systems theory and of cybernetics mediated via systems biology and systems neuroscience. Specifically, the neural mechanisms of the generation and pharmacological control are in the center of the finding biomarkers.

Abnormal neural oscillations have been identified as at least neural correlates of pathological neurological and psychiatric syndromes. Specifically, it has been demonstrated that the frequency of hippocampal theta rhythm measured in the synthetic local field potential of CA1 pyramidal cells can be reduced by modulating synaptic and intrinsic cellular parameters in a biophysically realistic model of the septo-hippocampal system. The reduction of mean theta frequency in this paradigm predicts the clinical efficacy of anxiolytic drugs.

In a recent combined physiological and computational studies [35] we gave some suggestion for the mechanisms by which anxiolytics produce their characteristic effects on the slope and intercept of the stimulus-frequency relationship of hippocampal theta.

Altered hippocampal theta might be related to other disorders, as well. Our general hypothesis to Alzheimer's disease is that early biochemical events could be connected to changes in oscillatory electric signals. We are building a biophysically realistic neural network model of the CA1 hippocampus and medial septum to account for mechanisms to relate $A\beta$ toxicity and changes in theta-related patterns.

We consider our model studies as early examples for the emerging field of quantitative systems pharmacology. Much work should be done to have well-founded and validated biophysically detailed computational models for offering new pharmacotherapies against anxiety and AD.

Acknowledgements PE thanks the Henry Luce Foundation for letting him serve as a Henry R. Luce Professor. TJ thanks the Heyl Science Scholarship Fund and the Barry Goldwater Scholarship Program for their support in attending Kalamazoo College. TK is a full time employee and shareholder of Pfizer Inc and is on sabbatical leave from the Wigner Research Centre for Physics of the Hungarian Academy of Sciences. Some research reported in this paper was supported by a BBSRC New Investigator grant (BB/G01342X/1 and X/2) to CL.

References

1. Aradi I, Érdi P. Computational neuropharmacology: dynamical approaches in drug discovery. Trends Pharmacol Sci. 2006;27(5):240–3.
2. Aranda L, Santín LJ, Begega A, Aguirre JA, Arias JL. Supramammillary and adjacent nuclei lesions impair spatial working memory and induce anxiolitic-like behavior. Behav Brain Res. 2006;167(1):156–64.
3. Arbib MA, Érdi P, Szentágothai J. Neural organization. Cambridge: MIT Press; 1997.
4. Brown P. Abnormal oscillatory synchronisation in the motor system leads to impaired movement. Curr Opin Neurobiol. 2007;17:656–64.
5. Burgess N. Grid cells and theta as oscillatory interference: theory and predictions. Hippocampus 2008;18(12):1157–74.
6. Buzsaki G. Theta oscillations in the hippocampus. Neuron 2002;33:325–40.
7. Buzsaki G. Rhythms of the brain. New York: Oxford University Press; 2006.
8. Calderone DJ, Lakatos P, Butler PD, Castellanos FD. Entrainment of neural oscillations as a modifiable substrate of attention. Trends Cognitive Sci. 2014;18:300–9.
9. Coombes S, Terry JR. The dynamics of neurological disease: integrating computational, experimental and clinical neuroscience. Eur J Neurosci. 2012;36:2118–20.
10. Crick F. Of molecules and men. Seattle: University of Washington Press; 1966.
11. Crick FH, Watson JD. Molecular structure of nucleic acids; a structure for deoxyribose nucleic acid. Nature 1953;171:737–738.
12. de Hemptinne C, Ryapolova-Webb ES, Air EL, et al. Exaggerated phase-amplitude coupling in the primary motor cortex in Parkinson disease. Proc Natl Acad Sci U S A. 2013;110:4780–5.
13. Denham MJ, Borisyuk RM. A model of theta rhythm production in the septal-hippocampal system and its modulation by ascending brain stem pathways. Hippocampus 2000;10(6):698–716.
14. Dupuy J-P. The mechanization of the mind: on the origins of cognitive science. Princeton: Princeton University Press; 2000.
15. Erdi P. Complexity explained. Heidelberg: Springer; 2007.
16. Érdi P, Kiss T, Tóth J, Ujfalussy B, Zalányi L. From systems biology to dynamical neuropharmacology: proposal for a new methodology. IEE Proc Syst Biol 2006;153(4):299–308.
17. Farmer S. Neural rhythms in Parkinson disease. Brain 2002;125(6):1175–6.
18. Ferguson KA, Skinner FK. Modeling oscillatory dynamics in brain microcircuits as a way to help uncover neurological disease mechanisms: a proposal. Chaos. 2013;23:046108.
19. Finkel LH. Neuroengineering models of brain disease. Annu Rev Biomed Eng. 2000;2:577–606.
20. Fries P. A mechanism for cognitive dynamics: neuronal communication through neuronal coherence. Trends Cogn Sci. 2005;9:474–80.
21. Geerts H, Spiros A, Roberts P, Carr R. Quantitative systems pharmacology as an extension of pk/pd modeling in cns research and development. J Pharmacokinet Pharmacodyn. 2013;40(3):257–65.
22. Gray JA, McNaughton N. The neuropsychology of anxiety. Oxford: Oxford university press; 2000.
23. Hájos N, Freund TF, Nusser Z, Rancz EA, Mody I. Cell type- and synapse-specific variability in synaptic GABAA receptor occupancy. Eur J Neurosci 2000;12(3):810–8.
24. Hájos M, Hoffmann WE, Orbán G, Kiss T, Erdi P. Modulation of septo-hippocampal theta activity by GABAA receptors: an experimental and computational approach. Neuroscience 2004;126(3):599–610.
25. Hájos M, Morozova E, Siok C, Atchison K, Nolan CE, Riddell D, Kiss T, Hajos-Korcsok E. Effects of the γ-secretase inhibitor semagacestat on hippocampal neuronal network oscillation. Vol 4. 2013.

26. Hardy J, Allsop D. Amyloid deposition as the central event in the aetiology of Alzheimer's disease. Trends Pharmacol Sci. 1991;12(10):383–8.
27. Harris JA, Devidze N, Verret L, Ho K, Halabisky B, Thwin MT, Kim D, et al. Transsynaptic progression of amyloid-β-induced neuronal dysfunction within the entorhinal-hippocampal network. Neuron 2010;68(3):428–41.
28. Huh CYL, Goutagny R, Williams S. Glutamatergic neurons of the mouse medial septum and diagonal band of Broca synaptically drive hippocampal pyramidal cells: relevance for hippocampal theta rhythm. J Neurosci Off J Soc Neurosci 2010;30(47):15951–61.
29. Hulbert S, Adeli H. EEG/MEG- and imaging-based diagnosis of Alzheimer's disease. Rev Neurosci. 2013;24:563–76.
30. Ingalls BP. Mathematical modeling in systems biology. Boston: The MIT Press; 2013.
31. Green JD, Arduini AA. Hippocampal electrical activity in arousal. J Neurophysiol. 1954;17:553–7.
32. Jeewajee A, Lever C, Burton S, OKeefe J, Burgess N. Environmental novelty is signalled by reduction of the hippocampal theta frequency. Hippocampus 2008;18(4):340–8.
33. Jensen O, Colgin LL. Cross-frequency coupling between neuronal oscillations. Trends Cogn Sci. 2007;11:267–9.
34. Jeong J. EEG dynamics in patients with Alzheimer's disease. Clin Neurophysiol. 2004;115:1490–505.
35. John T, Kiss T, Lever C, Érdi P. Anxiolytic drugs and altered hippocampal theta rhythms: the quantitative systems pharmacological approach. Network (Bristol, England). 2014;25(1–2):20–37.
36. Kinney GG, Patino P, Mermet-Bouvier Y, Starrett JE Jr, Gribkoff VK. Cognition-enhancing drugs increase stimulated hippocampal theta rhythm amplitude in the urethane-anesthetized rat. J Pharmacol Exp Ther. 1999;291:99–106.
37. Kirk IJ. Frequency modulation of hippocampal theta by the supramammillary nucleus, and other hypothalamo-hippocampal interactions: mechanisms and functional implications. Neurosci Biobehav Rev. 1998;22(2):291–302.
38. Kirk IJ, McNaughton N. Mapping the differential effects of procaine on frequency and amplitude of reticularly elicited hippocampal rhythmical slow activity. Hippocampus 1993;3(4):517–25.
39. Kiss T, Hoffmann WE, Hajós M. Delta oscillation and short-term plasticity in the rat medial prefrontal cortex: modelling nmda hypofunction of schizophrenia. Int J Neuropsychopharmacol. 2011;14:29–42.
40. Kocsis B, Li S. In vivo contribution of h-channels in the septal pacemaker to theta rhythm generation. Eur J Neurosci. 2004;20:2149–58.
41. Kramis R, Vanderwolf CH, Bland BH. Two types of hippocampal rhythmical slow activity in both the rabbit and the rat: relations to behaviour and effects of atropine, diethyl ether, urethane, and pentobarbital. Exp Neurol. 1975;49:58–85.
42. Lakatos P, Karmos G, Mehta AD, Ulbert I, Schroeder CE. Entrainment of neuronal oscillations as a mechanism of attentional selection. Science 2010;320:110–11.
43. Lopes da Silva FH, Blanes W, Kalitzin SN, Parra J, Suffczynski P, Velis DN. Dynamical diseases of brain systems: different routes to epileptic seizures. IEEE Trans Biomed Eng. 2003;50:540–8.
44. McCarthy MM, Moore-Kochlacs C, Gu X, Boyden ES, Han X, Kopell N. Striatal origin of the pathologic beta oscillations in Parkinson's disease. Proc Nat Acad Sci U S A. 2011;108:11620–5.
45. McNaughton N, Kocsis B, Hajós M. Elicited hippocampal theta rhythm: a screen for anxiolytic and procognitive drugs through changes in hippocampal function? Behav Pharmacol. 2007;18:329–46.
46. McNaughton N, Kocsis B, Hajós M. Elicited hippocampal theta rhythm: a screen for anxiolytic and procognitive drugs through changes in hippocampal function? Behav Pharmacol. 2007;18:329–46.

47. Moran LV, Hong E. High vs low frequency neural oscillations in schizophrenia. Schizophr Bull. 2011;37:659–63.
48. Morange M. A history of molecular biology (Trans. M Cobb). Cambridge: Harvard University Press; 1998.
49. Mormann F, Kreuz T, Rieke C, Andrzejak RG, Kraskov A, David P, Elger CE, Lehnertz K. On the predictability of epileptic seizures. Clin Neurophysiol. 2005;116:569–87.
50. Mudher A, Lovestone S. Alzheimer's disease - do tauists and baptists finally shake hands? Trends Neurosci. 2002;25(1):22–26.
51. Nakazawa K, Zsiros V, Jiang Z, Nakao K, Kolata S, Zhang S, Belforte JE. Gabaergic interneuron origin of schizophrenia pathophysiology. Neuropharmacology 2012;62:1574–83.
52. Noble D. Mind over molecule: systems biology for neuroscience and psychiatry. In: Tretter F, Winterer G, Gebicke-Haerter PJ, Mendoza RE, editors. Systems biology in psychiatric research. Hoboken: Wiley-Blackwell; 2010.
53. O'Keefe J, Nadel L. The hippocampus as a cognitive map. Oxford: Oxford University Press; 1978.
54. Palop JJ, Chin J, Roberson ED, Wang J, Thwin MT, Bien-Ly N, Yoo J, et al. Aberrant excitatory neuronal activity and compensatory remodeling of inhibitory hippocampal circuits in mouse models of Alzheimer's disease. Neuron 2007;55(5):697–711.
55. Pitts W, McCulloch W. A logical calculus of the ideas immanent in nervous activity. Bull Math Biophys. 1943;7:115–33.
56. Richardson MP. New observations may inform seizure models: very fast and very slow oscillations. Prog Biophys Mol Biol. 2011;105:5–13.
57. Richardson RM. Abnormal cortical brain rhythms in Parkinson disease. Neurosurgery 2013;72:N23–4.
58. Rosen R. Anticipatory systems: philosophical, mathematical and methodological foundations. Pergamon Press; 1985.
59. Rosen R. Life itself: a comprehensive inquiry into the nature, origin, and fabrication of life. Columbia Univ. Press; 1991.
60. Rotstein HG, Kaper TJ, Kramer MA. Introduction to focus issue: rhythms and dynamic transitions in neurological disease: modeling, computation, and experiment. Chaos 2013;23:046001.
61. Rykiel EJ Jr. Testing ecological models: the meaning of validation. Ecol Model. 1996;90(3):229–44.
62. Sainsbury RS. Hippocampal theta: a sensory-inhibition theory of function. Neurosci Biobehav Rev. 1998;22(2):237–41.
63. Sargent RG. Verification and validation of simulation models. J Simul. 2012;7(1):12–24.
64. Scott L, Feng J, Kiss T, Needle E, Atchison K, Kawabe TT, Milici AJ, Hajós-Korcsok E, Riddell D, Hajós M. Age-dependent disruption in hippocampal θ oscillation in amyloid-β overproducing transgenic mice. Neurobiol Aging. 2012;33(7):13–23.
65. Seidenbecher T, Laxmi TR, Stork O, Pape HC. Amygdalar and hippocampal theta rhythm synchronization during fear memory retrieval. Science 2003;301(5634):846–50.
66. Siok CJ, Taylor CP, Hajos M. Anxiolytic profile of pregabalin on elicited hippocampal theta oscillation. Neuropharmacology 2009;56:379–85.
67. Sorger PK, Allerheiligen SRB, et al. Quantitative and systems pharmacology in the post-genomic era: new approaches to discovering drugs and understanding therapeutic mechanisms. An NIH White Paper by the QSP Workshop Group; 2011.
68. Sperling RA, Laviolette PS, OKeefe K, OBrien J, Rentz DM, Pihlajamaki M, Marshall G, et al. Amyloid deposition is associated with impaired default network function in older persons without dementia. Neuron 2009;63(2):178–88.
69. Traub RD, Michelson-Law H, Bibbig AE, Buhl EH, Whittington MA. Gap junctions, fast oscillations and the initiation of seizures. Adv Exp Med Biol. 2004;548:110–22.
70. Tretter F, Winterer G, Gebicke-Haerter PJ, Mendoza RE, editors. Systems biology in psychiatric research. Wiley-Blackwell; 2010.

71. Uhlhaas PJ, Singer W. Abnormal neural oscillations and synchrony in schizophrenia. Nat Rev Neurosci. 2010;11:100–13.
72. Ujfalussy B, Érdi P, Kiss T. Multi-level models. In: Cutsuridis VI, Graham B, Cobb S, editor. Hippocampal microcircuits. A computational modeler's resource book. Springer; 2010. pp. 527–554.
73. Van Regenmortel MHV. Biological complexity emerges from the ashes of genetic reductionism. J Mol Recognit. 2004;17:145–8.
74. Vanderwolf CH. Hippocampal electrical activity and voluntary movement in the rat. Electroencephalogr Clin Neurophysiol. 1969;26:407–18.
75. Vecchio F, Babiloni C, Lizio R, Fallani FV, Blinowska K, Verrienti G, Frisoni G, Rossini PM. Resting state cortical EEG rhythms in Alzheimer's disease: toward EEG markers for clinical applications: a review. Suppl Clin Neurophysiol. 2013;62:223–36.
76. Verret L, Mann EO, Hang GB, Barth AMI, Cobos I, Ho K, Devidze N, et al. Inhibitory interneuron deficit links altered network activity and cognitive dysfunction in Alzheimer model. Cell. 2012;149(3):708–21.
77. Vertes RP. Hippocampal theta rhythm: a tag for short-term memory. Hippocampus 2005;15(7):923–35.
78. Villette V, Poindessous-Jazat F, Simon A, Léna C, Roullot E, Bellessort B, Epelbaum J, Dutar P, Stéphan A. Decreased rhythmic GABAergic septal activity and memory-associated theta oscillations after hippocampal amyloid-beta pathology in the rat. J Neurosci Off J Soc Neurosci. 2010;30(33):10991–1003.
79. Volman V, Behrens MM, Sejnowski TJ. Downregulation of parvalbumin at cortical gaba synapses reduces network gamma oscillatory activity. J Neurosci. 2011;31(49):18137–48.
80. von Bertalanff L. An outline of general system theory. Br J Philos Sci. 1950;1:114–29.
81. von der Malsburg C. Binding in models of perception and brain function. Curr Opin Neurobiol. 1995;5:520–26.
82. Wang XJ. Pacemaker neurons for the theta rhythm and their synchronization in the septohippocampal reciprocal loop. J Neurophysiol. 2002;87(2):889–900.
83. Wang XJ. Neurophysiological and computational principles of cortical rhythms in cognition. Physiol Rev. 2010;90:1195–268.
84. Wang XJ, Buzsáki G. Gamma oscillation by synaptic inhibition in a hippocampal interneuronal network model. J Neurosci. 1996,16:6402–13.
85. Wells CE, Amos DP, Jeewajee A, Douchamps V, Rodgers J, O'Keefe J, Burgess N, Lever C. Novelty and anxiolytic drugs dissociate two components of hippocampal theta in behaving rats. J Neurosci. 2013;33(20):8650–67.
86. Wiener N. Cybernetics or control and communication in the animal and the machine. Cambridge: MIT Press; 1948.
87. Wolkenhauer O. Systems biology: the reincarnation of systems theory applied in biology? Brief Bioinform. 2001;2:258–70.
88. Woodnorth M-A, McNaughton N. Similar effects of medial supramammillary or systemic injection of chlordiazepoxide on both theta frequency and fixed-interval responding. Cognit Affect Behav Neurosci. 2002;2(1):76–83.
89. Yamazaki CK, Shirao T, Sasagawa Y, Maruyama Y, Akita H, Saji M, Sekino Y. Lesions of the supramammillary nucleus decrease self-grooming behavior of rats placed in an open field. Kitakanto Med J. 2011;61(2):287–92.
90. Yeung M, Dickson CT, Treit D. Intrahippocampal infusion of the ih blocker ZD7288 slows evoked theta rhythm and produces anxiolytic-like effects in the elevated plus maze. Hippocampus 2013;23(4):278–86.

Chapter 3
DCM, Conductance Based Models and Clinical Applications

A. C. Marreiros, D. A Pinotsis, P. Brown and K. J. Friston

Abstract This chapter reviews some recent advances in dynamic causal modelling (DCM) of electrophysiology, in particular with respect to conductance based models and clinical applications. DCM addresses observed responses of complex neuronal systems by looking at the neuronal interactions that generate them and how these responses reflect the underlying neurobiology. DCM is a technique for inferring the biophysical properties of cortical sources and their directed connectivity based on distinct neuronal and observation models. The DCM framework uses mathematical formalisms of neural masses, neural fields and mean-fields as forward or generative models for observed neuronal activity. We here consider conductance based neural mass, mean-field and field models—and review their latest technical developments. We use dynamically rich conductance based models to generate responses in laminar-specific populations of excitatory and inhibitory cells. These models allow for the evaluation of neuronal connections and high-order statistics of neuronal states, using Bayesian estimation and inference. We also discuss recent clinical applications of DCM for convolution based neural mass models, in particular for the study of Parkinson's disease. We present a study of data from Parkinsonian patients, and model the large-scale network changes underlying the pathological excess of beta oscillations that characterise the Parkinsonian state.

A. C. Marreiros (✉)
Max Planck Institute for Biological Cybernetics, 72076 Tübingen, Germany
e-mail: andre.marreiros@tuebingen.mpg.de

D. A. Pinotsis (✉) · P. Brown · K. J. Friston
The Wellcome Trust Centre for Neuroimaging, University College London,
Queen Square, London WC1N 3BG, UK
e-mail: d.pinotsis@ucl.ac.uk

P. Brown
e-mail: peter.brown@ndcn.ox.ac.uk

K. J. Friston
e-mail: k.friston@ucl.ac.uk

A. C. Marreiros · P. Brown
Nuffield Department of Clinical Neurosciences, John Radcliffe Hospital,
University of Oxford, Oxford, UK

Keywords Conductance based model · Neural mass model · Neural field model · Dynamical causal modelling · Model validation · Electrophysiology · EEG · MEG · LFP · Parkinson's disease · Neuromodulation

Abbreviations

BF	Bayes factor
BG	Basal ganglia nuclei
BMS	Bayesian model selection
DCM	Dynamic Causal Modelling
EEG	Electroencephalography
EM	Expectation-Maximization
ERP	Event-Related Potential
fMRI	Functional Magnetic Resonance Imaging
FN	FitzHugh-Nagumo
GLM	General linear model
GPe	Globus Pallidus externa
GPi	Globus Pallidus interna
JR	Jansen and Rit
HH	Hodgkin-Huxley
LFP	Local Field Potential
MEG	Magnetoencephalography
MFM	Mean field model
MM	Method of moments
MMN	Mismatch Negativity
MRI	Magnetic Resonance Imaging
NMM	Neural mass model
NFM	Neural field model
ODE	Ordinary differential equation
PD	Parkinson's disease
SDE	Stochastic differential equation
SEP	Somatosensory evoked potential
SPM	Statistical parametric mapping
SSR	Steady state responses
STN	Subthalamic nucleus

Introduction

System models can be extremely helpful in the mechanistic understanding of neural systems. Models of effective connectivity, i.e. the causal influences that system elements exert over another, are essential for studying the functional integration of neuronal populations—and for understanding the mechanisms that underlie neuronal dynamics. Dynamical Causal Modelling (DCM; [1]) is a general framework for inferring processes and mechanisms at the neuronal level from measurements of functional neuroimaging data, including fMRI [2], EEG/MEG [3] and local field potentials [4, 5]. In contrast to models of functional connectivity or statistical dependencies, DCM does not operate on the measured time-series directly. Instead, it combines a model of the hidden neuronal dynamics with a forward model (or generative model) that translates neuronal states into predicted measurements of how observed data were caused.

DCM is a technique for determining the effective connectivity in neural systems. It considers mean field models as generative models for ERP, ERF, and LFP data measured with EEG, MEG, or invasive electrodes. DCM can estimate effective coupling among brain regions and infer whether this coupling is modulated by experimental manipulations like language [6–10], learning [11, 12], perception and attention [13–16], motor function [17, 18], memory and pharmacology [19, 20]. Crucially it provides the basis of Bayesian model selection that enables different models or hypotheses to be compared in terms of their evidence, given some observed responses or data (BMS; [21, 22]). Since the original description of DCM [2], a number of methodological developments have improved and extended DCM for EEG/MEG [3], DCM for induced responses [23], DCM for neural-mass and mean-field models [24], DCM for steady-state responses [4], DCM for phase coupling [19], DCM for complex-valued data [25] and DCM for neural fields [5, 26, 27]. A review of developments for electrophysiological data can be found in [12], a critical review of the biophysical and statistical foundations in [28], and three more recent reviews on neural masses and fields in [29–31].

In general, modelling inference techniques must be shown to be reliable and valid when applied to a variety of well-characterized datasets. Along the years DCM has been submitted to various reliability and validity tests which are of paramount importance for a rigorous interpretation of findings. DCM *face validity* has been explored in terms of simulations (e.g., [4, 32, 33]), *construct validity* has been established in relation to large-scale models of spiking neurons (e.g.[34]), and *predictive validity* has match predictions from independent experimental measures such as microdialysis (e.g. Ref. [105]) and using mismatch negativity sensory learning paradigm (e.g. [35]).

The first part of this chapter reviews generative models of electrophysiological responses and their use in DCM. Our focus in this chapter is on the nature of the models employed by DCM—noting that their inversion or fitting uses (exactly the same) standard Bayesian techniques. In our survey of these models, we will introduce three cardinal distinctions; namely the distinction between *convolution* and *conductance* dynamics, the distinction between *neural mass* and *mean field* formulations and the distinction between *point sources* and *neural field* models. The first distinction pertains to the dynamics or equations of motion within a single population. Convolution models formulate synaptic dynamics in terms of a (linear) convolution operator, where as conductance based models consider the (non-linear) coupling between conductance and voltage. The second distinction is between the behaviour of a neuronal population or ensemble of neurons—as described with their mean or a *point probability mass* over state space. This contrasts with mean field approaches that model the ensemble density, where different ensemble densities are coupled through their expectations; hence mean field models. Finally, we will consider the distinction between spatial models of populations as point sources (c.f., equivalent current dipoles) and models that have an explicit spatial domain over (cortical) manifolds that call on neural fields. We will focus on conductance based neural mass and field models; largely because these embody key nonlinearities and are consequently capable of fitting observed neuronal responses with a high degree of detail. In the final section, we illustrate a clinical application of DCM, present a case study of data from Parkinsonian patients and discuss its validation.

Causal Models of Neuronal Interactions

DCM is a technique for determining effective connectivity on the basis of measured neuronal data [12, 24, 30]. By neural systems, we mean networks of cortical sources comprising regions or sources that interact with each other. The dynamics of these regions are represented by a single or several state variables. These state variables could refer to various neurophysiological properties, e.g. postsynaptic potentials, status of ion channels, etc. Critically, the state variables interact with each other, i.e. the evolution of each state variable depends on at least one other state variable. For example, the postsynaptic membrane potential depends on which and how many ion channels are open; similarly, the probability of voltage-dependent ion channels opening depends on the membrane potential. Such mutual functional dependencies between the state variables of the system can be expressed quite naturally by a set of ordinary or partial differential equations that operate on the state vector.

The nature of the causal relations between state variables requires a set of parameters θ that determine the form and strength of influences among state variables x. In neural systems, these parameters usually correspond to time constants or synaptic strengths of the connections between the system elements. The mathematical form of the dependencies $F = F(x,u,\theta)$ and the pattern of absent and present connections represent the *structure* of the system. Each element of the system or region is then driven by some endogenous or subcortical input u. We can therefore write down a general state equation for non-autonomous deterministic systems in the following manner,

$$\dot{x} = F(x,u,\theta) \qquad (1)$$

A model whose form follows this general state equation provides a causal description of how system dynamics result from system structure, because it describes (i) when and where external inputs enter the system; and (ii) how the state changes induced by these inputs evolve in time—depending on the system's structure. Given a particular temporal sequence of inputs $u(t)$ and an initial state $x(0)$, one obtains a complete description of how the dynamics of the system (i.e. the trajectory of its state vector in time) results from its structure by integration of the general state equation. It provides in this way, a general form for models of effective connectivity in neural systems.

In the DCMs considered in this chapter, we assume that all processes in the system are deterministic and occur instantaneously. Whether or not this assumption is valid depends on the particular system of interest. If necessary, random components (noise) and delays can be accounted for by using stochastic differential equations and delay differential equations, respectively. We also assume that we know the inputs that enter the system. This is a tenable assumption in neuroimaging because the inputs are experimentally controlled variables, e.g. changes in stimuli or instructions. It may also be helpful to point out that using time-invariant dependencies F and parameters θ does not exclude modelling time-dependent changes of the network behaviour. Although the mathematical form of F *per se* is static, the use of time-varying inputs u allows for dynamic changes in what components of F are 'activated'. Also, there is no principled distinction between states and time-invariant parameters. Therefore, estimating time-varying parameters can be treated as a state estimation problem.

This approach regards an experiment as a designed perturbation of neuronal dynamics that are promulgated and distributed throughout a system of coupled anatomical nodes to change region-specific neuronal activity. These changes engender, through a measurement-specific forward model, responses that are used to identify the architecture and time constants of the system at a neuronal level. An important conceptual aspect of dynamic causal models pertains to how the experimental inputs enter the model and cause neuronal responses.

DCM using Convolution Based Models

In this section, we provide a brief review of DCM with convolution based models. For more details about convolution-based models we refer the reader to [3, 12, 29–32]. In general, the aim of DCM is to estimate, and make inferences about (i) the coupling among brain areas, (ii) how that coupling is influenced by experimental changes (*e.g.* time or cognitive set) and (iii) what are the underlying neurobiological determinants that can account for the variability of observed activity. Crucially, one constructs a reasonably realistic neuronal model of interacting cortical regions or nodes and supplements this with a forward model of how neuronal or synaptic activity translates into a measured response. This enables the parameters of the neuronal model (*e.g.*, effective connectivity) to be estimated from observed data. This process is the same for both the convolution models discussed in this section and the conductance based models discussed in later sections. Electrophysiology has been used for decades as a measure of perceptual, cognitive operations, etc [36, 37]. However, much remains to be established about the exact neurobiological mechanisms underlying their generation [38–40]. DCM for ERPs was developed as a biologically plausible model to understand how event-related responses result from the dynamics in coupled neural ensembles. It rests on a neural mass model which uses established connectivity rules in hierarchical sensory systems to assemble a network of coupled cortical sources. This kind of neural-mass model has been widely used to model electrophysiological recordings (*e.g.*, [41–47] and has also been used as the basis of a generative model for event-related potentials and induced or steady-state responses that can be inverted using real data [3, 4, 23, 48–50].

The DCM developed in [3], uses the connectivity rules described in [51] to model a network of coupled sources. These rules are based on a partitioning of the cortical sheet into supra-, infra-granular layers and granular layer (layer IV). Generally speaking, bottom-up or forward connections originate in agranular layers and terminate in layer IV. Top-down or backward connections target agranular layers. Lateral connections originate in agranular layers and target all layers. These long-range or extrinsic cortico-cortical connections are excitatory and arise from pyramidal cells.

Each region or source is modelled using a neural mass model described in [32], based on the model of [44]. This model emulates the activity of a cortical area using three neuronal subpopulations, assigned to granular and agranular layers. A population of excitatory pyramidal (output) cells receives inputs from inhibitory and excitatory populations of interneurons, via intrinsic connections (intrinsic connections are confined to the cortical sheet). Within this model, excitatory interneurons can be regarded as spiny stellate cells found predominantly in layer IV and in receipt of forward connections. Excitatory pyramidal cells and inhibitory interneurons are considered to occupy agranular layers and receive backward and lateral inputs (see Fig. 3.1).

To model event-related responses, the network receives inputs via input connections. These connections are exactly the same as forward connections and deliver inputs to the spiny stellate cells in layer IV. The vector C controls the influence of the input on each source. The lower, upper and leading diagonal matrices A^F, A^B, A^L encode forward, backward and lateral connections respectively. The DCM here is specified in terms of the state equations shown in Fig. 3.1 and a linear output equation

$$\dot{x} = F(x,u,\theta) \cdot y = Lx_0 + \varepsilon \qquad (2)$$

where x_0 represents the trans-membrane potential of pyramidal cells and L is a lead field matrix coupling electrical sources to the EEG channels [52].

Within each subpopulation, the evolution of neuronal states rests on two operators. The first transforms the average density of pre-synaptic inputs into the average postsynaptic membrane potential. This is modelled by a linear transformation with excitatory and inhibitory kernels parameterised by $H_{e,i}$ and $\tau_{e,i}$. $H_{e,i}$ control the maximum post-synaptic potential and $\tau_{e,i}$ represent a lumped rate-constant. The second operator S transforms the average potential of each subpopulation into an average firing rate. This is assumed to be instantaneous and is a sigmoid function. Interactions, among the subpopulations, depend on constants $\gamma_{1,2,3,4}$, which control the strength of intrinsic connections and reflect the total number of synapses expressed by each subpopulation. Having specified the DCM in terms of these equations one can estimate the coupling parameters from empirical data using a standard variational Bayesian scheme, under a Laplace approximation to the true posterior [53]. This is known as Variational Laplace.

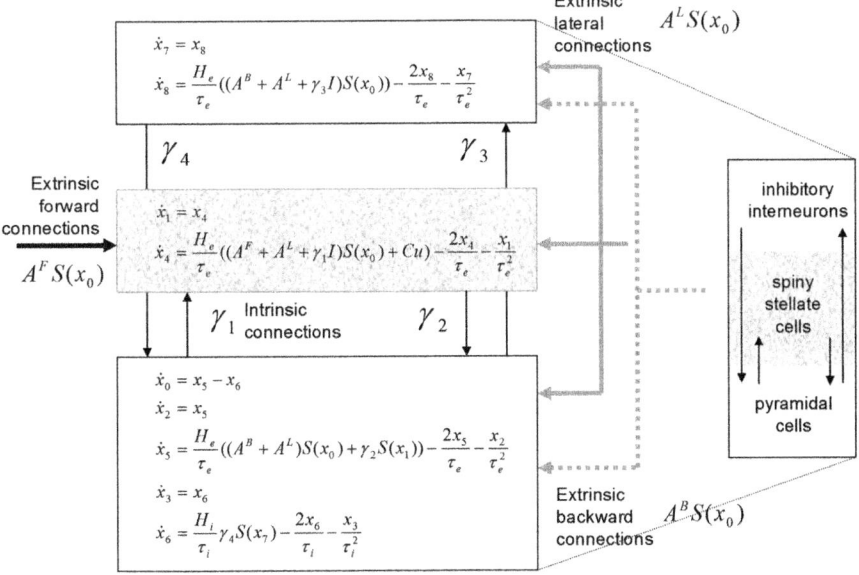

Neuronal model

Fig. 3.1 Convolution-based neural mass model. Schematic of the DCM used to model electrophysiological responses. This schematic shows the state equations describing the dynamics of sources or regions. Each source is modelled with three subpopulations (pyramidal, spiny stellate and inhibitory interneurons) as described in [32, 44]. These have been assigned to granular and agranular cortical layers which receive forward and backward connections respectively [3]

DCM Using Neural Mass and Mean Field Models of Ensemble Activity

The use of neuronal models to simulate large networks of neurons has enjoyed recent developments, involving both direct simulations of large numbers of neurons (which can be computationally expensive); e.g., [54] and probabilistic approaches; e.g. [55, 56]. Probabilistic approaches model the population density directly and bypass direct simulations of individual neurons. This (mean field) treatment of neuronal models exploits a probability mass approximation. This effectively replaces coupled Fokker-Planck equations describing population density dynamics, with equations of motion for expected neuronal states and their dispersion; that is, their first and second moments. These equations are formulated in terms of the mean and covariance of the population density over each neuronal state, as a function of time. In other words, an ensemble density on a high-dimensional phase-space is approximated with a series of low-dimension ensembles that are coupled through mean-field effects. The product of these marginal densities is then used to approximate the full density. Critically, the mean-field coupling induces nonlinear dependencies among the density dynamics of each ensemble. This typically requires a nonlinear Fokker-Planck equation for each ensemble. The Fokker-Planck equation prescribes the evolution of the ensemble dynamics, given any initial conditions and equations of motion that constitute the neuronal model. However, it does not specify how to encode or parameterize the density. There are several approaches to density parameterization [56–61]. These include binning the phase-space and using a discrete approximation to a free-form density. However, this can lead to a vast number of differential equations, especially if there are multiple states for each population. One solution is to reduce the dimension of the phase-space to render the integration of the Fokker-Planck more tracTable (e.g., [62]). Alternatively, one can assume the density has a fixed parametric form and deal only with its sufficient statistics [63–65]. The simplest form is a delta-function or point mass; under this assumption one obtains neural-mass models. In short, we replace the full ensemble density with a mass at a particular point and then summarize the density dynamics by the location of that mass. What we are left with is a set of non-linear differential equations describing the dynamic evolution of this mode. In the full nonlinear Fokker-Planck formulation, different phase-functions or probability density moments could couple to each other; both within and between ensembles. For example, the average depolarisation in one ensemble could be affected by the dispersion or variance of depolarisation in another, see [66]. In neural-mass models, one ignores this potential dependency because only the expectations or first moments are coupled. There are several devices that are used to compensate for this simplification. Perhaps the most ubiquitous is the use of a sigmoid function $\varsigma(V)$ relating expected depolarisation to expected firing-rate [67, 68]. This implicitly encodes variability in the post-synaptic depolarisation, relative to the potential at which the neuron would fire. This affords the considerable simplification of the dynamics and allows one to focus on the behaviour of a large number of ensembles, without having to worry about an explosion in the number of dimensions or differential equations one has to integrate. Important generalisations of neural-mass models, which allow for states that are function of position on the cortical sheet, are referred to as *neural-field models*, as we will see later.

Conductance Based Models

The neuronal dynamics here conform to a simplified [69] model, where the states $x^{(i)} = \{V(i), g_1^{(i)}, g_2^{(i)}, \cdots\}$ comprise transmembrane potential and a series of conductances corresponding to different types of ion channel. The dynamics are given by the stochastic differential equations

$$C\dot{V}^{(i)} = \sum_k g_k^{(i)}(V_k - V^{(i)}) + I + \Gamma_V$$
$$\dot{g}_k^{(i)} = \kappa_k^{(i)}(\varsigma_k^{(i)} - g_k^{(i)}) + \Gamma_k \quad (3)$$

Table 3.1 Comparison of ensemble, Mean-Field (MF) and Neural-Mass (NM) models. The evolution of the *Ensemble* dynamics decomposes into deterministic flow and diffusion. Those reduce to a simpler *MFM* form under Gaussian (Laplace) assumptions, where first order population dynamics are a function of flow and the curvature of the flow, and the second order statistics are a function of the gradients of flow. Those reduce to a simpler *NMM* form by fixing its second order statistics (dispersion) to constant values

Model	Description	Equation
Ensemble	Stochastic differential equation that describes how the states evolve as functions of each other and some random fluctuations	$dx = f(x,u)dt + \sigma dw$
MFM	Differential equation that describes how the mean and covariance of a neuronal populations evolve. This rests on mean-field and Laplace approximations of the ensemble dynamics	$\dot{\mu}_i^{(j)} = f_i^{(j)}(\mu, \Sigma, u) + \frac{1}{2}tr(\Sigma^{(j)}\partial_{xx}f_i^{(j)})$ $\dot{\Sigma}^{(j)} = \partial_x f^{(j)}\Sigma + \Sigma\partial_x f^{(j)T} + D^{(j)} + D^{(j)T}$
NMM	Differential equation that describes how the density evolves as a function of the mean. Obtained by fixing the covariance of the MFM	$\dot{\mu}_i^{(j)} = f_i^{(j)}(\mu, \Sigma, u) + \frac{1}{2}tr(\Sigma^{(j)}\partial_{xx}f_i^{(j)})$ $\dot{\Sigma}^{(j)} = 0$

They are effectively the governing equations for a parallel resistance-capacitance circuit; the first says that the rate of change of transmembrane potential (times capacitance, C) is equal to the sum of all currents across the membrane (plus exogenous current, $I = u$). These currents are, by Ohm's law, the product of potential difference between the voltage and reversal potential, V_k for each type of conductance. These currents will either hyperpolarise or depolarise the cell, depending on whether they are mediated by inhibitory or excitatory receptors respectively (*i.e.*, whether V_k is negative or positive). Conductances change dynamically with a characteristic rate constant κ_k and can be regarded as the number of open channels. Channels open in proportion to pre-synaptic input ς_k and close in proportion to the number open.

Ensemble models of neuronal populations can employ mean-field (MFM) and neural-mass (NMM) approximations. Ensemble models (Eq. 3.3) provide the trajectories of many neurons to form a sample density of population dynamics. The MFM is obtained by a mean-field and a Laplace approximation to these densities. The NMM is a special case of the mean-field model in which we ignore all but the first moment of the density (*i.e.*, the mean or mode). In other words, the NMM discounts dynamics of second-order statistics (*i.e.*, variance) of the neuronal states. The mean-field models allow us to model interactions between the mean of neuronal states (e.g., firing rates) and their dispersion or variance over each neuronal population modelled (c.f., [70]). The key behaviour we are interested in is the coupling between the mean and variance of the ensemble, which is lost in the NMM. The different models and their mathematical representations are summarised in Table 3.1.

The three populations of Fig. 3.2 below emulate a source and yield predictions for observed electromagnetic responses. Similarly to the network of Fig. 3.1, each source comprises two excitatory populations and an inhibitory population. These are taken to represent input cells (spiny stellate cells in the granular layer of cortex),

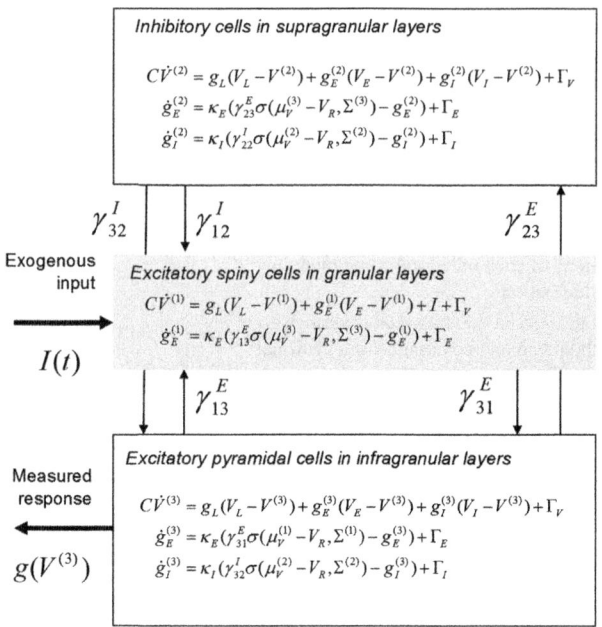

Fig. 3.2 Conductance-based neural mass model. Neuronal state-equations for a source model with a layered architecture comprising three interconnected populations (Spiny-stellate, Interneurons, and Pyramidal cells), each of which has three different states (Voltage, Excitatory and Inhibitory conductances)

inhibitory interneurons (allocated somewhat arbitrarily to the superficial layers) and output cells (pyramidal cells in the deep layers). The deployment and intrinsic connections among these populations are shown in Fig. 3.2.

In this model, we use three conductance types: leaky, excitatory and inhibitory conductance. This gives, for each population

$$\begin{aligned}
C\dot{V}^{(i)} &= g_L(V_L - V^{(i)}) + g_E^{(i)}(V_E - V^{(i)}) + g_I^{(i)}(V_I - V^{(i)}) + I + \Gamma_V \\
\dot{g}_E^{(i)} &= \kappa_E(\varsigma_E^{(i)} - g_E^{(i)}) + \Gamma_E \\
\dot{g}_I^{(i)} &= \kappa_I(\varsigma_I^{(i)} - g_I^{(i)}) + \Gamma_I \\
\varsigma_k^{(i)} &= \sum_j \gamma_{ij}^k \sigma(\mu_V^{(j)} - V_R, \Sigma^{(j)})
\end{aligned} \quad (4)$$

Notice that the leaky conductance does not change, which means the states reduce to $x^{(i)} = \{V^{(i)}, g_E^{(i)}, g_I^{(i)}\}$. Furthermore, for simplicity, we have assumed that the rate-constants, like the reversal potentials are the same for each population. The excitatory and inhibitory nature of each population is defined entirely by the specification of the non-zero intrinsic connections γ_{ij}^k, see Fig. 3.2.

This approach was first applied to FitzHugh-Nagumo (FN) neurons [63, 71] and later to Hodgkin-Huxley (HH) neurons [64, 72]. This approach assumes that the distributions of the variables are approximately Gaussian so that they can be charac-

3 DCM, Conductance Based Models and Clinical Applications

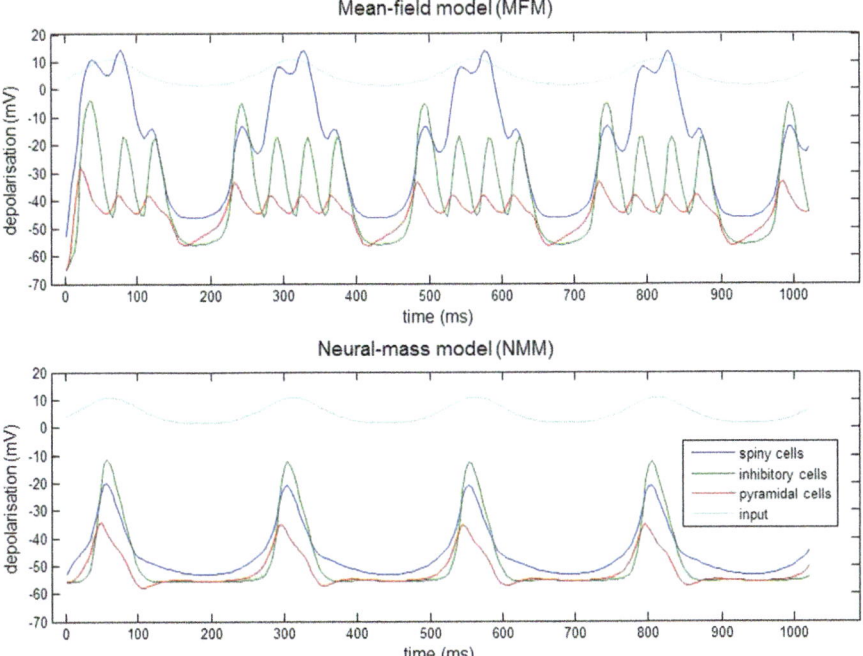

Fig. 3.3 Nested oscillations. Three-population source (Fig. 3.2) driven by slow sinusoidal input for both MFM and NMM. Input is shown in *light blue*, spiny interneuron depolarization in *dark blue*, inhibitory interneurons in green and pyramidal depolarization in *red*. The nonlinear interactions between voltage and conductance produces phase-amplitude coupling in the ensuing dynamics. The MFM shows deeper oscillatory responses during the nested oscillations. This simulation is an illustration of how small differences between models can have large effects on the nature of predicted neuronal responses

terized by their first and second order moments; i.e., the means and covariances. In related work, Hasegawa described a dynamical mean-field approximation (DMA) to simulate the activities of a neuronal network. This method allows for qualitative or semi-quantitative inference on the properties of ensembles or clusters of FN and HH neurons; see [65, 73].

Using the three population source from Fig. 3.2 it can be seen that the population responses of the MFM and NMM show clear qualitative differences in dynamic repertoire, with the MFM presenting limit-cycle attractors after bifurcations from a fixed-point [24], which could be useful for modelling nonlinear or quasi-periodic dynamics, like nested oscillations (Fig. 3.3). This produces phase-amplitude coupling, between the inhibitory population and the spiny population that is driven by the low-frequency input. The bursting and associated nested oscillations are caused by nonlinear interactions between voltage and conductance, which are augmented by coupling between their respective means and dispersions.

Conductance based models are inherently non-linear models that can explain a richer set of dynamic phenomena, like phase amplitude or cross-frequency coupling because of the multiplicative interactions between state variables. Mean field models also include a nonlinearity that follows from the interaction between first and second order moments (see equations in Table 3.1). This might be the reason that MFM captures faster population dynamics better than the equivalent NMM. In [74], it was shown that the NMM was the best model for explaining the MMN (mismatch negativity) data and in contrast the MFM was a better model for the SEP (somatosensory evoked potentials) data. This may be because MFM evokes a more profound depolarisation in neuronal populations as compared to the NMM. In fact, the role of higher moments can be assessed empirically in a Bayesian model selection framework, see below and [74]. In general, questions about density dynamics of this sort can be answered using Bayesian model comparison [21, 75].

The NMM and MFM models were extended in [76] through the inclusion of a third ligand-gated ion channel to model conductances mediated by the NMDA receptor. The introduction of the NMDA ion channels to pyramidal cells and inhibitory interneurons further constrained and specified the neuronal populations laminar specific responses [77]. This richly parameterized model was then used to analyse and recover underlying empirical neural activations in pharmacologically induced changes in receptor processing using MEG [76], during a visuo-spatial working memory task. Remarkably, the DCM parameter estimates disclosed an effect of L-Dopa on delay period activity and revealed the dual mechanisms of dopaminergic modulation of glutamatergic processing, in agreement with predictions from the animal and computational literature [78–82].

Conductance-based Neural Field Models

As with the NMM and MFM models reviewed above, conductance based neural field models are inherently nonlinear models. On top of the multiplicative nonlinearity involving the state variables, they also characterize nonlinear interactions (diffusion) between neighbouring populations on a cortical patch mediated by intrinsic or lateral connections.

Neural field models consider the cortical surface as a Euclidean manifold upon which spatiotemporally extended neural dynamics unfold. These models have been used extensively to predict brain activity, see [5, 29, 30, 83–88] and were obtained using mean-field techniques from statistical physics that allow one to average neural activity in both space and time [89]—and later generalised to considered delays in the propagation of spikes over space [90]. Recent work has considered the link between networks of stochastic neurons and neural field theory by using convolution models (with alpha type kernels) to characterize postsynaptic filtering: some studies have focused on the role of higher order correlations, starting from neural networks and obtaining neural field equations in a rigorous manner; e.g., [91, 92], while others have considered a chain of individual fast spiking neurons [93], communicating through *spike* fields [94]. These authors focused on the complementary nature of spiking and neural field models and on eliminating the need to track individual spikes [95]. We focus here on the relation of neural field models to the

3 DCM, Conductance Based Models and Clinical Applications

conductance based neural mass models we considered in the previous section. For convolution-based neural field approaches in DCM, we refer the reader to [5, 27, 29, 30].

This section considers the behaviour of neuronal populations, where conductance dynamics replace the convolution dynamics—and the input rate field is a function of both time and space. This allows us to integrate field models to predict responses and therefore, in principle, use these spatial models as generative or observation models of empirical data.

We describe below a model that is nonlinear in the neuronal states, as with single unit conductance models and the model of [87]. This model entails a multiplicative nonlinearity, involving membrane depolarization and presynaptic input and has successfully reproduced the known actions of anaesthetic agents on EEG spectra, see e.g. [96–100]. This model is distinguished by the fact that it incorporates distinct cell types with different sets of conductances and local conduction effects. Similarly to the models presented above, it comprises three biologically plausible populations, each endowed with excitatory and inhibitory receptors. We focus on the propagation of spike rate fluctuations over cortical patches and the effect this spatiotemporal dynamics has on membrane dynamics gated by ionotropic receptor proteins. In particular, we consider laminar specific connections among two-dimensional populations (layers) that conform to canonical cortical microcircuitry. The parameterization of each population or layer involves a receptor complement based on findings in cellular neuroscience. However, this model incorporates lateral propagation of neuronal spiking activity that is parameterized through an intrinsic (local) conduction velocity. This model is summarized in Fig. 3.4 below: This figure shows the evolution equations that specify a conductance-based field model of a single source. This model contains the same three populations as in previous figures and is an extension of the well-known Jansen and Rit (JR) model. As with earlier models, second-order differential equations mediate a linear convolution of presynaptic activity to produce postsynaptic depolarization. This depolarization gives rise to firing rates within each sub-population that provide inputs to other populations.

This model is a recent addition to conductance based models implemented in the DCM toolbox of the academic freeware Statistical Parametric Mapping (SPM) and has not been yet used for the analysis of empirical data. Here, as a first step, we will consider a single sensor and cortical source driven by white noise input (see also [88]) and illustrate the ability of this model to account for observed evoked responses of the sort recorded with e.g. local field potential electrodes. In particular, we generated synthetic electrophysiological responses by integrating the equations in Fig. 3.4 from their fixed points and characterised the responses to external (excitatory) impulses to spiny stellate cells, in the time domain. Electrophysiological signals (LFP or M/EEG data) were simulated by passing neuronal responses through a lead field that varies with location on the cortical patch. The resulting responses in sensor space (see Figs. 3.5 and 3.6) are given by a mixture of currents flowing in and out of pyramidal cells in Fig. 3.4:

$$y(t,\theta) = \int L(x,\theta) Q \dot{v}(x,t) dx \qquad (5)$$

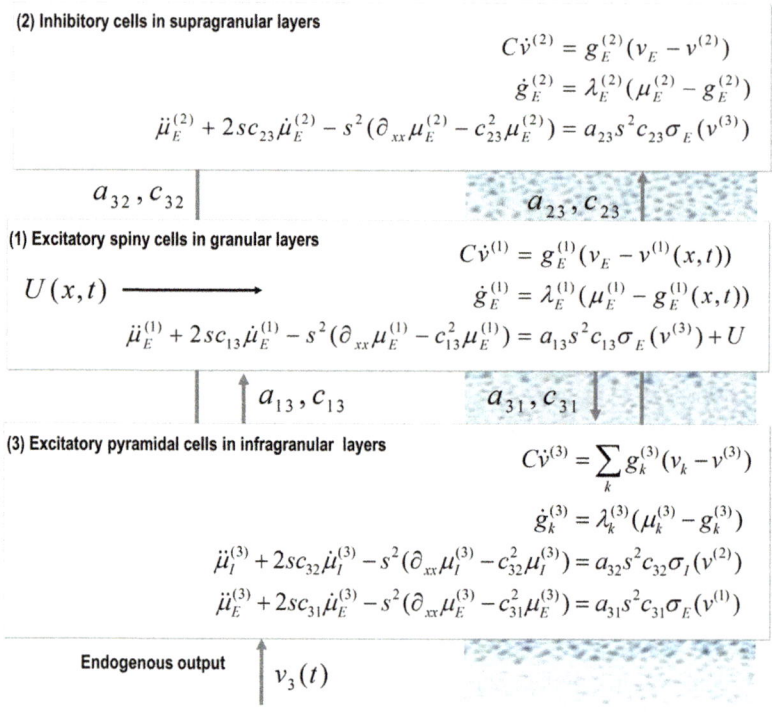

Fig. 3.4 A conductance-based neural field model. This schematic summarizes the equations of motion or state equations that specify a conductance based neural field model of a single source. This model contains three populations, each associated with a specific cortical layer. These equations describe changes in expected neuronal states (e.g., voltage or depolarization) that subtend observed local field potentials or EEG signals. These changes occur as a result of propagating presynaptic input through synaptic dynamics. Mean firing rates within each layer are then transformed through a nonlinear (sigmoid) voltage-firing rate function to provide (presynaptic) inputs to other populations. These inputs are weighted by connection strengths and are gated by the states of synaptic ion channels

In this equation, $Q \subset \theta$ is a vector of coefficients that weight the relative contributions of different populations to the observed signal and $L(x,\theta)$ is the lead field. This depends upon parameters θ and we assume it is a Gaussian function of location—as in previous models of LFP or MEG recordings, see [5]. This equation is analogous to the usual (electromagnetic) gain matrix for equivalent current dipoles. We assume here that these dipoles are created by pyramidal cells whose current is the primary source of an LFP signal. With spatially extended sources (patches), this equation integrates out the dependence on the source locations within a patch and provides a time series for each sensor.

Fig. 3.5 Impulse responses of conductance-based mass and field models. Responses to impulses of different amplitudes for mass (*top*) and field (*bottom*) conductance based models. The responses are normalized with respect to the amplitude of each input. The blue lines illustrate responses to small perturbations. The red lines illustrate responses to intermediate sized inputs, where conductance based models show an augmented response, due to their nonlinearity. The green lines show responses for larger inputs, where the saturation effects due to the sigmoid activation function are evident. Nonlinear effects are more pronounced in the field model– with attenuation of the response amplitude, even for intermediate input amplitudes

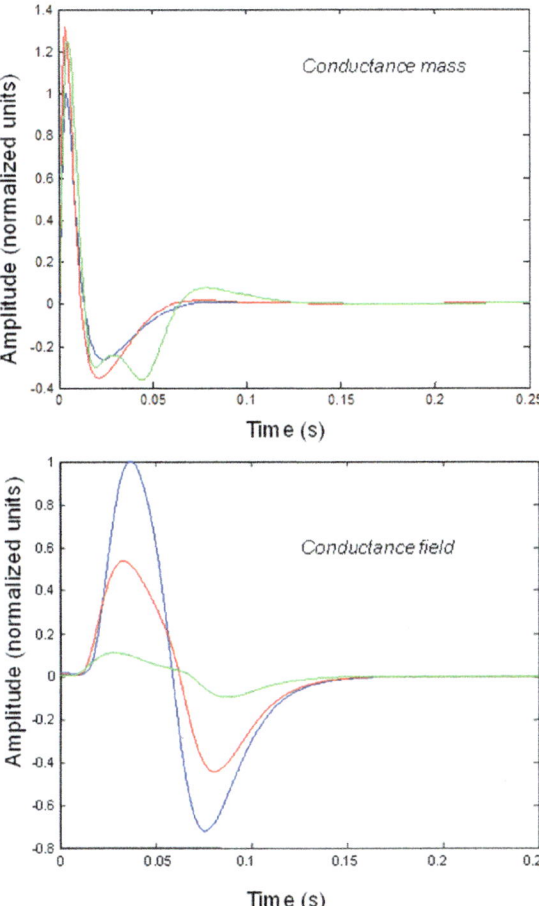

We modelled a cortical source (approximated with 11 grid points) and used the model equations (see Fig. 3.5) to generate impulse response functions. The parameters of this model are provided in Table 3.2. The results reported below were chosen to illustrate key behaviours in terms of ERPs, following changes in parameter values. We also consider the corresponding result for the mass variant of our field model, that is a simplified Morris-Lecar type model (that neglects fast voltage-dependent conductances) introduced in [24]. This model uses the same equations but assume that all neurons of a population are located at (approximately) the same point.

The resulting model is based on the Rall and Goldestein equations [101] and is formally related to Ermentrout's [102] reduction of the model described in [103]. Mass models have often been used to characterize pharmacological manipulations

Table 3.2 Parameters of conductance-based mass and field models. Prior expectation of the conductance-based neural mass and field model parameters

Parameter	Physiological interpretation	Value
g_L	Leakage conductance	1
$\alpha_{13}, \alpha_{23}, \alpha_{31}, \alpha_{32}$	Amplitude of intrinsic connectivity kernels	(1/10, 1, 1/2, 1)*3/10 (field) 1/2, 1, 1/2, 1 (mass)
c_{ij}	Intrinsic connectivity decay constant	1 (mm^{-1})
v_L, v_E, v_I	Reversal potential	−70, 60, −90 (mV)
v_R	Threshold potential	−40 (mV)
C	Membrane capacitance	8 (pFnS^{-1})
s	Conduction speed	3 m/s
$\lambda, \tilde{\lambda}$	Postsynaptic rate constants	1/4, 1/16 (ms^{-1})
ℓ	Radius of cortical patch	7 (mm)

and the action of sedative agents [87, 97, 99, 104–106]. This usually entails assuming that a neurotransmitter manipulation changes a particular parameter, whose effects are quantified using a contribution or structural stability analysis, where structural stability refers to how much the system changes with perturbations to the parameters.

We now focus on generic differences mediated by conductance based field and mass models. To do this, we integrated the corresponding equations for (impulse) inputs of different amplitudes and plotted temporal responses resulting from fixed point perturbations. Linear models are insensitive to the amplitude of the input, in the sense that the impulse responses scale linearly with amplitude. Our interest here was in departures from linearity—such as saturation—that belie the nonlinear aspects of the models. Figure 3.5 shows the responses of the mass and field models to an impulse delivered to stellate cells. Note that these responses have been renormalized with respect to the amplitude of each input. The red (green) curves depict responses to double (ten times) the input reported by the blue curves. We used the same parameters for both models: see Table 3.2.

It can be seen that there are marked differences between the model responses. The top panel depicts the response of the mass model and the lower panel shows the equivalent results for the field model. One can see that large inputs produce substantial sub-additive saturation effects (blue versus green lines in Fig. 3.4): for the mass model, the nonlinearities produce an inverted U relationship between the amplitude of the response, relative to the input. In other words, the form of the input-output amplitude relationship differs quantitatively for the mass (inverted U) and field (decreasing) model (see Fig. 3.5).

The above illustrations of system's predictions assume that spectral responses result from fixed point perturbations. For conductance models, a change in the parameters changes both the expansion point and the system's flow (provided the flow is non-zero). Figure 3.5 shows the dependence of the conductance model's fixed

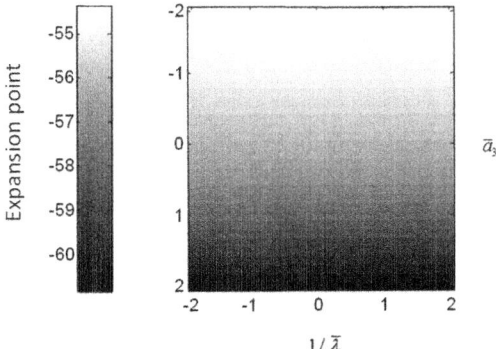

Fig. 3.6 Conductance mass model mean depolarization. Mean depolarization of the pyramidal population of the conductance neural mass model as a function of parameter changes. This corresponds to the fixed point around which the impulse responses of Fig. 3.5 were computed

points on parameter perturbations. The model parameterization used here renders the expansion point relatively insensitive to changes in the synaptic time constant. Figure 3.6 shows the results for the conductance mass model; results for its field variant were very similar.

Clinical Applications and Parkinson's Disease

DCM has contributed to a mechanistic understanding of brain function and drug mechanisms and could serve as an important diagnostic tool for diseases linked to abnormalities in connectivity and synaptic plasticity, like schizophrenia [107–109], Parkinson's [110–114], consciousness and drug effects [115–119] and epilepsy [120–128]. For some other important clinical applications of DCM the reader is referred to [129–135].

In the following, we discuss an application to Parkinson's disease focusing on pathological alterations of beta oscillations in Parkinsonian patients and evidence for an abnormal increase in the gain of the cortical drive to STN and the STN-GPe coupling [111]. In line with a previous study in the rat model [110], we used a DCM for steady state responses (SSR) to predict observed spectral densities [4]. The dynamics of these sources is specified by a set of first-order differential equations [110]. The ensemble firing of one population drives the average membrane potential of others through either glutamate (which produces postsynaptic depolarisation) or GABA (hyperpolarisation) as a neurotransmitter. These effects are mediated by a postsynaptic (alpha) kernel that is either positive or negative. The (excitatory or inhibitory) influence of one subpopulation on another is parameterised by extrinsic effective connectivity (between sources) or intrinsic connectivity (within sources). Effective connectivity is modelled as a gain factor that couples discharge rates in one subpopulation to depolarisation in another.

The validity of the approach used here (DCM for SSR) has been addressed previously ([4, 136]). These studies established that both the form of the model and

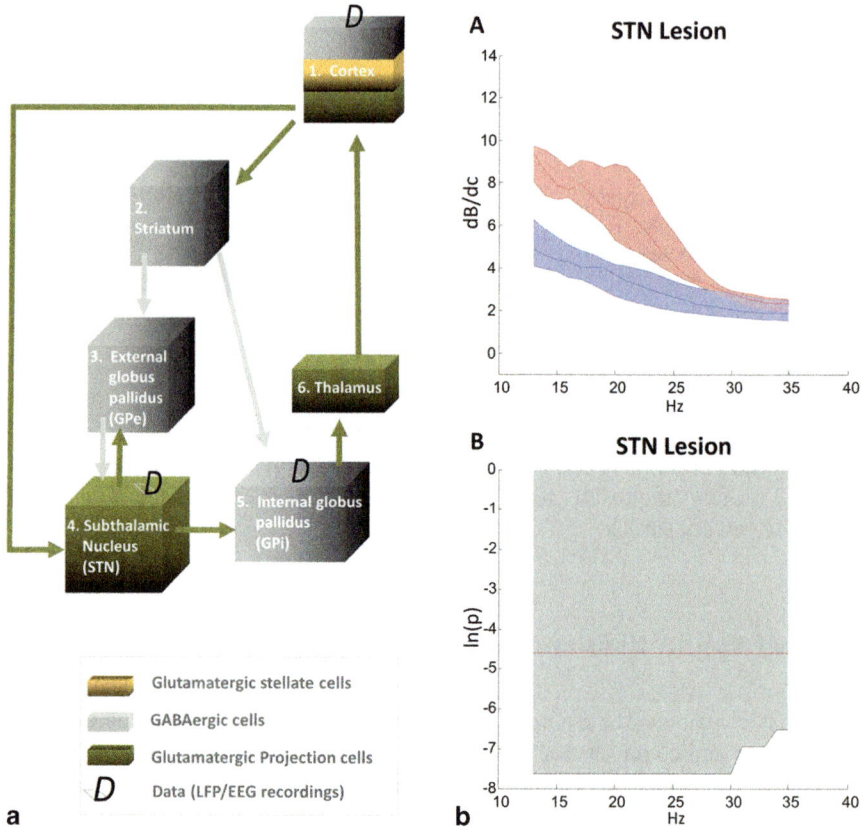

Fig. 3.7 The DCM model structure (**a**). The DCM comprised the principal nodes and connections in the human motor cortico-basal ganglia-thalamocortical loop: the nodes included motor cortex, modelled by a three layer cell ensemble comprising input, excitatory spiny stellate cells, projection (pyramidal) glutamatergic cells and inhibitory GABAergic interneurons. Excitatory projections from cortex innervated the Striatum, and STN (the hyperdirect pathway). The striatum comprised an inhibitory cell mass that projected to two other inhibitory cell masses, GPe (as part of the indirect pathway), and GPi (*via* the direct pathway). The GPe and STN expressed reciprocal connections, and signals from the hyperdirect and indirect pathways were conveyed via excitatory STN projections to the GPi. The thalamus, which excited the cortex, was itself inhibited by connections from GPi. Data, *D*, used for the model inversion were acquired from recordings in cortex, STN and GPi. Lesion Analysis (**b**). (*A*) Effect of lesioning all connections to and from STN on net beta activity in the Parkinsonian network (*red*—without lesion: *blue*—with lesion) the shaded areas correspond to lower and upper quartiles of the contribution results from the 12 DCM. (*B*) Wilcoxon signed-rank test of the effects on lesioning: beta activity was profoundly and significantly suppressed. Horizontal red line denotes $p<0.01$. Results were identical if the population of STN neurons was removed from the model

its key parameters can be recovered in terms of conditional probability densities. We first used synthetic datasets that included noisy data and tested for face validity and indentifiability. We ensured the inversion scheme was able to recover veridical estimates and that model comparison using the log-evidence was able to identify the correct model. We also established physiological validity of the model, using empirical LFP data. For more details, we refer the interested reader to ([4, 136]).

In the application of DCM to Parkinsonian circuits considered here, it is not possible to sample more than a few sites in patients. This calls for an assessment of robustness of the spectral DCM in face of 'hidden' neuronal areas. To address this, we generated synthetic data from a model where the underlying network of regions was known, and used this dataset to evaluate the model evidence of a family of DCM models that differed in architecture and, more significantly, in hidden neuronal area number. We also evaluated the precision of the connectivity parameter estimates for each model by permuting the amount of available data in each network. These results showed that DCM was able to identify the correct model and recover the true parameter values reliably in settings with different levels of observation noise.

The model architecture is depicted in Fig. 3.7a [111] and is similar to that used in a previous study in the 6-hydroxy-dopamine (6OHDA) midbrain lesioned rodent model of Parkinsonism [110]. Three (layered) populations were used to model the cortical source [4], and a single population of neurons, either glutamatergic (excitatory) or GABAergic (inhibitory) was used for BG nuclei (Fig. 3.7a). The major glutamatergic and GABAergic connections between six key components of the cortico–basal ganglia–thalamocortical circuit were thus incorporated into our standard model architecture [137]. In particular, it included the two elements that are thought to be crucial for the expression of exaggerated beta oscillations in Parkinsonism; the hyperdirect pathway [138, 139] and the reciprocal STN–GPe network [140, 141]. The effects of the experimental conditions, ON and OFF levodopa, were explained by the same model through changes in the extrinsic connections in the network. The DCM was focused on the frequency window from 13 to 35 Hz as this is the frequency band that has been most clearly implicated in Parkinsonism in both correlative [142] and intervention [138, 143, 144] studies.

We found significant and relatively consistent differences in effective connectivity between the treated and untreated Parkinsonian states. First, the effective connection strength of the cortical 'hyperdirect' input to the STN was increased in the OFF state, consistent with other findings [139, 145, 146]. Second, the GPe–STN connection was strengthened in the OFF state, consistent with the key role of overactivity in the indirect pathway in PD [147, 148]. Contribution analysis identified connections whose beta promoting potency is much greater when embedded in the 'OFF' network. Here, the tendency to promote beta activity was quantified by the derivative $d\beta/dc$ in response to small changes in connection strength. This approach revealed that all three connections that are increased in the OFF state are also promoting beta synchrony, as did the connection from STN to GPe in the OFF state circuit. This suggests that the strengthening of the three connections is primarily pathological and not compensatory. In addition, the reciprocal connections between GPe to STN emerge as having a particularly important role in promoting beta synchrony in the 'OFF' compared to the 'ON' state circuit. Finally, the face validity of the model was tested by mimicking the effects of a STN lesion (Fig. 3.7b), and confirmed our expectation that this would attenuate beta activity across the basal ganglia–thalamocortical system [149]. This provided a way of identifying pathology and potential therapeutic targets at the synaptic level. We confirmed the prediction that this would attenuate beta activity across the basal ganglia–thalamocortical system [149]. Lesioning, micro-lesioning or muscimol inactivation of this nucleus reduces beta oscillations and improves Parkinsonism in primates and humans [149, 150]. Crucially, the model was successful in making valid predictions regarding the consequence of lesions of the subthalamic nucleus and its connections.

Conclusions

This chapter has reviewed some recent results in dynamic causal modelling (DCM) of electrophysiology, in particular with respect to conductance based models and clinical applications. DCM allows one to address observed responses of complex neuronal systems by looking at the neuronal interactions that generate them and how these responses reflect the underlying neurobiology. We have considered conductance based neural mass, mean-field and field models and reviewed their latest technical developments. We used dynamically rich conductance based models to generate responses in laminar-specific populations of excitatory and inhibitory cells. These models allow for the evaluation of neuronal connections and high-order statistics of neuronal states, using Bayesian estimation and inference. We also discussed the latest clinical applications of DCM with convolution based models, in particular for the study of Parkinson's disease. We presented an analysis involving Parkinsonian patients, and modelled the large-scale network changes underlying the pathological excess of beta oscillations that characterise the Parkinsonian state.

Acknowledgments ACM is funded by the Wellcome Trust and the Max-Planck Society, Tübingen. KJF and DAP are funded by the Wellcome Trust. PB is funded by the Medical Research Council UK, Wellcome Trust, Rosetrees Trust and the National Institute for Health Research Oxford Bio-medical Research Centre.

References

1. Marreiros AC, Stephan KE, Friston KJ. Dynamic causal modeling. Scholarpedia. 2010;5(7):9568.
2. Friston KJ, Harrison L, Penny W. Dynamic causal modelling. Neuroimage. 2003;19:1273–302.
3. David O, Kiebel SJ, Harrison LM, Mattout J, Kilner JM, Friston KJ. Dynamic causal modeling of evoked responses in EEG and MEG. Neuroimage. 2006;30(4):1255–72. Epub 9 Feb 2006.
4. Moran RJ, Stephan KE, Seidenbecher T, Pape HC, Dolan RJ, Friston KJ. Dynamic causal models of steady-state responses. Neuroimage. 2009;44(3):796–811. doi: 10.1016/j.neuroimage.2008.09.048. Epub 17 Oct 2008.
5. Pinotsis DA, Moran RJ, Friston KJ. Dynamic causal modeling with neural fields. Neuroimage. 2012 Jan 16;59(2):1261–74. http://www.pubmedcentral.nih.gov/articlerender.fcgi?artid=3236998&tool=pmcentrez&rendertype=abstract. Accessed 9 Jan 2014. (Elsevier Inc.).

6. Noppeney U, Josephs O, Hocking J, Price CJ, Friston KJ. The effect of prior visual information on recognition of speech and sounds. Cereb Cortex. 2008;18(3):598–609. http://www.ncbi.nlm.nih.gov/pubmed/17617658. Accessed 9 Jan 2014.
7. Schofield TM, Iverson P, Kiebel SJ, Stephan KE, Kilner JM, Friston KJ, et al. Changing meaning causes coupling changes within higher levels of the cortical hierarchy. Proc Natl Acad Sci U S A. 2009;106:11765–70.
8. Hartwigsen G, Saur D, Price CJ, Ulmer S, Baumgaertner A. Perturbation of the left inferior frontal gyrus triggers adaptive plasticity in the right homologous area during speech production. Proc Natl Acad Sci U S A. 2013;110(41):16402–7.
9. Parkinson AL, Korzyukov O, Larson CR, Litvak V, Robin DA. Modulation of effective connectivity during vocalization with perturbed auditory feedback. Neuropsychologia. 2013;51(8):1471–80. http://www.ncbi.nlm.nih.gov/pubmed/23665378. Accessed 9 Jan 2014.
10. Woodhead ZVJ, Barnes GR, Penny W, Moran R, Teki S, Price CJ, et al. Reading front to back: MEG evidence for early feedback effects during word recognition. Cereb Cortex. 2014;24(3):817–25. doi: 10.1093/cercor/bhs365. Epub 21 Nov 2012.
11. Garrido MI, Friston KJ, Kiebel SJ, Stephan KE, Baldeweg T, Kilner JM. The functional anatomy of the MMN: a DCM study of the roving paradigm. Neuroimage. 2008;42(2):936–44. http://www.pubmedcentral.nih.gov/articlerender.fcgi?artid=2640481&tool=pmcentrez&rendertype=abstract. Accessed 9 Jan 2014.
12. Kiebel SJ, Garrido MI, Moran R, Chen C-C, Friston KJ. Dynamic causal modeling for EEG and MEG. Hum Brain Mapp. 2009;30(6):1866–76. http://www.ncbi.nlm.nih.gov/pubmed/19360734. Accessed 9 Jan 2014.
13. Auksztulewicz R, Spitzer B, Blankenburg F. Recurrent neural processing and somatosensory awareness. J Neurosci. 2012;32(3):799–805. http://www.ncbi.nlm.nih.gov/pubmed/22262878. Accessed 19 Jan 2014.
14. Auksztulewicz R, Blankenburg F. Subjective rating of weak tactile stimuli is parametrically encoded in event-related potentials. J Neurosci. 2013;33(29):11878–87. http://www.ncbi.nlm.nih.gov/pubmed/23864677. Accessed 19 Jan 2014.
15. Shigihara Y, Zeki S. Parallelism in the brain's visual form system. Eur J Neurosci. 2013;38(12):3712–20. http://www.ncbi.nlm.nih.gov/pubmed/24118503. Accessed 19 Jan 2014.
16. Vossel S, Weidner R, Driver J, Friston KJ, Fink GR. Deconstructing the architecture of dorsal and ventral attention systems with dynamic causal modeling. J Neurosci. 2012;32(31):10637–48. http://www.pubmedcentral.nih.gov/articlerender.fcgi?artid=3432566&tool=pmcentrez&rendertype=abstract. Accessed 17 Jan 2014.
17. Cárdenas-Morales L, Volz LJ, Michely J, Rehme AK, Pool E-M, Nettekoven C, et al. Network connectivity and individual responses to brain stimulation in the human motor system. Cereb Cortex. 2013;24(7):1697–707. doi: 10.1093/cercor/bht023. Epub 8 Feb 2013.
18. Van Wijk BCM, Litvak V, Friston KJ, Daffertshofer A. Nonlinear coupling between occipital and motor cortex during motor imagery: a dynamic causal modeling study. Neuroimage. 2013;71:104–13. http://www.ncbi.nlm.nih.gov/pubmed/23313570. Accessed 19 Jan 2014. (Elsevier Inc.)
19. Penny WD, Litvak V, Fuentemilla L, Duzel E, Friston K. Dynamic causal models for phase coupling. J Neurosci Methods. 2009;183(1):19–30. http://www.pubmedcentral.nih.gov/articlerender.fcgi?artid=2751835&tool=pmcentrez&rendertype=abstract. Accessed 16 Jan 2014.
20. Moran RJ, Symmonds M, Stephan KE, Friston KJ, Dolan RJ. An in vivo assay of synaptic function mediating human cognition. Curr Biol. 2011;21:1320–5.
21. Penny WD, Stephan KE, Mechelli A, Friston KJ. Comparing dynamic causal models. Neuroimage. 2004;22(3):1157–72.
22. Stephan KE, Penny WD, Daunizeau J, Moran RJ, Friston KJ. Bayesian model selection for group studies. Neuroimage. 2009;46:1004–17. doi: 10.1016/j.neuroimage.2009.03.025. Epub 20 Mar 2009
23. Chen CC, Kiebel SJ, Friston KJ. Dynamic causal modelling of induced responses. Neuroimage. 2008;41(4):1293–312. http://www.ncbi.nlm.nih.gov/pubmed/18485744. Accessed 9 Jan 2014.

24. Marreiros AC, Kiebel SJ, Daunizeau J, Harrison LM, Friston KJ. Population dynamics under the laplace assumption. Neuroimage. 2009;44:701–14.
25. Friston KJ, Bastos A, Litvak V, Stephan KE, Fries P, Moran RJ. DCM for complex-valued data: cross-spectra, coherence and phase-delays. Neuroimage. 2012;59(1):439–55. http://www.pubmedcentral.nih.gov/articlerender.fcgi?artid=3200431&tool=pmcentrez&rendertype=abstract. Accessed 9 Jan 2014. (Elsevier Inc.)
26. Pinotsis DA, Schwarzkopf DS, Litvak V, Rees G, Barnes G, Friston KJ. Dynamic causal modelling of lateral interactions in the visual cortex. Neuroimage. 2012;66C:563–76. http://www.pubmedcentral.nih.gov/articlerender.fcgi?artid=3547173&tool=pmcentrez&rendertype=abstract. Accessed 9 Jan 2014.
27. Pinotsis DA, Brunet N, Bastos A, Bosman CA, Litvak V, Fries P, et al. Contrast gain-control and horizontal interactions in V1: A DCM study. Neuroimage. 2014;92:143–55. doi: 10.1016/j.neuroimage.2014.01.047. Epub 2 Feb 2014.
28. Daunizeau J, David O, Stephan KE. Dynamic causal modelling: a critical review of the biophysical and statistical foundations. Neuroimage. 2011;58(2):312–22. doi: 10.1016/j.neuroimage.2009.11.062. Epub Dec 1 2009.
29. Moran R, Pinotsis DA, Friston K. Neural masses and fields in dynamic causal modeling. Front Comput Neurosci. 2013;7(May):57. http://www.pubmedcentral.nih.gov/articlerender.fcgi?artid=3664834&tool=pmcentrez&rendertype=abstract. Accessed 14 Jan 2014.
30. Pinotsis DA, Friston KJ. Neural fields, masses and bayesian modelling. In: Coombes S, Graben PB, Potthast R, Wright JJ, editors. Neural Field Theory. Mathematical Neuroscience Series. Springer; 2014. doi: 10.13140/2.1.2100.3209.
31. Pinotsis DA, Friston KJ. Extracting novel information from neuroimaging data using neural fields. EPJ Nonlin Biomed Phys. 2014;2: 5.
32. David O, Friston KJ. A neural-mass model for MEG/EEG: coupling and neuronal dynamics. NeuroImage. 2003;20(3):1743–55.
33. David O, Cosmelli D, Friston KJ. Evaluation of different measures of functional connectivity using a neural-mass model. NeuroImage. 2004;21(2):659–73.
34. Lee L., Friston K., Horwitz B. Large-scale neural models and dynamic causal modelling. NeuroImage. 2006;30:1243–1254.
35. Garrido MI, Friston KJ, Kiebel SJ, Stephan KE, Baldeweg T, Kilner JM. The functional anatomy of the MMN: a DCM study of the roving paradigm. NeuroImage. 2008;42(2):936–44.
36. Regan, D., Spekreijse H. Electrophysiological correlate of binocular depth perception in man. Nature. 1970;225(5227):92–4. http://dx.doi.org/10.1038/225092a0. Accessed 30 March 2014.
37. Kutas M, Hillyard SA. Electrophysiology of cognitive processing. Annu Rev Psychol. 1983;34:33–61. http://www.annualreviews.org/doi/pdf/10.1146/annurev.ps.34.020183.000341. Accessed 30 March 2014.
38. Baillet S, Mosher JC, Leahy RM. Electromagnetic brain mapping. IEEE Signal Process Mag. 2001;18(6):14–30. http://ieeexplore.ieee.org/lpdocs/epic03/wrapper.htm?arnumber=962275. Accessed 19 March 2014.
39. Friston K. A theory of cortical responses. Philos Trans R Soc Lond B Biol Sci. 2005;360(1456):815–36. http://rstb.royalsocietypublishing.org/content/360/1456/815.short. Accessed 19 March 2014.
40. Friston K. The free-energy principle: a unified brain theory? Nat Rev Neurosci. 2010;11(2):127–38. http://dx.doi.org/10.1038/nrn2787. Accessed 19 March 2014. (Nat Publishing Group).
41. Lopes da Silva FH, Hoeks A, Smits H, Zetterberg LH. Model of brain rhythmic activity. The alpha-rhythm of the thalamus. Kybernetik. 1974;15:27–37.
42. Zetterberg LH, Kristiansson L, Mossberg K. Performance of a model for a local neuron population. Biol Cybern. 1978;31:15–26.
43. Elbert T, Ray WJ, Kowalik ZJ, Skinner JE, Graf KE, Birbaumer N. Chaos and physiology: deterministic chaos in excitable cell assemblies. Physiol Rev. 1994;74:1–47.

44. Jansen BH, Rit VG. Electroencephalogram and visual evoked potential generation in a mathematical model of coupled cortical columns. Biol Cybern. 1995;73:357–66.
45. Kincses WE, Braun C, Kaiser S, Elbert T. Modeling extended sources of event-related potentials using anatomical and physiological constraints. Hum Brain Mapp. 1999;8:182–93.
46. Wendling F, Bellanger JJ, Bartolomei F, Chauvel P. Relevance of nonlinear lumped-parameter models in the analysis of depth-EEG epileptic signals. Biol Cybern. 2000;83:367–78.
47. David O, Harrison L, Friston KJ. Modelling event-related responses in the brain. Neuroimage. 2005;25(3):756–70. http://www.ncbi.nlm.nih.gov/pubmed/15808977. (Elsevier).
48. Valdes PA, Jimenez JC, Riera J, Biscay R, Ozaki T. Nonlinear EEG analysis based on a neural mass model. Biol Cybern. 1999;81(5–6):415–24. http://www.ncbi.nlm.nih.gov/pubmed/10592017. Accessed 30 March 2014.
49. Jansen BH, Kavaipatti AB, Markusson O. Evoked potential enhancement using a neurophysiologically-based model. Methods Inf Med. 2001;40(4):338–45. http://www.ncbi.nlm.nih.gov/pubmed/11552347. Accessed 30 March 2014.
50. Brown HR, Friston KJ. The functional anatomy of attention: a DCM study. Front Hum Neurosci. 2013;7(December):784. http://www.pubmedcentral.nih.gov/articlerender.fcgi?artid=3845206&tool=pmcentrez&rendertype=abstract. Accessed 9 Jan 2014.
51. Felleman DJ, Van Essen DC. Distributed hierarchical processing in the primate cerebral cortex. Cereb Cortex. 1991;1:1–47.
52. Kiebel S, David O, Friston K. Dynamic causal modelling of evoked responses in EEG/MEG with lead field parameterization. Neuroimage. 2006;30(4):1273–84. Epub 21 Feb 2006.
53. Friston KJ. Bayesian estimation of dynamical systems: an application to fMRI. Neuroimage. 2002;16:513–30.
54. Izhikevich EM. Which model to use for cortical spiking neurons? IEEE Trans Neural Netw. 2004;15:1063–70.
55. De Groff D, Neelakanta PS, Sudhakar R, Aalo V. Stochastical aspects of neuronal dynamics: fokker-planck approach. Biol Cybern. 1993;69:155–64.
56. Nykamp DQ, Tranchina D. A population density approach that facilitates large-scale modeling of neural networks: extension to slow inhibitory synapses. Neural Comput. 2001;13:511–46.
57. Casti ARR, Omurtag A, Sornborger A, Kaplan E, Knight B, Victor J, et al. A population study of integrate-and-fire-or-burst neurons. Neural Comput. 2002;14:957–86.
58. Haskell E, Nykamp DQ, Tranchina D. Population density methods for large-scale modelling of neuronal networks with realistic synaptic kinetics: cutting the dimension down to size. Network. 2001;12:141–74.
59. Knight BW. Dynamics of encoding in neuron populations: some general mathematical features. Neural Comput. 2000;12:473–518.
60. Omurtag A, Knight BW, Sirovich L. On the simulation of large populations of neurons. J Comput Neurosci. 2000;8:51–63.
61. Sirovich L. Dynamics of neuronal populations: eigenfunction theory; some solvable cases. Network. 2003;14:249–72.
62. Chizhov AV, Graham LJ. Population model of hippocampal pyramidal neurons, linking a refractory density approach to conductance-based neurons. Phys Rev E Stat Nonlin Soft Matter Phys. 2007;75:011924.
63. Rodriguez R, Tuckwell HC. Statistical properties of stochastic nonlinear dynamical models of single spiking neurons and neural networks. Phys Rev E Stat Phys Plasmas Fluids Relat Interdiscip Topics. 1996;54(5):5585–90.
64. Rodriguez RC, Tuckwell H. Noisy spiking neurons and networks: useful approximations for firing probabilities and global behavior. Bio Systems. 1998;48:187–94.
65. Hasegawa H. Dynamical mean-field theory of spiking neuron ensembles: response to a single spike with independent noises. Phys Rev E Stat Nonlin Soft Matter Phys. 2003;67:041903.
66. Deco G, Jirsa VK, Robinson PA, Breakspear M, Friston K. The dynamic brain: from spiking neurons to neural masses and cortical fields. PLoS Comput Biol. 2008;4:e1000092.
67. Freeman WJ. Mass action in the nervous system. New York: Academic;1975.

68. Marreiros AC, Daunizeau J, Kiebel SJ, Friston KJ. Population dynamics: variance and the sigmoid activation function. Neuroimage. 2008;42:147–57.
69. Morris C, Lecar H. Voltage oscillations in the barnacle giant muscle fiber. Biophys J. 1981;35:193–213.
70. Harrison LM, David O, Friston KJ. Stochastic models of neuronal dynamics. Philos Trans R Soc Lond B Biol Sci. 2005;360:1075–91.
71. Tuckwell HC, Rodriguez R. Analytical and simulation results for stochastic fitzhugh-nagumo neurons and neural networks. J Comput Neurosci. 1998;5:91–113.
72. Rodriguez R, Tuckwel HC. A dynamical system for the approximate moments of nonlinear stochastic models of spiking neurons and networks. Math Comput Model 2000;31:175–80.
73. Hasegawa H. Dynamical mean-field theory of noisy spiking neuron ensembles: application to the hodgkin-huxley model. Phys Rev E Stat Nonlin Soft Matter Phys. 2003;68:041909.
74. Marreiros AC, Kiebel SJ, Friston KJ. A dynamic causal model study of neuronal population dynamics. Neuroimage. 2010;51:91–101.
75. Stephan KE, Penny WD, Daunizeau J, Moran RJ, Friston KJ. Bayesian model selection for group studies. Neuroimage. 2009;46(4):1004–17. http://www.pubmedcentral.nih.gov/articlerender.fcgi?artid=2703732&tool=pmcentrez&rendertype=abstract. Accessed 21 March 2014. (Elsevier Inc.).
76. Moran RJ, Stephan KE, Dolan RJ, Friston KJ. Consistent spectral predictors for dynamic causal models of steady-state responses. Neuroimage. 2011;55(4):1694–708. http://www.pubmedcentral.nih.gov/articlerender.fcgi?artid=3093618&tool=pmcentrez&rendertype=abstract. Accessed 9 Jan 2014. (Elsevier Inc.).
77. Brunel N, Wang XJ. Effects of neuromodulation in a cortical network model of object working memory dominated by recurrent inhibition. J Comput Neurosci. 2001;11(1):63–85. http://www.ncbi.nlm.nih.gov/pubmed/11524578. Accessed 28 March 2014.
78. Goldman-Rakic PS. Regional and cellular fractionation of working memory. Proc Natl Acad Sci U S A. 1996;93:13473–80.
79. Durstewitz D, Seamans JK, Sejnowski TJ. Neurocomputational models of working memory. Nat Neurosci. 2000;3 Suppl:1184–91.
80. Gorelova NA, Yang CR. Dopamine D1/D5 receptor activation modulates a persistent sodium current in rat prefrontal cortical neurons in vitro. J Neurophysiol. 2000;84:75–87.
81. Gonzalez-Islas C, Hablitz JJ. Dopamine enhances EPSCs in layer II-III pyramidal neurons in rat prefrontal cortex. J Neurosci. 2003;23:867–75.
82. Durstewitz D, Seamans JK. The dual-state theory of prefrontal cortex dopamine function with relevance to catechol-O-methyltransferase genotypes and schizophrenia. Biolo Psychiatry. 2008;64:739–49.
83. Nunez PL. The brain wave equation: a model for the EEG. Math Biosci. 1974;21:279–97.
84. Amari S. Homogeneous nets of neuron-like elements. Biol Cybern. 1975;17(4):211–20.
85. Amari S. Dynamics of pattern formation in lateral-inhibition type neural fields. Biol Cybern. 1977;27(2):77–87.
86. Breakspear M, Terry JR, Friston KJ. Modulation of excitatory synaptic coupling facilitates synchronization and complex dynamics in a biophysical model of neuronal dynamics. Network. 2003;14(4):703–32.
87. Liley DTJ, Bojak I. Understanding the transition to seizure by modeling the epileptiform activity of general anesthetic agents. J Clin Neurophysiol. 2005;22(5):300–13.
88. Pinotsis DA, Leite M, Friston KJ. On conductance-based neural field models. Front Comput Neurosci. 2013;7(November):158. http://www.pubmedcentral.nih.gov/articlerender.fcgi?artid=3824089&tool=pmcentrez&rendertype=abstract. Accessed 14 Jan 2014.
89. Wilson HR, Cowan JD. Excitatory and inhibitory interactions in localized populations of model neurons. Biophys J. 1972;12(1):1–24. http://www.pubmedcentral.nih.gov/articlerender.fcgi?artid=1484078&tool=pmcentrez&rendertype=abstract. Accessed 30 March 2014.
90. Jirsa V, Haken H. Field theory of electromagnetic brain activity. Phys Rev Lett. 1996;77:960–3.

91. Buice MA, Cowan JD, Chow CC. Systematic fluctuation expansion for neural network activity equations. Neural Comput. 2010;22:377–426.
92. Touboul JD, Ermentrout GB. Finite-size and correlation-induced effects in mean-field dynamics. J Comput Neurosci. 2011;31(3):453–84. http://www.ncbi.nlm.nih.gov/pubmed/21384156. Accessed 30 March 2014.
93. Rose RM, Hindmarsh JL. The assembly of ionic currents in a thalamic neuron. I. The three-dimensional model. Proc R Soc Lond B Biol Sci. 1989;237(1288):267–88. http://www.ncbi.nlm.nih.gov/pubmed/2571154. Accessed 30 March 2014.
94. Wilson MT, Robinson PA, O'Neill B, Steyn-Ross DA. Complementarity of spike- and rate-based dynamics of neural systems. PLoS Comput Biol. 2012;8(6):e1002560. http://www.pubmedcentral.nih.gov/articlerender.fcgi?artid=3380910&tool=pmcentrez&rendertype=abstract. Accessed 30 March 2014.
95. Robinson PA, Kim JW. Spike, rate, field, and hybrid methods for treating neuronal dynamics and interactions. J Neurosci Methods. 2012;205(2):283–94. http://www.ncbi.nlm.nih.gov/pubmed/22330795. Accessed 30 March 2014.
96. Liley DTJ, Cadusch PJ, Gray M, Nathan PJ. Drug-induced modification of the system properties associated with spontaneous human electroencephalographic activity. Phys Rev E Stat Nonlin Soft Matter Phys. 2003;68(5 Pt 1):051906. http://www.ncbi.nlm.nih.gov/pubmed/14682819. Accessed 30 March 2014.
97. Bojak I, Liley DTJ. Modeling the effects of anesthesia on the electroencephalogram. Phys Rev E Stat Nonlin Soft Matter Phys. 2005;71(4 Pt 1):041902. http://www.ncbi.nlm.nih.gov/pubmed/15903696. Accessed 30 March 2014.
98. Steyn-Ross ML, Steyn-Ross DA, Sleigh JW, Wilcocks LC. Toward a theory of the general-anesthetic-induced phase transition of the cerebral cortex. I. A thermodynamics analogy. Phys Rev E Stat Nonlin Soft Matter Phys. 2001;64(1 Pt 1):011917. http://www.ncbi.nlm.nih.gov/pubmed/11461298. Accessed 30 March 2014.
99. Steyn-Ross DA, Steyn-Ross ML, Sleigh JW, Wilson MT. Progress in modeling EEG effects of general anesthesia: biphasic response and hysteresis. New York: Springer; 2011.
100. Wilson MT, Sleigh JW, Steyn-Ross DA, Steyn-Ross ML. General anesthetic-induced seizures can be explained by a mean-field model of cortical dynamics. Anesthesiology. 2006;104(3):588–93. http://www.ncbi.nlm.nih.gov/pubmed/16508406. Accessed 30 March 2014.
101. Goldstein SS, Rall W. Changes of action potential shape and velocity for changing core conductor geometry. Biophys J. 1974;14(10):731–57. http://www.pubmedcentral.nih.gov/articlerender.fcgi?artid=1334570&tool=pmcentrez&rendertype=abstract. Accessed 27 March 2014.
102. Ermentrout B. Neural networks as spatio-temporal pattern-forming systems. Rep Prog Phys. 1998;61:353–430. http://iopscience.iop.org/0034-4885/61/4/002/pdf/0034-4885_61_4_002.pdf. Accessed 30 March 2014.
103. Somers DC, Nelson SB, Sur M. An emergent model of orientation selectivity in cat visual cortical simple cells. J Neurosci. 1995;15(8):5448–65. http://www.ncbi.nlm.nih.gov/pubmed/7643194. Accessed 30 March 2014.
104. Hutt A, Longtin A. Effects of the anesthetic agent propofol on neural populations. Cogn Neurodyn. 2010;4(1):37–59. http://www.pubmedcentral.nih.gov/articlerender.fcgi?artid=2837528&tool=pmcentrez&rendertype=abstract. Accessed 30 March 2014.
105. Moran RJ, Stephan KE, Kiebel SJ, Rombach N, O'Connor WT, Murphy KJ, et al. Bayesian estimation of synaptic physiology from the spectral responses of neural masses. Neuroimage. 2008;42:272–84.
106. Traub RD. Fast oscillations in cortical circuits. Cereb Cortex. 2014;24(11):2873–83. doi: 10.1093/cercor/bht140. Epub 2 Jun 2013.
107. Roiser JP, Wigton R, Kilner JM, Mendez MA, Hon N, Friston KJ, et al. Dysconnectivity in the frontoparietal attention network in schizophrenia. Front psychiatry. 2013;4(December):176. http://www.ncbi.nlm.nih.gov/pubmed/24399975. Accessed 14 Jan 2014.

108. Dima D, Dietrich DE, Dillo W, Emrich HM. Impaired top-down processes in schizophrenia: a DCM study of ERPs. Neuroimage. 2010;52(3):824–32. http://www.ncbi.nlm.nih.gov/pubmed/20056155. Accessed 9 Jan 2014. (Elsevier Inc).
109. Dima D, Frangou S, Burge L, Braeutigam S, James AC. Abnormal intrinsic and extrinsic connectivity within the magnetic mismatch negativity brain network in schizophrenia: a preliminary study. Schizophr Res. 2012;135(1–3):23–7. http://www.ncbi.nlm.nih.gov/pubmed/22264684. Accessed 16 Jan 2014. (Elsevier B.V.).
110. Moran RJ, Mallet N, Litvak V, Dolan RJ, Magill PJ, Friston KJ, et al. Alterations in brain connectivity underlying Beta oscillations in parkinsonism. PLoS Comput Biol. 2011;7(8):e1002124. doi: 10.1371/journal.pcbi.1002124. Epub 11 Aug 2011.
111. Marreiros AC, Cagnan H, Moran RJ, Friston KJ, Brown P. Basal ganglia-cortical interactions in parkinsonian patients. Neuroimage. 2012;66C:301–10. http://www.pubmedcentral.nih.gov/articlerender.fcgi?artid=3573233&tool=pmcentrez&rendertype=abstract. Accessed 16 Jan 2014.
112. Herz DM, Florin E, Christensen MS, Reck C, Barbe MT, Tscheuschler MK, et al. Dopamine replacement modulates oscillatory coupling between premotor and motor cortical areas in parkinson's disease. Cereb Cortex. 2014;24(11):2873–83. doi: 10.1093/cercor/bht140. Epub 2 Jun 2013.
113. Herz DM, Siebner HR, Hulme OJ, Florin E, Christensen MS, Timmermann L. Levodopa reinstates connectivity from prefrontal to premotor cortex during externally paced movement in Parkinson's disease. Neuroimage. 2014;90:15–23. doi: 10.1016/j.neuroimage.2013.11.023. Epub 22 Nov 2013.
114. Herz DM, Christensen MS, Reck C, Florin E, Barbe MT, Stahlhut C, et al. Task-specific modulation of effective connectivity during two simple unimanual motor tasks: a 122-channel EEG study. Neuroimage. 2012;59(4):3187–93. http://www.ncbi.nlm.nih.gov/pubmed/22146753. Accessed 19 Jan 2014. (Elsevier Inc).
115. Boly M, Garrido MI, Gosseries O, Bruno M-A, Boveroux P, Schnakers C, et al. Preserved feedforward but impaired top-down processes in the vegetative state. Science. 2011;332:858–62.
116. Boly M, Moran R, Murphy M, Boveroux P, Bruno M-A, Noirhomme Q, et al. Connectivity changes underlying spectral EEG changes during propofol-induced loss of consciousness. J Neurosci. 2012;32(20):7082–90. http://www.pubmedcentral.nih.gov/articlerender.fcgi?artid=3366913&tool=pmcentrez&rendertype=abstract. Accessed 16 Jan 2014.
117. Moran RJ, Jung F, Kumagai T, Endepols H, Graf R, Dolan RJ, et al. Dynamic causal models and physiological inference: a validation study using isoflurane anaesthesia in rodents. PLoS One. 2011;6(8):e22790. http://www.pubmedcentral.nih.gov/articlerender.fcgi?artid=3149050&tool=pmcentrez&rendertype=abstract. Accessed 19 Jan 2014.
118. Muthukumaraswamy SD, Carhart-Harris RL, Moran RJ, Brookes MJ, Williams TM, Errtizoe D, et al. Broadband cortical desynchronization underlies the human psychedelic state. J Neurosci. 2013;33(38):15171–83. http://www.ncbi.nlm.nih.gov/pubmed/24048847. Accessed 14 Jan 2014.
119. Schmidt A, Diaconescu AO, Kometer M, Friston KJ, Stephan KE, Vollenweider FX. Modeling ketamine effects on synaptic plasticity during the mismatch negativity. Cereb Cortex. 2013;23(10):2394–406. http://www.ncbi.nlm.nih.gov/pubmed/22875863. Accessed 19 Jan 2014.
120. David O, Guillemain I, Saillet S, Reyt S, Deransart C, Segebarth C, et al. Identifying neural drivers with functional MRI: an electrophysiological validation. PLoS Biol. 2008;6(12):2683–97. http://www.pubmedcentral.nih.gov/articlerender.fcgi?artid=2605917&tool=pmcentrez&rendertype=abstract. Accessed 11 Jan 2014.
121. Hamandi K, Powell HWR, Laufs H, Symms MR, Barker GJ, Parker GJM, et al. Combined EEG-fMRI and tractography to visualise propagation of epileptic activity. J Neurol Neuro-

surg Psychiatry. 2008;79(5):594–7. http://www.pubmedcentral.nih.gov/articlerender.fcgi?artid=2571962&tool=pmcentrez&rendertype=abstract. Accessed 10 Jan 2014.
122. Vaudano AE, Laufs H, Kiebel SJ, Carmichael DW, Hamandi K, Guye M, et al. Causal hierarchy within the thalamo-cortical network in spike and wave discharges. PLoS One. 2009;4(8):e6475. http://www.pubmedcentral.nih.gov/articlerender.fcgi?artid=2715100&tool=pmcentrez&rendertype=abstract. Accessed 16 Jan 2014.
123. Airaksinen AM, Niskanen J-P, Chamberlain R, Huttunen JK, Nissinen J, Garwood M, et al. Simultaneous fMRI and local field potential measurements during epileptic seizures in medetomidine-sedated rats using raser pulse sequence. Magn Reson Med. 2010;64(4):1191–9. http://www.pubmedcentral.nih.gov/articlerender.fcgi?artid=2946452&tool=pmcentrez&rendertype=abstract. Accessed 14 Jan 2014.
124. Airaksinen AM, Hekmatyar SK, Jerome N, Niskanen J-P, Huttunen JK, Pitkänen A, et al. Simultaneous BOLD fMRI and local field potential measurements during kainic acid-induced seizures. Epilepsia. 2012;53(7):1245–53. http://www.ncbi.nlm.nih.gov/pubmed/22690801. Accessed 14 Jan 2014.
125. Murta T, Leal A, Garrido MI, Figueiredo P. Dynamic causal modelling of epileptic seizure propagation pathways: a combined EEG-fMRI study. Neuroimage. 2012;62(3):1634–42. http://www.pubmedcentral.nih.gov/articlerender.fcgi?artid=3778869&tool=pmcentrez&rendertype=abstract. Accessed 16 Jan 2014. (Elsevier Inc.).
126. Vaudano AE, Carmichael DW, Salek-Haddadi A, Rampp S, Stefan H, Lemieux L, et al. Networks involved in seizure initiation. A reading epilepsy case studied with EEG-fMRI and MEG. Neurology. 2012;79(3):249–53. http://www.pubmedcentral.nih.gov/articlerender.fcgi?artid=3398433&tool=pmcentrez&rendertype=abstract.
127. Vaudano AE, Avanzini P, Tassi L, Ruggieri A, Cantalupo G, Benuzzi F, et al. Causality within the epileptic network: an EEG-fMRI study validated by intracranial EEG. Front Neurol. 2013;4(November):185. http://www.pubmedcentral.nih.gov/articlerender.fcgi?artid=3827676&tool=pmcentrez&rendertype=abstract. Accessed 17 Jan 2014.
128. Campo P, Garrido MI, Moran RJ, García-Morales I, Poch C, Toledano R, et al. Network reconfiguration and working memory impairment in mesial temporal lobe epilepsy. Neuroimage. 2013;72:48–54. http://www.pubmedcentral.nih.gov/articlerender.fcgi?artid=3610031&tool=pmcentrez&rendertype=abstract. Accessed 9 Jan 2014. (Elsevier Inc.).
129. Albouy P, Mattout J, Bouet R, Maby E, Sanchez G, Aguera P-E, et al. Impaired pitch perception and memory in congenital amusia: the deficit starts in the auditory cortex. Brain. 2013;136(Pt 5):1639–61. http://www.ncbi.nlm.nih.gov/pubmed/23616587. Accessed 15 Jan 2014.
130. Babajani-Feremi A, Gumenyuk V, Roth T, Drake CL, Soltanian-Zadeh H. Connectivity analysis of novelty process in habitual short sleepers. Neuroimage. 2012;63(3):1001–10. http://www.ncbi.nlm.nih.gov/pubmed/22906789. Accessed 13 Jan 2014. (Elsevier Inc.).
131. Campo P, Poch C, Toledano R, Igoa JM, Belinchón M, García-Morales I, et al. Anterobasal temporal lobe lesions alter recurrent functional connectivity within the ventral pathway during naming. J Neurosci. 2013;33(31):12679–88. http://www.ncbi.nlm.nih.gov/pubmed/23904604. Accessed 19 Jan 2014.
132. Hughes LE, Ghosh BCP, Rowe JB. Reorganisation of brain networks in frontotemporal dementia and progressive supranuclear palsy. NeuroImage Clin. 2013;2:459–68. http://www.pubmedcentral.nih.gov/articlerender.fcgi?artid=3708296&tool=pmcentrez&rendertype=abstract. Accessed 14 Jan 2014. (The Authors).
133. Kumar S, Sedley W, Nourski KV, Kawasaki H, Oya H, Patterson RD, et al. Predictive coding and pitch processing in the auditory cortex. J Cogn Neurosci. 2011;23(10):3084–94. http://www.pubmedcentral.nih.gov/articlerender.fcgi?artid=3821983&tool=pmcentrez&rendertype=abstract.
134. Silchenko AN, Adamchic I, Hauptmann C, Tass PA. Impact of acoustic coordinated reset neuromodulation on effective connectivity in a neural network of phantom sound. Neuro-

image. 2013;77:133–47. http://www.ncbi.nlm.nih.gov/pubmed/23528923. Accessed 14 Jan 2014. (Elsevier Inc.).
135. Teki S, Barnes GR, Penny WD, Iverson P, Woodhead ZVJ, Griffiths TD, et al. The right hemisphere supports but does not replace left hemisphere auditory function in patients with persisting aphasia. Brain. 2013;136(Pt 6):1901–12. http://www.ncbi.nlm.nih.gov/pubmed/23715097. Accessed 15 Jan 2014.
136. Pinotsis DA, Moran RJ, Friston KJ. Dynamic causal modeling with neural fields. Neuroimage. 2012 Jan 16;59(2):1261–74.
137. Smith Y, Bevan MD, Shink E, Bolam JP. Microcircuitry of the direct and indirect pathways of the basal ganglia. Neuroscience. 1998;86(2):353–87. http://www.ncbi.nlm.nih.gov/pubmed/9881853. Accessed 30 March 2014.
138. Gradinaru V, Mogri M, Thompson KR, Henderson JM, Deisseroth K. Optical deconstruction of parkinsonian neural circuitry. Science. 2009;324(5925):354–9. http://www.ncbi.nlm.nih.gov/pubmed/19299587. Accessed 21 March 2014.
139. Magill PJ, Bolam JP, Bevan MD. Dopamine regulates the impact of the cerebral cortex on the subthalamic nucleus-globus pallidus network. Neuroscience. 2001;106(2):313–30. http://www.ncbi.nlm.nih.gov/pubmed/11566503. Accessed 30 March 2014.
140. Cruz A V, Mallet N, Magill PJ, Brown P, Averbeck BB. Effects of dopamine depletion on information flow between the subthalamic nucleus and external globus pallidus. J Neurophysiol. 2011;106(4):2012–23. http://www.pubmedcentral.nih.gov/articlerender.fcgi?artid=3191831&tool=pmcentrez&rendertype=abstract. Accessed 19 March 2014.
141. Holgado AJN, Terry JR, Bogacz R. Conditions for the generation of beta oscillations in the subthalamic nucleus-globus pallidus network. J Neurosci. 2010;30(37):12340–52. http://www.ncbi.nlm.nih.gov/pubmed/20844130. Accessed 30 March 2014.
142. Jenkinson N, Brown P. New insights into the relationship between dopamine, beta oscillations and motor function. Trends Neurosci. 2011;34(12):611–8. http://www.ncbi.nlm.nih.gov/pubmed/22018805. Accessed 22 March 2014.
143. Chen CC, Lin WY, Chan HL, Hsu YT, Tu PH, Lee ST, et al. Stimulation of the subthalamic region at 20 Hz slows the development of grip force in Parkinson's disease. Exp Neurol. 2011;231(1):91–6. http://www.ncbi.nlm.nih.gov/pubmed/21683700. Accessed 30 March 2014.
144. Eusebio A, Chen CC, Lu CS, Lee ST, Tsai CH, Limousin P, et al. Effects of low-frequency stimulation of the subthalamic nucleus on movement in Parkinson's disease. Exp Neurol. 2008;209(1):125–30. Epub 18 Sep 2007.
145. Baudrexel S, Witte T, Seifried C, von Wegner F, Beissner F, Klein JC, et al. Resting state fMRI reveals increased subthalamic nucleus-motor cortex connectivity in parkinson's disease. Neuroimage. 2011;55(4):1728–38. http://www.ncbi.nlm.nih.gov/pubmed/21255661. Accessed 30 March 2014.
146. Dejean C, Gross CE, Bioulac B, Boraud T. Dynamic changes in the cortex-basal ganglia network after dopamine depletion in the rat. J Neurophysiol. 2008;100(1):385–96. http://www.ncbi.nlm.nih.gov/pubmed/18497362. Accessed 26 March 2014.
147. Bergman H, Wichmann T, DeLong MR. Reversal of experimental parkinsonism by lesions of the subthalamic nucleus. Science. 1990;249(4975):1436–8.
148. Kravitz AV, Freeze BS, Parker PRL, Kay K, Thwin MT, Deisseroth K, et al. Regulation of parkinsonian motor behaviours by optogenetic control of basal ganglia circuitry. Nature. 2010;466(7306):622–6. doi: 10.1038/nature09159. Epub 7 Jul 2010.
149. Chen CC, Pogosyan A, Zrinzo LU, Tisch S, Limousin P, Ashkan K, et al. Intra-operative recordings of local field potentials can help localize the subthalamic nucleus in Parkinson's disease surgery. Exp Neurol. 2006;198(1):214–21. http://www.ncbi.nlm.nih.gov/pubmed/16403500. Accessed 30 March 2014.
150. Tachibana Y, Iwamuro H, Kita H, Takada M, Nambu A. Subthalamo-pallidal interactions underlying parkinsonian neuronal oscillations in the primate basal ganglia. Eur J Neurosci. 2011;34(9):1470–84. http://www.ncbi.nlm.nih.gov/pubmed/22034978. Accessed 19 March 2014.

Chapter 4
Computational Models of Closed–Loop Deep Brain Stimulation

Yixin Guo and Kelly Toppin

Abstract Deep Brain Stimulation (DBS) is a neurosurgical intervention that sends electrical signals to the brain to effectively alleviate the symptoms of neurological disorders such as Parkinson's disease. Although the conventional high frequency DBS shows remarkable therapeutic success, it is desirable to overcome the downsides of such a form of open–loop DBS. Using computational models, we explore a closed–loop DBS paradigm, multi–site delayed feedback stimulation (MDFS), that may potentially overcome the drawbacks of constant high frequency DBS. We first develop a biological-faithful computational network model of basal ganglia and thalamus in parkinsonian conditions. The model mimics the pathological neuronal activity observed in the basal ganglia in parkinsonian conditions, such as increased firing rate, bursting patterns, and synchronization. We then evaluate the outcome of closed–loop MDFS being applied to the parkinsonian network by examining both quantitative measures of neurons in the basal ganglia and the relay error of thalamocortical (TC) neurons. Our computational results show that closed–loop MDFS significantly diminish TC relay error by breaking the bursting pattern and desynchronizing the synchronized clusters in the basal ganglia. The design of MDFS suggests that it is superior to open–loop stimulation in that not only the stimulation signal is guided by changes in neuronal activities specific to disorders being treated, but also MDFS shows much lower energy consumption compared with the conventional high frequency DBS. To support the computational results and feasibility of closed–loop DBS, we further review some previous work that validates the evaluation measure of TC relay error and some recent experimental studies that validate the on–demand type of DBS.

Keywords Stimulation · Model · Parkinsonian network · Patterns · Basal ganglia · Neuron

---

Y. Guo (✉) · K. Toppin
Department of Mathematics, Drexel University, Philadelphia, PA 19104, USA
e-mail: yixin@math.drexel.edu

Introduction

Deep Brain Stimulation (DBS)

Deep Brain Stimulation (DBS) is a neurosurgical procedure involving implantation of a stimulator that sends electrical impulses to nuclei in the brain. DBS can effectively alleviate the symptoms of neurological movement disorders such as Parkinson's disease (PD). As these diseases progress, neurologists often need to hand–tune the standard DBS parameters, such as the frequency, the duration, and the amplitude of a pulse train, for a given patient to sustain the treatment effect. Although DBS has shown remarkable therapeutic success, neither the mechanisms for the effectiveness of DBS nor the possible improvement on drawbacks of the conventional DBS are fully addressed. In the following subsections, we will briefly introduce the unsolved issues of the conventional DBS, the pathology of PD, the current biophysical computational models in studying the underlying mechanism of DBS, and the existing studies in finding better alternatives to the conventional DBS protocol using such models.

PD Pathology

Basal ganglia dysfunction has been implied in Parkinson's disease. Both experimental model parkinsonism of non–human primate and PD patients are shown to be associated with changes in neural activity patterns, including increases in synchrony, firing rates, and bursts in the subthalamic nucleus (STN) and the internal segment of the globus pallidus (GPi) [4, 8, 9, 38, 43, 47, 54, 61, 82, 84].

In the basal ganglia circuit, the GPi is the major output structure of the basal ganglia motor territory to the thalamus [1]. The mean firing rate, the firing pattern such as oscillations or bursts, and the synchrony of GPi population are proven to be biologically important neuronal features to the downstream information processing of thalamocortical relay [4, 8, 9, 38, 43, 47, 54, 61, 73, 82, 84]. First, using single–unit recording, researchers have shown that the mean GPi firing rate (95.2 ± 2.3 Hz) of PD is significantly higher [73]. Second, the symptom severity may be more relevant to the neuronal firing patterns than to the firing rates of GPi [4, 8, 9, 15, 38, 43, 47, 54, 56, 61, 66, 76, 82, 84]. In particular, numerous experimental studies have demonstrated that neurons within both STN and GPi show an increased level of bursting activity during parkinsonian state [4, 9, 47, 54, 61, 73]. Third, measuring synchronized activity across many neuronal elements of GPi using local field potential (LFP) recording [71] showed that the untreated PD is associated with an increased tendency to synchronization.

Since motor outputs from GPi target the anterior ventrolateral nucleus of the thalamus (VLa) [1, 17, 40, 50, 85], which serves to relay signals between cortical areas [22, 23, 24, 28], it has been hypothesized that pathological GPi outputs may induce parkinsonian signs by changing thalamic activity patterns [16, 30, 31, 49, 72], in particular, by compromising thalamocortical (TC) relay [11, 25, 26, 58, 67].

Drawback of Conventional DBS

The apparatus of the conventional DBS consists of (1) an electrode that is approximately 1.25 mm in diameter, (2) a stimulator including battery, and (3) an extension cord connecting the two. The conventional DBS delivers an ongoing stream of high frequency current pulses to the stimulation target. The electrode is placed into the stimulation target such as STN and GPi. In particular, both STN or GPi stimulation sites are commonly practiced for treating PD [33, 81, 82]. Deep brain stimulation has significant advantages compared with ablative surgery pallidotomy (destroy part of GPi) or thalamotomy (destroy part of thalamus). Considering DBS does not require making a destructive lesion in the brain, DBS is, in principle, reversible and does not preclude the use of possible future therapies in PD requiring integrity of the basal ganglia circuitry.

Although DBS has achieved remarkable success, it is desirable to improve DBS when considering some of its significant **drawbacks**:

- "Dumb" stimulation [3]: The conventional DBS employs a stimulation with *constant* high frequency pulses that relies on external force determined by parameters such as type of stimulation (monopolar or bipolar), voltage, frequency (Hz), and pulse width (in ms). Such form of stimulation is considered "dumb" because the external force is not guided by the changes in the brain's electrical activity relevant to the disorder being treated.
- Laborious DBS parameter tuning: For a given patient, tuning the DBS parameters to gain optimal treatment efficacy for a given stage of the disease could be a laborious and difficult task, especially for those patients with movement disorders that would take months to see the therapeutic effect of DBS.
- High energy cost and invasive surgery to replace the battery [2]: It requires surgery of opening chest wall to replace the battery of the pulse generator. The cost of applying chronic DBS for patients of some other movement disorders, such as dystonia, is even higher than those for PD patients due to the younger age of onset [80]. Hence, more frequent invasive surgeries for battery replacements is needed [41, 46]. If the stimulation can be energy efficient, the battery's life will be prolonged, and fewer invasive surgeries are needed.

Computational Study of Conventional and Closed–Loop DBS

Multi–site Delayed Feedback Stimulation (MDFS) was first suggested as an alternative to the conventional DBS in [34, 35, 36, 64, 77]. In their MDFS protocol, the LFP of the stimulation target population is measured, filtered, delayed and fed back into the same ensemble through multiple stimulation sites that have different time delays. Compared with the conventional DBS, MDFS has at least two advantages. First, it is noticeably "smarter" in that it can be adjusted to brain's own electrical signals, whereas the conventional DBS entirely relies on external forces. Second, the energy required to administer such stimulation can be maintained at lower level. Although these earlier works contributed toward an excellent idea, the outcomes of MDFS are limited by the choice of model and the focus of study.

Hauptmann et al. [34, 35, 36] used both non–biophysical phase models and the Morris–Lecar model to describe the activity of excitatory neurons. Other work by Rosenblum and collaborators [64, 77] used the Hindmarsh–Rose equations. Both phase models and the Hindmarsh–Rose model are not biophysical models. The Morris–Lecar model is a simplified 2–dimensional reduced model in which several of the ionic currents are not presented in the tracking of real membrane potential. These models are very limited in describing the neuronal behavior of the basal ganglia.

In [34, 35, 36, 64, 77], only one neuron population that is representative of the excitatory STN neurons in the basal ganglia were considered. The authors introduced local synaptic coupling among the excitatory neurons that is responsible for the synchronization of the stimulated population. In general, it is straightforward to induce synchronization by strong excitatory coupling; however, this synchronization mechanism is unfaithful to the anatomy of basal ganglia and the pathology of PD [32].

The computational work in [34, 35, 36, 64, 77] reported desynchronizing effect on abnormal synchrony of neuron ensembles. The stimulation neither breaks the bursting pattern nor reduces the burst rate which are important characteristics of GPi and STN activity in PD. None of the work in [34, 35, 36, 64, 77] incorporated any criteria to evaluate the downstream neuronal behavior in the thalamus that is relevant to the clinical effectiveness of MDFS.

Rubin and Terman's (RT) Basal–Ganglia Thalamocortical Network Model

Rubin and Terman [66] laid the ground work on the theoretical analysis of DBS using a conductance–based network model of basal ganglia and thalamus. They set up the first Hodgkin–Huxley type of computational model including the important nuclei, STN, GPe (external segment of globus pallidus), GPi, and thalamocortical (TC) cell related to PD. Another major contribution of this work is that they defined a criteria of DBS effectiveness; that is, the fidelity of TC relaying sensorimotor signals. In their computational model of parkinsonian conditions, rhythmic inhibition from GPi to the thalamus compromises the TC relay ability to respond faithfully to sensorimotor signals arriving at the thalamus via corticothalamic projections. Under the GPi inhibition altered by the conventional DBS, TC neurons respond faithfully to incoming sensorimotor signals, and then the conventional DBS achieves clinical effectiveness. This criteria later was further validated by incorporating experimental and human data into a model of TC relay neurons by Guo et al. [26, 27]. The scope of Rubin and Terman's work [66] was to probe the working mechanism of the conventional DBS. It is a logical next step to build upon their network model and further develop better alternatives to the conventional DBS [66].

MDFS Using a Biophysical Model

Guo and Rubin [25] reported the first work of STN stimulation in the form of MDFS with a biophysical–detailed basal ganglia network model based on Rubin and Terman's model [66]. In [25], both inhibitory population GPe, GPi, and excitatory population STN are included. There is no self–coupling among the excitatory STN neurons. Therefore the synchronized bursting clusters in STN are not induced by strong excitatory self–coupling such as previous work [34, 35, 36, 64, 77]. The synchronized clusters exist because of the topological structure and coupling between different neuronal groups. Such a setup is consistent with the basal ganglia anatomy that there is not synaptic coupling among excitatory STN neurons [32]. Guo and Rubin demonstrated that MDFS applied in STN population not only breaks the pathological synchrony, but also eliminates the bursting patterns presented in STN neurons. The reduction of the average firing rate is a natural result from the burst elimination. They further evaluated the outcome of MDFS by looking at the TC relay fidelity. Their results show that MDFS restores the TC relay ability by desynchronization and burst elimination in a parkinsonian basal–ganglia thalamocortical network. Even though they have some success in this first attempt, the network they consider in their model is too small and the spatial structure of the neurons is too rigid to be representative of the real basal ganglia circuit in [25]. There are only tens of neurons in the basal–ganglia network including GPe, STN, and GPi nuclei (16 in each type). Guo and Rubin placed 16 STN neurons spatially in a perfect symmetric square grid. They measured LFP from the center of the STN neurons and fed the filtered LFP signal back to STN through four symmetric stimulation sites. Little is known about the geometric structure of the STN neurons. However, the assumption of perfect symmetry of these neurons is an unrealistic extreme. Moreover, the symmetry in STN neurons and stimulation sites eliminates many possible variations in the administration of MDFS. One natural question that follows is whether MDFS may still be able to restore TC relay fidelity if the network is larger with non-symmetric stimulation sites among randomly placed neurons.

Other Computational Study on DBS

Dovzhenok and others [18] studied the action of DBS on partially synchronized oscillatory dynamics. Their basal ganglia model, based on the RT model, consisted of two populations: an array of 20 GPe neurons and an array of 10 STN neurons that was tuned to reproduce experimentally recorded data in parkinsonian conditions. They generated the LFP feedback signal by considering only the stimulation current, the synaptic current flowing into the neuron and the synaptic currents flowing into the two closest STN neighbors. They excludes all the ionic currents of STN neurons in the calculation of LFP signal, unlike other computational work including the ionic currents [26, 34, 35, 36, 45, 64, 77]. They concluded that DBS is more likely to increase the synchrony in the basal ganglia thereby increasing PD symptoms. Their findings, opposite to many other computation results [25, 26, 34, 35, 36, 45, 64, 77] and experimental results [44], may be due to the exclusion of ionic currents in the LFP calculation.

Lourens [45] augmented the model by Terman et al. [76], which is the prelude of the RT model, with spike timing dependent plasticity (STDP) for the inhibitory connections within the GPe. He showed that a STDP rule that regulates the synaptic weights between GPe cells stabilizes the two possible coexisting network states, a healthy state (desynchronized) and a PD state (synchronized). Since STDP can learn the good (desynchronized) dynamics or bad (synchronized) dynamics by changing its synaptic connections between GPe neurons, DBS can teach the network with STDP to transition to more healthy activity by forcing the network in a healthy activity region. He showed that constant DBS has a counterproductive effect on PD since the STDP moves the network to a more synchronized state once the DBS is removed. But a short duration DBS with sufficiently high amplitude can use STDP to move the PD network to a healthier state.

Gorzelic and others [20] explored the use of classical feedback control methods in the application of DBS. Gorzelic et al. used the model from [67] consisting of 8 STN neurons, 8 GPe neurons, 8 GPi neurons and 2 TC cells to develop and test model–based rational feedback controller design. They tested two high level control strategies: one of which was driven by online estimate of thalamic reliability (reliability-based control) and another that acts to eliminate substantial decreases in the inhibition from GPi to the thalamus (GPi synaptic conductance control). They designed six control laws that are inspired by the proportional–integral–derivative (PID) methodology. They concluded that best controller based on performance was the amplitude proportional with derivative control and integral bias, which is a full PID controller. Their findings suggested that model–based rational design of feedback controllers for PD can play an important role in the development of closed loop DBS.

Scope of the Current Work

We study a form of closed–loop DBS that has been suggested as an alternative to the conventional DBS, namely multi–site STN stimulation (MDFS) with delays between stimulation periods at different stimulation sites [34, 36, 75]. We consider such stimulation with the current injection based on a local field potential signal recorded from the STN population [25, 34, 35, 64, 77]. To perform our investigation, we introduce this stimulation paradigms into a computational model based on earlier work [25, 26, 67, 76]. This model consists of a neural network of conductance–based STN, GPe, GPi, and TC neurons that, by design, generates parkinsonian activity patterns in the absence of stimulation. Our goal is to explore the impact of closed–loop MDFS on TC relay fidelity. We will further compare TC relay performance with basal ganglia modulation across PD state and PD state with MDFS and other types of DBS, such as the conventional DBS, coordinated reset periodic DBS.

The main goal of the current work is to explore whether the closed–loop MDFS can improve thalamic relay responses in parkinsonian conditions modeled by a network of conductance–based, single–compartment computational neurons. In section "Computational Model of Basal Ganglia Thalamic Network", as the first step, we will introduce such a network model of parkinsonian conditions and the measure of the TC relay fidelity, both adopted from earlier work [25, 66, 76]. In section

"Multi–site Delayed Feedback Stimulation", we will give details on how to administrate closed–loop MDFS and do a comparison study between open–loop and closed–loop DBS. In the final section, we will review results on TC relay fidelity with experimental and human recording data of GPi neurons to validate our evaluation criteria of DBS, based on TC relay error index.

Computational Model of Basal Ganglia Thalamic Network

To develop a biologically faithful PD network model, we will adopt the same Hodgkin–Huxley model of basal ganglia thalamic network as in [25, 66, 76] with modifications to incorporate MDFS stimulation. Neurons in the basal ganglia and the thalamus communicate through various excitatory and inhibitory synaptic connections and receive certain external inputs. The basal ganglia circuit consists of GPe, GPi, STN neurons, and striatal input as shown on Fig. 4.1. All straight lines with arrows represent synaptic connections or inputs. Both GPi and GPe receive excitatory inputs from STN. GPe and GPi are subject to a constant, inhibitory striatal input. There is synaptic coupling among inhibitory GPe neurons, and there is no coupling within the STN population and the GPi population. The thalamocortical (TC) cell is a relay station whose role is to respond under the GPi inhibition to incoming sensorimotor excitation via corticothalamic projections.

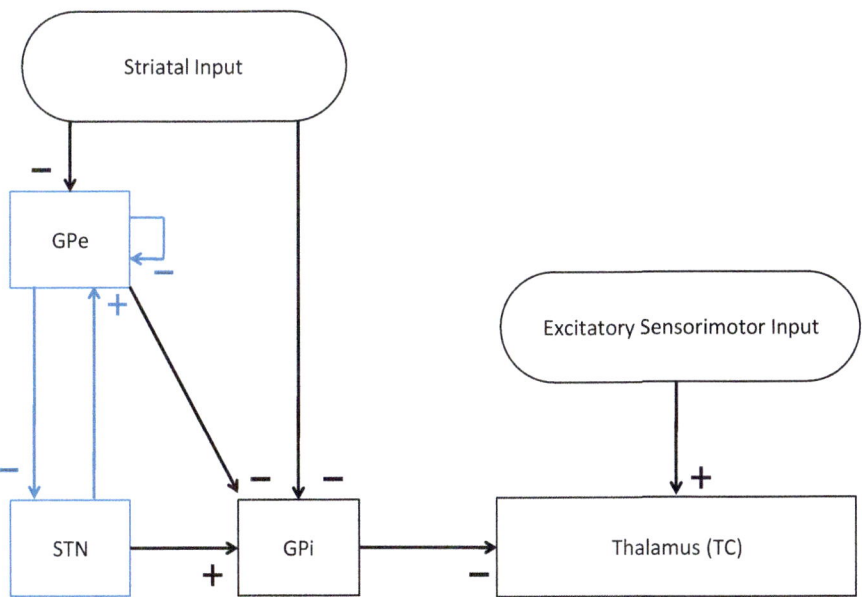

Fig. 4.1 Neuronal structure in the network model. Arrows labeled with a '−' sign represent inhibitory synaptic connections or inputs. Arrows labeled with a '+' sign are excitatory synaptic connections or inputs

Model Equations for STN, GPe and GPi Neurons

The network model consists of TC, STN, GPe, and GPi neurons. We will first describe the equations of STN, GPe and GPi neuron in the model network [25, 66, 76]. Then we give the details of the conductance–based model equations for the TC neuron and the error index to measure TC relay [25, 26, 66], which we use to evaluate the DBS effectiveness in later sections. All specifics of the functions and parameter values used for each type of neuron in the model are given in the Appendix.

STN Neurons
The STN voltage equation that we use, of the form

$$C_m v'_{Sn} = -I_L - I_{Na} - I_K - I_T - I_{Ca} - I_{AHP} - I_{Ge \to Sn} + I^{stim},$$

was introduced in [76]. All the currents and corresponding kinetics are the same except that we make some parameter adjustments so that STN firing patterns are more similar to those reported *in vivo* [7, 78, 79]. $I_{Ge \to Sn}$ is the inhibitory input current from GPe to STN. I^{stim} is the external stimulation applied to STN, which is multi–site coordinated reset stimulation (CRS), multi–site stimulation based on the local field potential (LFP), or conventional high frequency stimulation. The details of these types of stimulation will be discussed further in sections "Multi–site Delayed Feedback Stimulation".

GPe Neurons
The voltage of each model GPe neuron obeys the equation

$$C_m v'_{Ge} = -I_L - I_{Na} - I_K - I_T - I_{Ca} - I_{AHP} - I_{Ge \to Ge} - I_{Sn \to Ge} + I_{app},$$

where $I_{Ge \to Ge}$ is the inhibitory input from other GPe cells, $I_{Sn \to Ge}$ is the excitatory input from STN cells, and I_{app} is a constant external current that represents hyperpolarizing striatal input to all GPe cells.

GPi Neurons
The voltage equation for each model GPi neuron is similar to that for the GPe neurons, namely

$$C_m v'_{Gi} = -I_L - I_{Na} - I_K - I_T - I_{Ca} - I_{AHP} - I_{Sn \to Gi} + I_{Ge \to Gi} + I_{appi},$$

where $I_{Sn \to Gi}$ represents the excitatory input from STN to GPi, $I_{Ge \to Gi}$ is the inhibitory input from GPe to GPi, and I_{appi} are constant external inputs that represent hyperpolarizing currents from striatum to all GPi cells.

TC Neuron and Its Relay Responses

The model for each TC neuron takes the form

$$C_m v' = -I_L - I_{Na} - I_K - I_T - I_{Gi \to TC} + I_E \qquad (4.1)$$

$$h' = (h_\infty(v) - h)/\tau_h(v)$$

$$r' = (r_\infty(v) - r)/\tau(v)$$

In system (1), $I_L = g_L(v - v_L)$, $I_{Na} = g_{Na} m_\infty^3(v) h(v - v_{Na})$, and $I_K = g_K (0.75 (1-h))^4 (v - v_K)$ are leak, sodium, and potassium currents, respectively. We apply a standard reduction in our expression for the potassium current to decrease the dimensionality of the model by one variable [62]. The current $I_T = g_T p_\infty^2(v) r(v - v_T)$ is a low–threshold calcium current, where r is the inactivation and $p_\infty^2(v)$ is the activation. The membrane capacitance C_m is normalized to 1 $\mu F/cm^2$ in all the neural models included in the current work.

Additional terms in system (4.1) are inputs that the model TC neuron receives. One is the inhibitory input current from the GPi, $I_{Gi \to TC}$, such that

$$I_{Gi \to TC} = g_{Gi} s_{Gi} (v - V_{Gi}), \qquad (4.2)$$

where g_{Gi} is the constant maximum conductance and V_{Gi} is the synaptic reversal potential. s_{Gi} satisfies the equation

$$s_{Gi}' = \alpha_{Gi}(1 - s_{Gi}) S_\infty(v) - \beta_{Gi} s_{Gi}, \qquad (4.3)$$

where

$$S_\infty(x) = (1 + e^{-(x+57)/2})^{-1}.$$

The other input to the model TC neuron, I_E, represents simulated excitatory sensorimotor signals to the TC neuron. We assume that these are sufficiently strong to induce a spike in the absence of inhibition and therefore may represent synchronized inputs from multiple presynaptic cells. We tune the parameters so that the TC cell yields spontaneous spikes at rate of roughly 12 Hz in the absence of both inhibitory GPi and excitatory synaptic inputs. The parameter values chosen place the model TC neuron near transition from silence to spontaneous oscillations. In the model, $I_E = g_E s(v - v_E)$, where $g_E = 0.018 mS/cm^2$, and s satisfies equation

$$s' = \alpha(1-s)exc(t) - \beta s$$

where $\alpha = 0.8 ms^{-1}$ and $\beta = 0.25 ms^{-1}$. The function $exc(t)$ controls the onset and offset of the excitation: $exc(t) = 1$ during each excitatory input, whereas $exc(t) = 0$ between excitatory inputs. The periodic $exc(t)$ takes the following form:

$$exc(t) = H(sin(2\pi t/p))(1 - H(sin(2\pi(t+d)/p))),$$

where the period $p = 50$ ms and duration $d = 5$ ms, and $H(x)$ is the Heaviside step function, such that $H(x) = 0$ if $x < 0$ and $H(x) = 1$ if $x > 0$. Hence, $exc(t) = 1$ from time 0 up to time d, from time p up to time $p+d$, from time $2p$ up to time $2p+d$, and so on. A similar periodic function was used in previous work [25, 26, 66]. A baseline input frequency of 20 Hz is consistent with the high–pass filtering of corticothalamic inputs observed in vivo [13]; at this input rate, the model TC cells rarely fire spontaneous spikes between inputs.

We evaluate the TC relay fidelity in the parkinsonian network and the performance of all stimulation protocols through calculating the TC relay error index defined as

$$EI = \frac{(miss\ TC\ responses + bad\ TC\ responses)}{number\ of\ excitatory\ stimuli}. \qquad (4.4)$$

We will use an existing algorithm by Guo et al. [25, 26, 27] to identify good, miss and bad TC responses and count the number of each type of TC responses. A good TC response (see example on Fig. 4.6) refers to one single TC spike corresponding to one sensorimotor input (the small bumps of the middle trace on Fig. 4.6) within a designated time window and before the next input arrives. In a "normal" state, a TC relay cell generates good TC responses to most of excitatory sensorimotor inputs. In a parkinsonian state, the thalamus is no longer able to transmit sensorimotor signals faithfully. Two types of TC relay error occur: one is bad TC relay response, the other is miss TC relay response (see examples in Fig. 4.6). The algorithm to determine a bad or miss response counts at most one error per excitatory sensorimotor input [25, 26, 66]. Therefore, EI should be a value between 0 and 1.

Model Design of a Parkinsonian Network

We design a model of parkinsonian network which consists STN, GPe, GPi and TC neurons introduced in sections "Model Equations for STN, GPe and GPi Neurons" and "TC Neuron and Its Relay Responses". Each STN, GPe, and GPi group includes 112 neurons. We incorporated two relay TC neurons into the parkinsonian network to evaluate the performance of DBS. The current parkinsonian network model is significantly advanced from the previous toy network that only has 16 neurons in each group [25, 66]. The network model mimics the pathological neuronal activity observed in the basal ganglia in parkinsonian conditions, such as increased firing rate, bursting patterns, and synchronization in STN and GPi neurons [4, 8, 9, 38, 43, 47, 54, 61, 83, 84]. We consider this rhythmic clustered regime in STN and GPi as the parkinsonian state and refer to the network in this state as the parkinsonian network. In the following subsections, we focus on three aspects of the PD network model: the coupling structure in the STN–GPe loop, the averaged GPi synaptic input going into a TC relay neuron, and the TC relay error index.

Coupling Within the STN–GPe Loop

One important feature of the PD network depicted on Fig. 4.1 is its capability of inducing rhythmic bursting patterns in GPi cells to mimic those seen experimentally in parkinsonian conditions [25, 26]. Since each GPi neuron receives synaptic inputs from a single corresponding STN neuron and a single GPe neuron, the rhythmic, bursting, and synchronized activity in a group of GPi neurons is really induced by the upstream structure of STN-GPe subnetwork that forms a loop. (See GPe and STN neurons with arrows between them and the arrow starting from GPe and back to itself on Fig. 4.1.) Therefore, it is crucial to set up appropriate coupling among the STN–GPe loop.

As shown previously [76], the STN and GPe subnetwork can generate both irregular asynchronous and synchronous activity [5, 59, 76]. In [25], Guo and Rubin designed a small STN–GPe network with specific symmetry and asymmetry in the coupling of sub–populations of STN and GPe, each of which has only 16 neurons. Each STN cells receives a designated inhibition signal from a GPe cell. And the STN cells are segregated into two groups and provide excitation via the design of strong and weak coupling back to GPe cells which are also segregated into two clusters.

Based on previous work [25, 66, 76], we successfully design a parkinsonian network relying on the coupling structure of the STN/GPe loop in a much larger network with many more neurons. We demonstrate that the STN cells in the two segregated groups can form two rhythmically bursting clusters. Figure 4.2 shows the synchronized clusters in the STN population of the PD network model. Half of the 112 STN neurons are in one synchronized bursting cluster, and the other half forms the second synchronized bursting cluster.

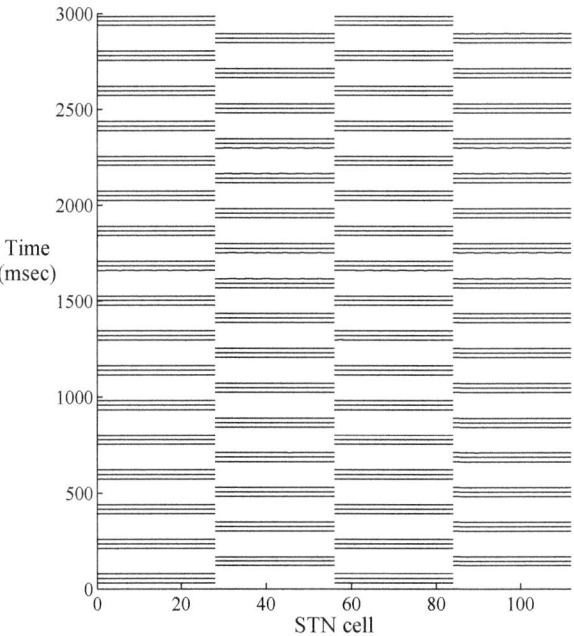

Fig. 4.2 Two synchronized STN clusters. The horizontal axis is the STN cell index. STN cell 1–28 and cell 57–84 form one synchronized cluster. Cell 29–56 and 85–112 form the second synchronized cluster. The *horizontal bars* represent spike times. Since STN cell 1–28 are in perfect synchrony, we only see one long bar across cell 1–28

The detailed structure of connections between STN and GPe neurons is depicted on Fig. 4.3. The synaptic connections within the STN–GPe loop are structured sparsely according to [66]. The STN–GPe loop consists of two subpopulations of 56 STN neurons and 56 GPe neurons each. The neurons in one subpopulation, mainly the STN neurons are connected to the GPe neurons in the other subpopulation via weak synaptic connections (all dashed lines with an arrow on Fig. 4.3). Each subpopulation is further divided into four groups of 14 neurons, (four blocks in each row on Fig. 4.3. Each group of 14 STN are connected to two groups of 14 GPe neurons through strong excitation (all black lines with an arrow on Fig. 4.3). Each group of 14 GPe cells sends strong inhibition to two groups of 14 STN neurons (all black lines with a solid circle on Fig. 4.3). There is also self coupling among GPe neurons. With the connections demonstrated on Fig. 4.3, the four groups of STN neurons (total 56) in the second row of Fig. 4.3, with 14 in each group, form one bursting cluster, and the fourth row of total 56 STN neurons forms the second bursting cluster. Then the STN–GPe loop sends excitation (from the STN neurons)

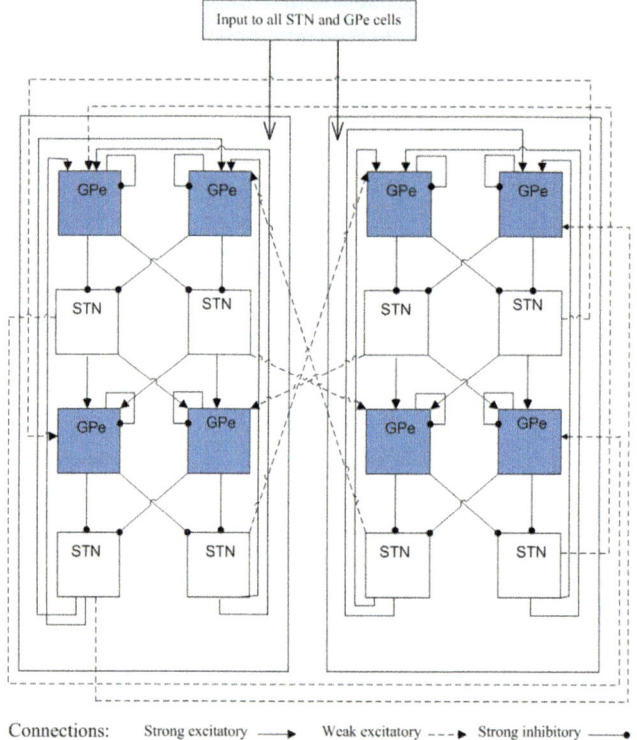

Fig. 4.3 Coupling structure in the STN–GPe loop. Top row of GPe neurons with total 56, 14 in each block, are in one cluster. The second row includes 56 synchronized bursting STN neurons with 14 in each block. Similar structures in the third and fourth row. *Solid line* with an *arrow* and *solid line* with a *solid circle* represent strong excitatory and inhibitory connections, respectively. *Dashed line* with an *arrow* is weak excitatory coupling

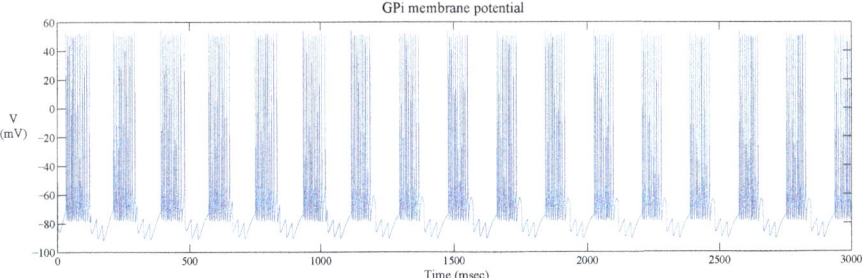

Fig. 4.4 The Membrane potential of a GPi neuron in a parkinsonian network model

and inhibition (from the GPe neurons) to the downstream GPi neurons, which cause the GPi model neurons to imitate the activity seen experimentally in monkeys with drug induced PD or humans with PD.

Averaged GPi Synaptic Input to a TC Cell

The network model includes 112 GPi neurons in the downstream of the STN–GPe loop. Each GPi cell receives input from a single corresponding STN neuron. Thus, the synchronized, rhythmic, bursting outputs of each STN cluster induce rhythmic, bursting and synchronized activity in a corresponding group of GPi neurons. Figure 4.4 shows the bursting pattern of one GPi neuron in the PD network we designed.

To further correlate the population effect of the synchronized GPi bursts to a TC neuron, we compute averaged GPi synaptic input over half of the GPi population who fire synchronized bursts in the PD network. In our network, the synaptic input from GPi to a TC neuron, $I_{Gi \to TC}$, comes from a subgroup of GPi neurons. This input takes the form

$$I_{Gi_j \to TC_j} = g_{Gi} \sum_{k \in \Omega_j} s_{Gi_j}^k (v - V_{Gi}) \qquad (4.5)$$

where $s_{Gi_j}^k$ satisfies Eq. (4.3). We define

$$sg_1 \equiv \sum_{k \in \Omega_1} s_{Gi_1}^k, \text{ and } sg_2 \equiv \sum_{k \in \Omega_2} s_{Gi_2}^k,$$

where Ω_1 is one synchronized subgroup which consists of 56 GPi cells, Ω_2 is the other synchronized subgroup with the same number of GPi cells. sg_1, the total synaptic input from GPi neurons in group Ω_1 going into TC cell 1, is the top trace shown on Fig. 4.6. sg_2 is the total synaptic input from GPi neurons in group Ω_2 going into TC cell 2 (not shown).

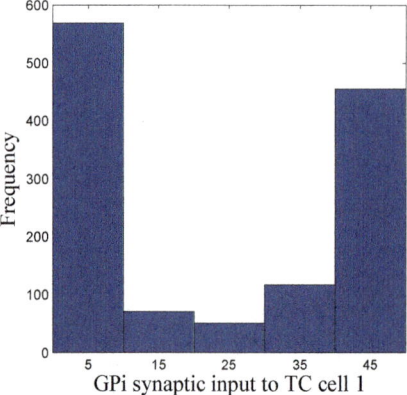 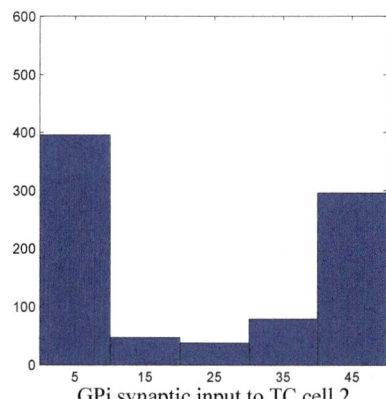

Fig. 4.5 Histograms of time-averaged sg_1 (*left*) and sg_2 (*right*) in the parkinsonian network. Both histograms include two dominant bins, centered at 5 and 45, due to the quiescent and bursting phases, respectively, of GPi activity

Based on the form of Eq. (4.3), each $s_{Gi_j}^k$ is between 0 and 1, and hence $sg_i \in [0, 56]$, $i = 1, 2$ for each i. In our simulations, we use the variability of the time–average of each sg_i as an indicator of GPi rhythmicity. Specifically, we form histograms based on the frequency with which each sg_i time course, averaged over 5 ms time windows, takes different values in bins that cover the range [0,56]. We display five bins centered at 5 through 45 with increment 10. In the parkinsonian network without stimulation, the average sg_i values mostly fall into the first bin centered at 5 and the fifth bin centered at 45, as displayed on Fig. 4.5. This result occurs because GPi firing is rhythmic and bursting (see the top traces on Fig. 4.6), such that GPi synaptic output is high during each burst and low between bursts. A few values do fall into the middle bins, due to transition between bursting and quiescent phases.

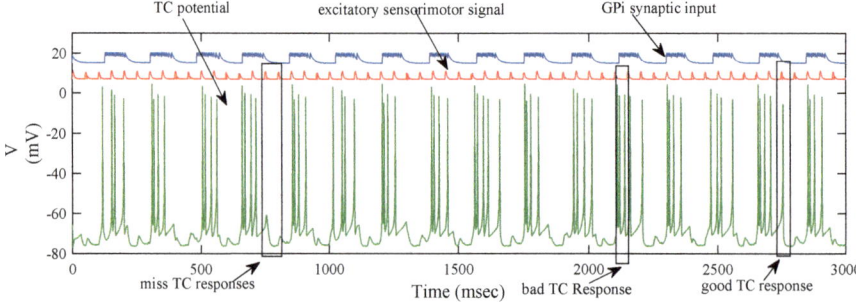

Fig. 4.6 The membrane potentials of a TC neuron responding to excitatory sensorimotor signals (*middle trace*), along with the total synaptic input the neuron receives from 56 GPi neurons (*top trace*). Note that the vertical axis label applies to the voltage trace, while the latter two traces are placed and scaled arbitrarily. Also, observe that the total GPi synaptic input is rhythmic and bursting, representing parkinsonian conditions

However, very different results emerge when stimulation is applied to the STN neurons (Fig. 4.10 in section "Results" and Fig. 4.14 in section "Comparison Between Open–Loop and Closed–Loop Stimulations").

TC Relay in the PD Network

In the parkinsonian network we designed, the synaptic input from GPi to TC (the top trace on Fig. 4.6) appears to be rhythmic and bursting. Although the TC neuron responds with a single spike (a good TC response shown on Fig. 4.6) to some of the excitatory inputs that it receives, others elicit either no spikes, or multiple spikes, which are miss TC responses and bad TC responses, respectively (both are shown on Fig. 4.6). Overall, the TC relay error is apparently at the higher end towards 1 in the parkinsonian network.

Multi–site Delayed Feedback Stimulation

We focus on investigating MDFS in a large biophysical network model of basal ganglia and thalamus. The objective is to gain understanding on how the MDFS form of closed–loop DBS might achieve its therapeutic effectiveness by the proper selection of stimulation target sites and parameters. In particular, using a scaled up basal ganglia thalamocortical parkinsonian network with hundreds of neurons, we develop a closed–loop feedback stimulation protocol applied on STN that can restore TC relay ability. The MDFS closed–loop stimulation is based on the LFP signal of the STN population, then is applied to the same STN neurons through multiple stimulation sites. See the schematic demonstration of MDFS being applied to the STN population on Fig. 4.7. In the following sections, we describe the choice of the stimulation sites, the generation of the stimulation currents and the improvement of TC relay during the closed–loop stimulation.

Layout Structure and Stimulation Sites of the Target Population

We use two geometric structures for the stimulation target population in the current study. In prior work [25], Guo and Rubin placed 16 STN neurons on a simple four by four square grid where the LFP was collected at the center, and four stimulation sites were chosen symmetrically around the center. This square grid structure is far from the real topology of STN neurons. To improve in this aspect, we use a layout structure of target population on a circle neuronal sheet where all STN neurons are placed both symmetrically (left figure of Fig. 4.8) and randomly (right figure of Fig. 4.8). In the random layout, the positions of neurons are sampled by imposing a minimum distance between any pair of neurons (e.g. excluding two neurons to appear at the same location.) The center of the circle where LFP of the target population is measured and calculated is at the origin of the 2–D plane.

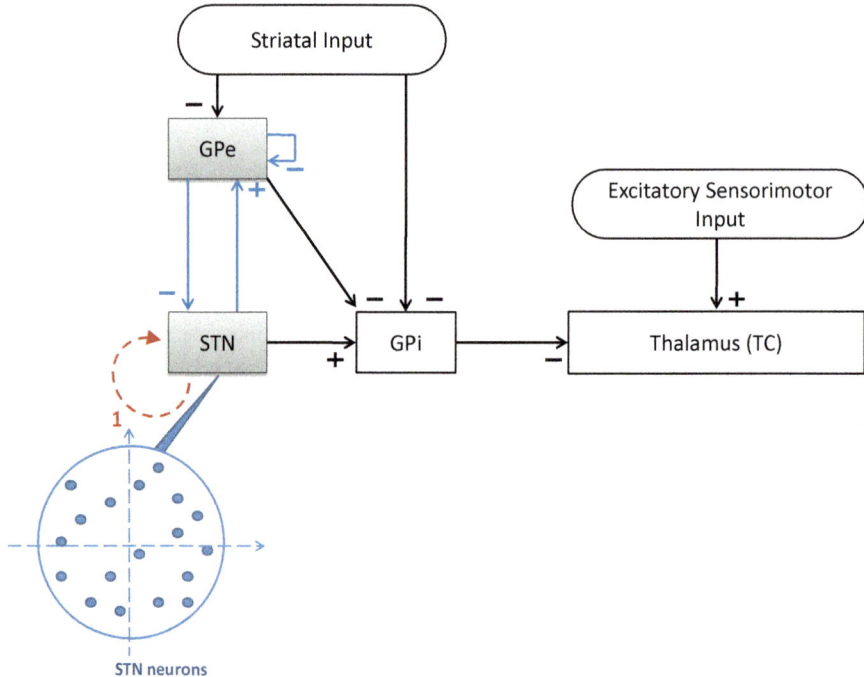

Fig. 4.7 Neuronal structure in the network model. *Arrows* labeled with a '−' sign represent inhibitory synaptic connections or inputs. *Arrows* labeled with a '+' sign are excitatory synaptic connections or inputs. The *dash arc* with an *arrow* represents the stimulation current. The tail of the *dashed arc* is where the stimulation current is gathered. The head of the *dashed arrow* points to the stimulation target population. The *blowup* shows the random topological structures of stimulation targets–STN with *blue arrows* representing the STN–GPe loop

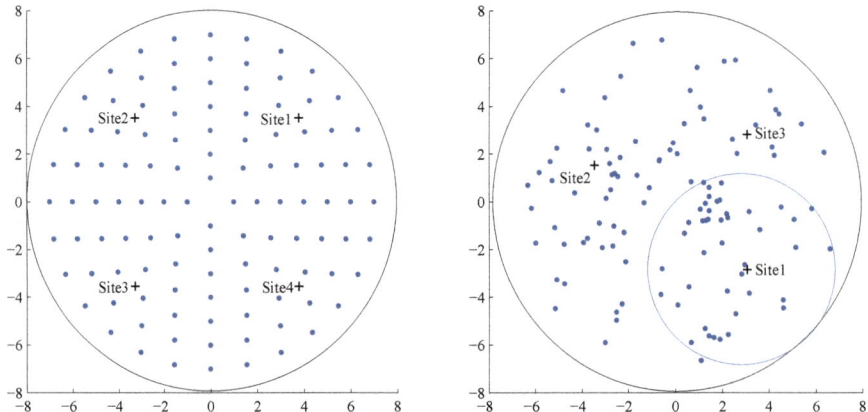

Fig. 4.8 *Left* symmetric layout of 112 STN neurons on a circle sheet. Four stimulation sites are marked with '+' signs. *Right* random layout of 112 STN neurons on a circle sheet. Three stimulation sites (*marked with* '+') are chosen using reference circles. One reference circle centered at site 1, with radius as a half of the neuron circle sheet is shown. LFP is computed by (6) at the center of the neuron sheet in both layouts

Choosing appropriate multiple stimulation sites is a crucial first step to the application of MDFS. Previous study of STN MDFS [25] and work by Hahn and McIntyre [29] have found that the stimulation becomes more effective when it reaches a larger portion of the STN population. In theory, more stimulation sites lead to better outcome of MDFS; however, it is not desirable to have too many stimulation sites as it needs more electrode contact points. In the symmetric layout structure of STN, we choose four stimulation sites symmetrically in the four quarters of the circle (shown on the left figure on Fig. 4.8). In the random layout of STN neurons, to choose proper number and locations of stimulation sites, we define a *reference circle* with radius as half of the neuronal sheet circle. The *reference circle* is inside tangent to the circle of neuronal sheet and passes through the center of the neuronal sheet circle. Then we slide the *reference circle* through the neuronal sheet to identify the most populated region. We place the first stimulation site at the center of the *reference circle* that contains the most populated region of STN neurons. Then we use a second *reference circle* to slide through the region not covered by the first *reference circle* to identify the next most populated region. Overall, we may use up to four *reference circles* to identify at most four sites to cover at least 85% of the circle neuronal sheet.

Generation of Stimulation Current from Feedback

We use LFP signal measured and calculated from the center of the circle STN neuronal sheet as the base for the stimulation current. According to current–source density analysis [37, 42, 51, 55], the field potential depends on the linear sum of potentials from sources (currents injected into the extracellular medium) and sinks (current removed from the extracellular medium). For a single point source in a homogeneous medium, the extracellular field potential Φ is given by $\Phi = \dfrac{R_e I_v}{4\pi r}$ [55], where r is the distance from the point source, I_v is the current from that point source, and R_e is the constant extracellular resistance. This translates to the following formula for the computation of LFP:

$$V_{LFP}(t) = \frac{R_e}{4\pi} \sum_{j=1}^{N} \frac{I_j}{r_j}. \tag{4.6}$$

In Eq. (4.6), the extracellular resistance R_e is set to 1, and r_j is the distance between neuron j and the electrode measuring LFP at the center of the circle of the neuronal sheet. N is the number of neurons in the circle. I_j is the total current from neuron j, which consists of ionic currents I_j^{ion} and external currents $I_j^{external}$, including both the stimulation and the synaptic currents from the presynaptic GPe neurons. Hence, we have $I_j = I_j^{ion} + I_j^{external}$ where $I_j^{ion} = I_j^l + I_j^{Ca} + I_j^{Na} + I_j + I_j^{AHP} + I_j^T$. $I_j^{external}$ is the following calculation:

$$I_j^{external} = -I_j^{Gi \to Sn} + \lambda I_j^{stim \to Sn} \text{ for STN}$$

Here $\lambda = 1$ if the stimulation target is the population where LFP is measured, and $\lambda = 0$ if the stimulation target is different from the neuron population where LFP is measured. The $\lambda = 0$ case is not considered in the current work.

We notice that there is a major difference between our method of calculating LFP signal and that of computing LFP signal by [18]. They did not include any of the ionic currents in the calculation of LFP. After a thorough study on computing the LFP of biophysical model neurons, we conclude that all the ionic currents should be accounted for in computing LFP, which is the same as the previous work [34, 35, 36, 37, 42, 51, 55]. In the review paper [10], the authors provide insight on how the ionic currents contribute to the extracellular signal such as LFP. Based on their study, the signs of all the ionic currents included in our network model depend on whether these currents act like sinks or sources. Sink is defined as a site on the neuronal membrane where positive charges enter the neuron and source as a site on the neuronal membrane where negative charges enter the neuron. Using the principles outlined in [10], currents, Na, Ca, Cl and T are sinks in our LFP calculation. While k, AHP, and the synaptic currents are sources.

The LFP signal $V_{LFP}(t)$ is re–scaled and filtered by a low–pass harmonic oscillator to generate the stimulation current. The low–pass filtering of $V_{LFP}(t)$ is implemented by the equation

$$x'' + ax' + bx = \mu V_{LFP}(t), \; a^2 < 4b. \tag{4.7}$$

There are multiple reasons we choose Eq. (4.7) as the feedback controller. First, it can smoothen the LFP signal and only let the low frequency component pass through, which fits our goal to find an alternative to the high frequency conventional DBS. The filtering is also consistent with the experimental design by Little et. al. [44]. Second, parameters a and b determine the period of the harmonic oscillator. We can set the period of the filter roughly the same as that of the bursting cycle of the population where LFP is collected. Since LFP reflects the population activity [14, 86], strong synchronization of the target population results in large variation of the LFP, then resulting in larger amplitude of signal x, which consequently leads to stronger stimulation signals. When neurons in the target population is desynchronized, the LFP has small variations, and, therefore, the stimulation current is small. This is a feedback control that relies on the neurons' own activity to make adjustments. Third, the parameter μ in the forcing term provides another degree of freedom to tune MDFS according to its outcome. We can scale LFP up or down through μ depending on the expected outcome of MDFS.

We will then apply the stimulation signal x via multiple sites with time delays [25, 34, 35, 36, 60, 63, 64]. The stimulation that the jth neuron in the target receives from multiple sites is given by

$$I_j^{stim \to STN} = \frac{h(t)}{N} \sum_{k=1}^{M} e^{-2 dist(j,k)} x^k (t - (k-1)\tau), \tag{4.8}$$

where $h(t)$ defines the stimulus onset and offset times; N is the number of neurons in the stimulation target; M is the number of stimulation sites; $dist(j,k)$ is the distance between the jth neuron and the kth stimulation site; and $x^k(t-(k-1)\tau)$ is the time delayed signal from Eq. (4.7) that is delivered at the kth stimulation site.

Results

We apply the stimulation current by Eq. (4.8) to each STN neuron in the parkinsonian network. Our computation shows that STN MDFS based on LFP can effectively break the synchronized bursting clusters of the STN population. Figure 4.9 shows that under MDFS, the STN neurons no longer burst or form synchronized clusters.

We further look at how the disappearance of bursting clusters in the STN population changes the downstream GPi firing. During the LFP stimulation, the averaged values of sg_i, $i=1,2$, defined in section "Averaged GPi Synaptic Input to a TC Cell", are low. Histograms of these values show that these GPi output measures mainly lie in the bin centered at 5 (Fig. 4.10) due to the elimination of synchronized bursting in the STN population and hence the reduction of synaptic excitation from STN to GPi. During the stimulation period, from 800 to 2500 ms. TC relay errors are correspondingly reduced dramatically. Both TC cells show much improved TC relay performance as in Fig. 4.11.

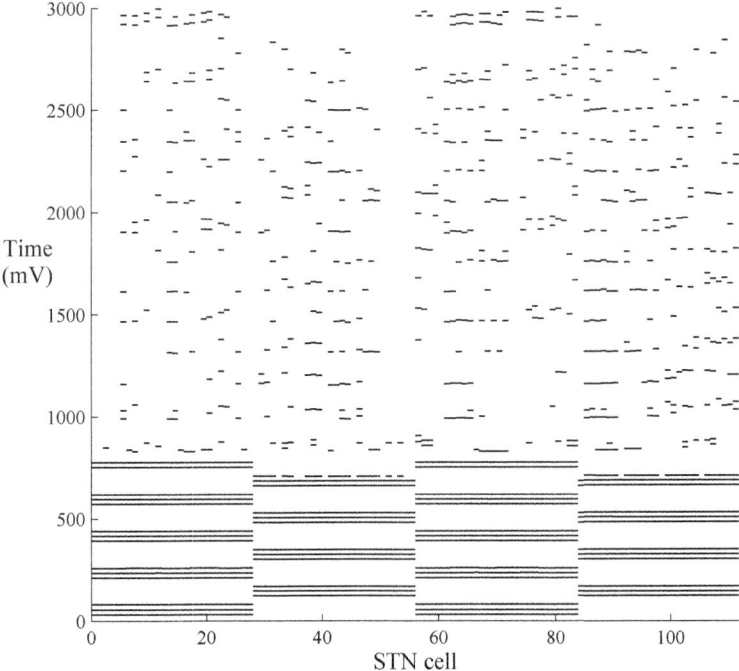

Fig. 4.9 Activity of STN neurons in the parkinsonian network with MDFS. The 112 STN neurons form two synchronized clusters before the multi–site stimulation is turned on at 800 ms. During the stimulation period from 800 to 2500 ms, synchronized STN bursting clusters no longer exist

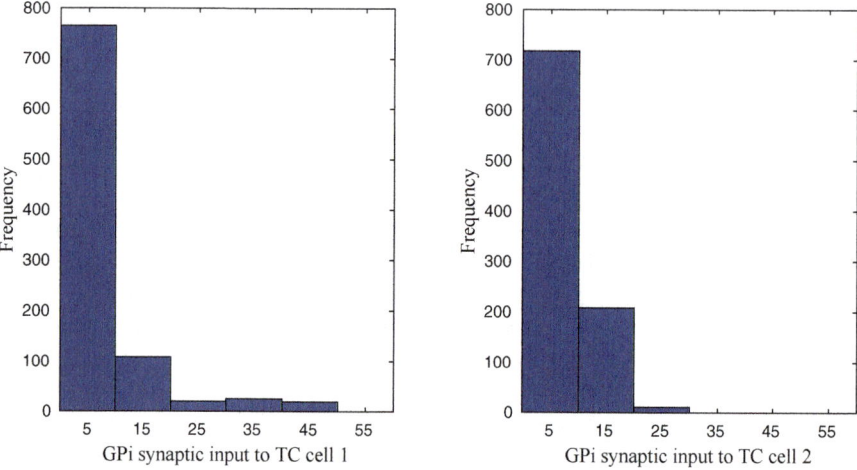

Fig. 4.10 Histograms of sg_i values, ($i = 1, 2$), representing the average strength of GPi synaptic input to each TC cell, during MDFS stimulation. *Left figure* is sg_1, the *right figure* is sg_2, and they are defined in section "Averaged GPi Synaptic Input to a TC Cell"

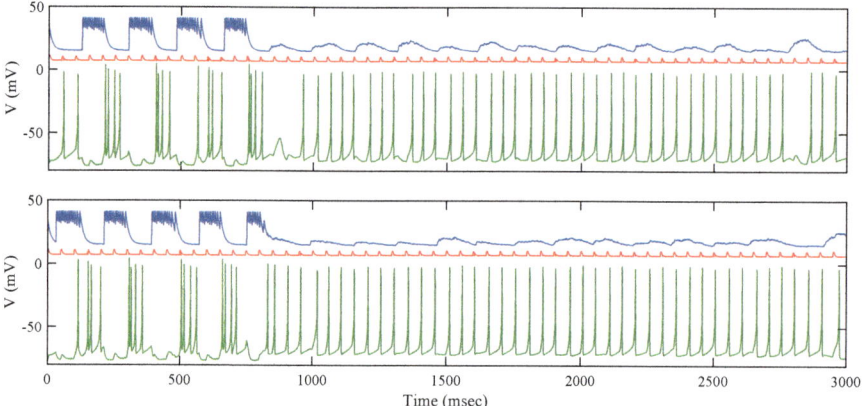

Fig. 4.11 The two TC cells respond to excitatory sensorimotor signals (*middle trace* on each figure) faithfully between the stimulation onset (800ms) and offset (2500 ms) due to the change of total GPi synaptic input (*top trace* on each figure). Parameter values for the stimulation are $a = 0.0025, b = 0.00136, \tau = 42.5$ms, $\mu = 0.0002$

Comparison Between Open–Loop and Closed–Loop Stimulations

We compared our closed–loop MDFS with the open–loop high frequency conventional DBS and the open–loop CRS that was experimentally tested on PD monkeys recently [75]. We apply all three stimulation protocols to our PD network to examine which one can produce better TC relay performance.

The high frequency stimulation (HF) is given by the following formula [25, 66],

$$I_k^{stim} = a_0 h(t) hvf(sin(\rho * t) - a_1) \quad (4.9)$$

4 Computational Models of Closed–Loop Deep Brain Stimulation

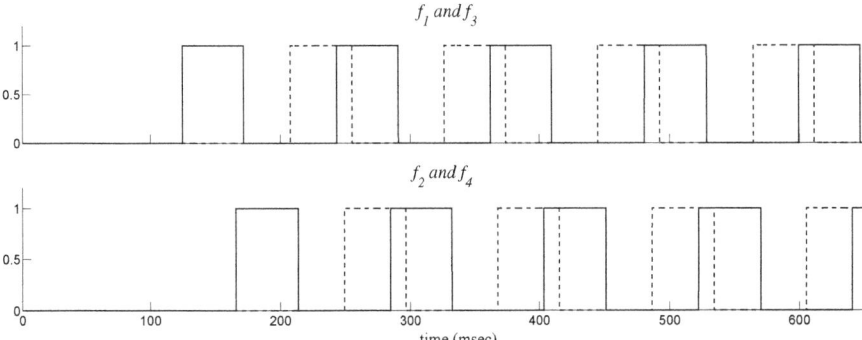

Fig. 4.12 The periodic step functions $f_1, f_2, f_3,$ and f_4 used to administer the four stimulation signals at four sites. f_1 and f_2 are shown with *solid lines*, while f_3 and f_4 are given by *dashed lines*. Note that between any pair of $f_1, f_2, f_3,$ and f_4 there is a temporal overlap. Within stimulation ON periods during which $f_k(t) = 1$, the stimulation signal consists of a train of high frequency pulses, given by $hvf(sin(\rho * t) - a_1)$, where $hvf(x) = 1/(1+exp(-x/0.001))$, (high frequency pulses not shown here). (Fig. 4.12 is taken and modified from Ref. [25])

where $a_0 = 140, a_1 = 0.9, \rho = 0.7,$ and $h(t) = H(t-l_1)(1-H(t-l_2))$. Since H denotes the Heaviside function, h(t) equals 1 on the time interval (l_1, l_2) and 0 outside this interval, where l_1 is the stimulation on time, and l_2 is the stimulation off time. $hvf(x) = 1/(1+exp(-x/0.001))$, with $x = sin(\rho * t) - a_1$, gives the high frequency pulses.

The coordinated reset stimulation (CRS) is given by the following formula [25],

$$I_k^{stim} = a_0 h(t) hvf(sin(\rho * t) - a_1) f_k,$$

where all parameters are given in the Appendix. Compared with the conventional high frequency stimulation (4.9), CRS incorporates additional periodic step functions f_k, $k = 1, 2, 3, 4$, where k is the stimulation site. $f_k, k = 1, 2, 3, 4$, are basically the same copy of one step function with phase shift τ_0 through four sites, as plotted in Fig. 4.12.

CRS with particular choices of period, phase shift among multiple sites and stimulation amplitude can reduce TC relay error dramatically. When CRS is on from 800 to 2500 ms (Fig. 4.13), the two STN synchronized bursting clusters (presented before 800 ms) are no longer there. Once the stimulation is turned off again, the STN neurons return to a two cluster firing pattern. Although there is still some structure of four or more small sub–clusters to the STN firing pattern during CRS, the stimulation outcomes, illustrated by Figs. 4.14 and 4.15, suggest that CRS can restore TC relay fidelity by changing the firing pattern of STN neurons. Compared with the PD condition where the GPi activity mainly has its strength at the bins centered at 5 and 45 due to GPi rhythmicity, GPi strength mostly concentrates on the bins centered at 15 and 25.

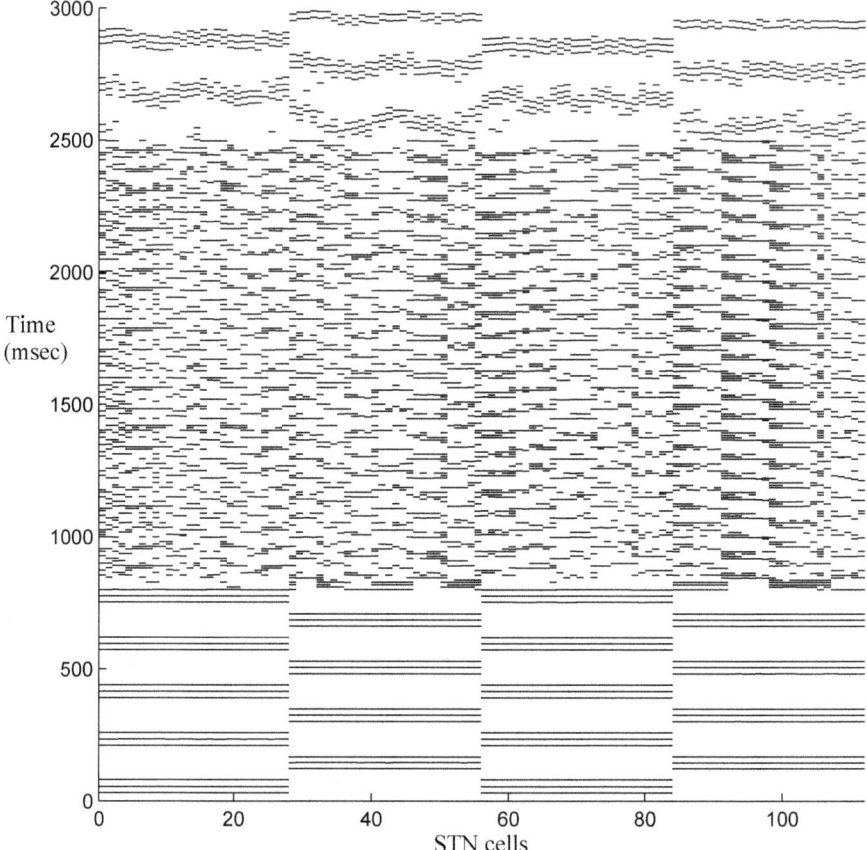

Fig. 4.13 Spikes times of 112 STN cells. CRS is on from 800 to 2500 ms with a $0=58$ and the phase shift in f_k is 38 ms. Before stimulation, there are two highly synchronized clusters. During CRS, the STN population develops into four or more small sub–clusters and once CRS is removed the two clusters start to reappear again

In addition to the fidelity of TC relay and the averaged GPi synaptic input, we further compare the energy consumption between the conventional high frequency DBS and closed–loop MDFS. Using the energy expenditure formula of DBS stimulation, previously introduced in Feng et al. [19], later used in [20], we find that the energy expenditure of conventional DBS is much higher than that of MDFS.

Principal Component Analysis

Analyzing the level of synchrony in the activity of the network is important in our work. Principal component analysis (PCA) provides a framework for our analysis of synchronization levels in our networks. The central idea of PCA is to reduce the

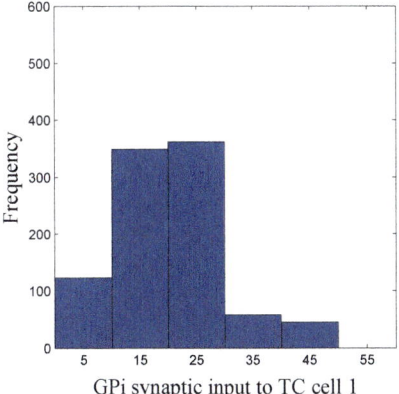
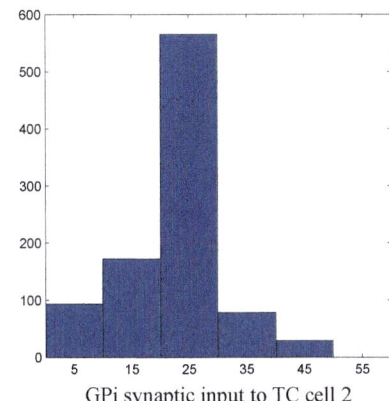

Fig. 4.14 Histograms of (*left*) and (*right*) of the PD network with CRS. The GPi activity mostly concentrates on the central bins

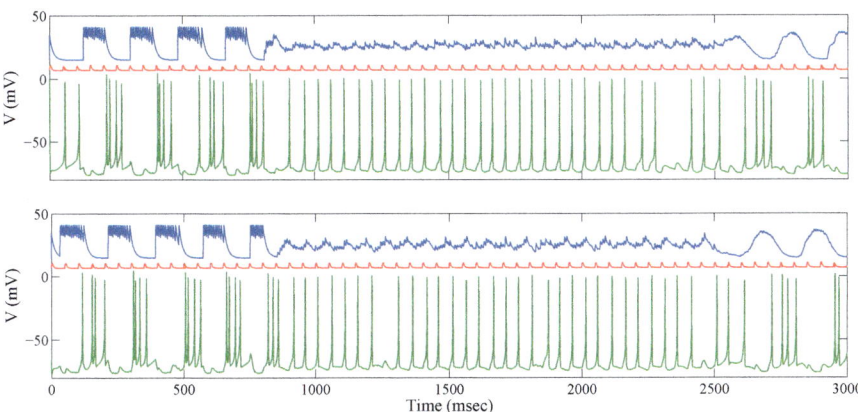

Fig. 4.15 TC relay performance of TC cell 1 (*top figure*) and TC cell 2 (*bottom figure*) in the PD network with CRS. The two TC cells respond to excitatory sensorimotor signals (*middle trace* on each figure) faithfully between the stimulation onset and offset due to the change of total GPi synaptic input (*top trace* on each figure). Note that the GPi synaptic input does not reduce to a level as low as in the MDFS case

dimensionality of a data set in which there is a large number of possibly correlated variables, while retaining as much as possible of the variation present in the data set. The reduction is achieved by using an orthogonal transformation to convert a set of p observation variables into a new set of values of linearly uncorrelated variables called principal components (PC). These new uncorrelated variables, PC, are ordered such that the first few retain most of the variation present in the original data. A few necessary steps are involved in finding the PCs. The first step is to organize the data set of p variables with n observations into a matrix of p columns and n rows. The next step is to find the mean of each column and subtract the mean from each observation within that column. This produces a data set of mean zero which is essential in the calculation of the PCs. Then we should calculate the $p \times p$ covariance matrix, find the eigenvalues of this covariance matrix and sort the eigenvalues in a decreasing order. In the final step, we need to select m number of PCs to account for most of the variation [39]. The methods for choosing the m number of PCs include: cumulative percentage of total variation, size of variances of PC and log–eigenvalue diagram. Further details of each method can be found in [39].

It was suggested in [5, 39, 57, 70] that PCA can be used as a method for quantitatively measuring network synchrony. Since PCA sequentially finds a specified number of PCs in which most of the variation is captured, a lack of variation in a set of p variables in the original dataset will result in a small number of m PCs capturing most of its variation. Thus the smaller m is the more synchronized the network is. The number of PCs, m, is a measure of the level of synchrony in the activity of the neuronal network.

We perform PCA on our parkinsonian network, using the STN neuron population data, with three different stimulations, HF, CRS and MDFS. We capture the 112 STN voltage traces from our network. Then we calculate the eigenvalues and eigenvectors of its covariance matrix. Next based on the cumulative percentage of total variation method, we select m PCs that can account for a predetermined amount, 99%, of the total variation in the original STN voltage data. In our study m is the number of PCs needed to account for at least 99% of the total variation in the STN population. For our parkinsonian network without any stimulation, we need $m = 2$ principal components (PCs) to capture 99% of the variation. Thus the STN population, as we know, is highly synchronized. After applying MDFS stimulation to our network, we need $m = 108$ PCs to account for the total variation in the set of 112 STN voltages. With HF stimulation, $m = 1$, meaning one PC is enough to capture 99% of the variation, and, for CRS, $m = 101$. Clearly HF produces a highly synchronized network, while CRS and MDFS desynchronize the network. The computations needed to generate PCs are completed using Matlab.

MDFS for Heterogeneous TC Cells

We take a further step to verify that the effectiveness of MDFS at restoring TC neurons relay fidelity is not specific to our baseline parameter values in our TC model [25]. We generate a population of 80 model TC neurons with heterogeneity by perturbing conductance values, g_L, g_{Na} and g_T through normal distributions around the baseline parameters with standard deviations 0.01, 0.05 and 0.08, respectively. We independently select g_L, g_{Na} and g_T from these distributions. Then we run the stimulations with these g_L, g_{na} and g_T values in all four scenarios: PD network without any stimulation, PD networks with conventional high frequency DBS, CRS, and MDFS. Our computation results suggest that all the members of the TC population with the three stimulations showed decreases in their error index compared to PD condition without any stimulation, as seen in Fig. 4.16. Overall, on average, MDFS is most effective in reducing the error index in the PD network as

4 Computational Models of Closed–Loop Deep Brain Stimulation

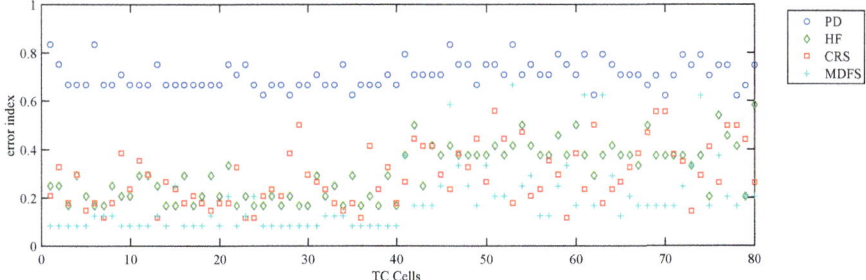

Fig. 4.16 Error index values for 80 model TC neurons with heterogeneous parameter values

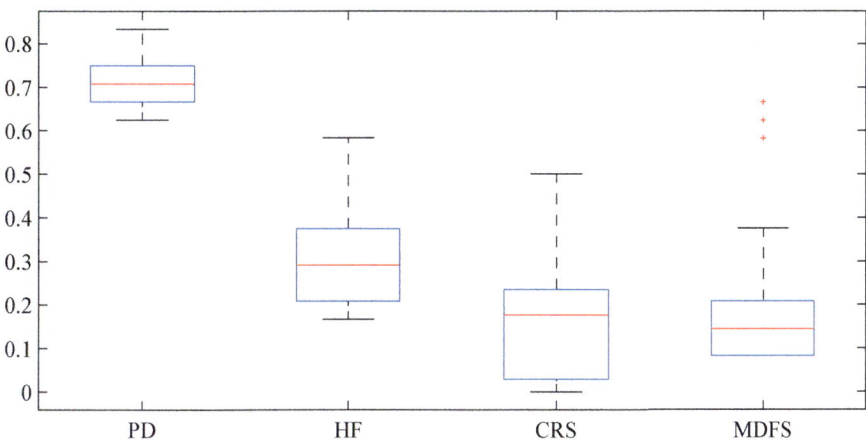

Fig. 4.17 Error index values for 80 model TC neurons with heterogeneous parameter values for conductances g_L, g_{Na} and g_T. Whisker plots show mean (*red line*), 25–75 percentile range (*blue box*), 95 % confidence interval (*black*) and outliers (*red plus signs*)

shown in Fig 4.17. There are three outliers from MDFS in the 80 TC cells, which is a lower percentage out of the 80 TC cells. We think the conductance combined with all the stimulation parameters create large perturbation outside of the parameter space for effective MDFS design. We will further investigate the parameter space of the administration of MDFS in another study.

Model Validation and Experimental Validation

Validation of TC Relay Fidelity: A Data–Driven Model

The justification of using TC relay to distinguish the parkinsonian state from a normal state and evaluate the performance of DBS is a difficult task. There seems no way to directly quantify TC relay fidelity in an experimental setting or real life [69]. However, one can take advantage of computational simulations to test the hypothesis that TC relay is faithful in normal condition, is compromised in PD condition, and is improved in PD condition with DBS. We review some of the work on data–driven model related to the validation of TC relay using recording data from monkeys and human patients [26, 27]. The model equation of a TC neuron is same as (4.1) except that the GPi inhibition term $I_{Gi \to TC}$ is no longer from a simulated network model as described in section "Computational Model of Basal Ganglia Thalamic Network". The GPi signal is based on spike timing from experimental recording of monkeys or clinical recording of human patients. The synaptic current from GPi to TC, $I_{Gi \to TC}$, is continuously generated by computation using spike times of the recording. Further details are given in [26, 27].

Guo et al. [26] examined TC relay responses under a variety of inhibitory signals obtained from single–unit GPi recordings from normal monkeys and PD monkeys (1-methyl-4-phenyl-1,2,3,6-tetrahydropyridine (MPTP) induced). The GPi recording were obtained in four conditions: the first was collected in normal control monkeys, the second was collected in PD monkeys without the conventional STN DBS, and the third and fourth were collected in PD monkeys under sub–therapeutic and the conventional therapeutic STN DBS. Using the data–driven TC model neuron, we were able to test how biologically observed changes in GPi neuronal activity affect TC signal transmission in a single TC model neuron and in a heterogeneous population of model TC cells. We calculated the error index of the model TC cell in four cases, normal control, PD without DBS, PD sub–therapeutic DBS, and, PD with therapeutic DBS. In each case, we examined TC relay responses to both periodic and random excitatory sensorimotor inputs. We found that the TC relay performance depends strongly on inhibitory GPi recording. The error index of the model TC cell responses to both periodic and random excitatory inputs is much higher in both PD monkeys without DBS or with sub–therapeutic DBS, and is consistently lower in normal and therapeutic DBS cases. Moreover, we found that the mean TC performances across all pairs of normal, PD without DBS, PD with therapeutic DBS are statistically different except no significant difference between normal and therapeutic DBS (see Fig. 4.2 in Ref. [26]). The computational results in Ref. [26] support the hypothesis that TC relay is compromised and STN–DBS may improve TC relay fidelity in parkinsonian conditions.

In a recent paper Guo et al. [27] used the same data–driven computational model of a TC cell to study the differences and similarities of thalamic relay in dystonia and Parkinson's disease. In this work, the GPi data that modulates TC relay was from recordings of several human patients with either Parkinson's disease or dystonia. Since we did not have recordings from a normal human without PD and PD patients with DBS, we could not distinguish the TC relay between normal, PD, PD with DBS states. However, this work shows another example in which TC relay response has higher errors in parkinsonian conditions.

Experimental Validation

Closing the loop of deep brain stimulation has gained more and more attention in recent years due to its feasibility to improve the therapeutic efficacy and the drawbacks in the conventional DBS (cDBS) [6, 12, 20, 21, 25, 44, 45, 48, 53, 52, 65, 66, 68, 75]. There are a few reports, toward closing the loop, on testing either adaptive or coordinated reset simulation [12, 65, 74]. The two most relevant to our current work are by Little et al. [44] and Tass et al. [74].

Little et al. report an on–demand type of adaptive DBS (aDBS), using brain–computer interfaces to interact with pathological brain signals, on eight human PD patients [44]. They record the LFP directly from the stimulating electrode in the STN, then they band–pass filter and amplify the LFP signal to extract beta oscillation in the LFP. They further rectify and smooth the filtered signal to produce a real time ongoing beta amplitude that is passed to the stimulator. Once the beta amplitude exceeds a user–defined threshold, the stimulator, empowered by a custom–built battery, starts to deliver pulse waveform to the stimulation target. Partial setup of their aDBS is in line with our computational study [25] except that we use the filtered signal to stimulate the target population, instead of a stimulator. Little et al. clinically test stimulations in three cases: cDBS, their aDBS and random intermittent stimulation not triggered by the beta amplitude [44]. They conclude that the improvement in patient's UPDRS (United Parkinson's Disease Rating Scale) with aDBS is much more significant than with cDBS while both aDBS and cDBS improved patient's UPDRS. They find that the improvement in UPDRS using the random stimulation is also significantly less than using aDBS.

Another exceptional work by Tass et al. [74] provides experimental evidence on the efficacy of CRS. They periodically apply brief high–frequency pulse trains with phase shifts in multi–sites of the stimulation electrode to MPTP–treated macaque monkeys. They find that coordinated reset type of stimulation reduces pathological synchronization presented in those PD monkeys. More importantly, the effect of the coordinated reset stimulation is not only acute but also are long–lasting beyond the CRS modulation on motor function in parkinsonian nonhuman primates. The long–lasting aftereffect has never been observed in the conventional DBS. Thus CRS may be superior to the conventional DBS.

Considering the extraordinary difficulty to design a closed–loop method in both experimental and clinical settings, the experimental and clinical work by Little et al. [44] and by Tass et al. [74] are remarkably interesting and encouraging to the community of computational study in the direction of closed–loop DBS.

Discussion

This current work, unlike previous works [25, 45], builds a larger biological–faithful computational network of PD that consists of hundreds of neurons in a more realistic geometric spatial structure in the stimulation target population. We develop and test the open–loop CRS and the closed–loop protocol that applies MDFS stimulation on STN population based on the feedback LFP signals collected within STN. The larger scale computational model of PD enables better understanding of the neuronal characteristics of PD. The plausible MDFS administration with appropriate choices of multiple stimulation sites and parameters introduces patient–specific stimulation. Our computational results suggest that MDFS may be able to overcome downsides of the conventional DBS and also may enable more effective and efficient DBS treatments on PD patients. This current framework can be extended to a large–scale network model of basal ganglia and thalamus with tens of thousands of neurons. Using such a network model, we can further investigate the possible outcomes of different stimulation protocols that may be more close to experimental or clinical settings.

Using computational models to explore the closed–loop stimulation paradigm is a safe and necessary first step, as experimental investigations of new stimulation protocols may be prohibited on humans due to its invasiveness and the associated uncertainty in its clinical effectiveness Computational models have been one step ahead in making meaningful predictions and in probing the possibility of maximizing DBS treatment efficacy while minimizing stimulation current requirements [25, 34, 35, 36, 60, 75].

Some of the computational studies were already validated by experimental work [6, 21, 44, 48, 74], which encourages the community of computational study to move forward in various directions. Based on the framework given in this paper, one immediate study we need to carry out is how to apply the MDFS strategy to the GPi population and examine the outcome on reducing TC relay error. Another important problem is how to optimize the administration of MDFS. In such a computational study, the optimization goal is to reduce the TC relay error in a parkinsonian network. However, the optimization process depends on many factors, such as determining appropriate stimulation target, number of sites, stimulation strength and the delay across the multiple sites. Not only will all of these factors affect the outcome of MDFS, they are also correlated with the measures of synchrony and burst occurrence rate in the basal ganglia. All these factors are key components in desynchronizing and breaking the bursting clusters in the stimulation target population. Hence, we need multiple constraints on these factors and the optimization is a multi–step process with several challenging subproblems associated with closed–loop DBS.

Appendix

In the following text, we use g_i to denote conductance in mS/cm^2 and v_i or V_i to denote reversal potentials in mV, where the subscript i are from the set $\{L, Na, K, Ca, AHP, T, E, Gi, Ge \to Ge, Ge \to Sn, Ge \to Gi, Sn \to Ge, Sn \to Gi\}$. τ with a subscript or both a superscript and a subscript is a time constant in units of msec. All α and β with subscripts are rate constants in units of ms^{-1}. Other parameters either are constants without units or with units given in the following text.

Functions for **TC** neurons in system (4.1):

$$m_\infty(v) = 1/(1+e^{-(v+37)/7}),\ p_\infty(v) = 1/(1+e^{-(v+60)/6.2}),$$

$$h_\infty(v) = 1/(1+e^{(v+41)/4}),\ r_\infty(v) = 1/(1+e^{(v+84)/4}),$$

$$\tau_h(v) = 1/(a_h(v)+b_h(v)),\ \tau_r(v) = 0.4(28+e^{-(v+25)/10.5}),$$

$$a_h(v) = 0.128e^{-(46+v)/18},\ b_h(v) = 4/(1+e^{-(23+v)/5}),$$

Parameters for **TC** neurons:

$$g_L = 0.14,\ g_{Na} = 3,\ g_K = 5,\ g_T = 5,\ g_E = 0.018,\ g_{Gi} = .009,$$

$$v_L = -72,\ v_{Na} = 50,\ v_K = -90,\ v_T = 90,\ v_E = 0,\ V_{Gi} = -85,$$

$$p = 50\text{ms},\ d = 5\text{ms},\ win_{off} = 12\text{ms}.$$

GPi currents:

$$I_L(v) = g_L(v-v_L),\ I_{Na} = g_{Na}(m_\infty^3(v))h(v-v_{Na}),\ I_K = g_K n^4(v-v_K),$$

$$I_T = g_T a_\infty^3(v)r(v-v_{Ca}),\ I_{Ca} = g_{Ca}s_\infty^2(v)(v-v_{Ca})$$

$$I_{AHP} = g_{AHP}(v-v_K)([Ca]/([Ca]+k)),$$

$$I_{Sn \to Gi} = g_{Sn \to Gi} s_{Sn \to Gi}(v-v_{Sn \to Gi}),$$

where $s_{Sn \to Gi}$ is listed under STN equations,
and

$$I_{Ge \to Gi} = g_{Ge \to Gi} s_{Ge \to Gi}(v-v_{Ge \to Gi}),$$

where $s_{Ge \to Gi}$ is listed under GPe equations.
$I_{appi} = -1\mu A$ is a constant applied current.
GPi equations and functions:

$$n' = \varphi_n(n_\infty(v) - n)/\tau_n(v),\ h' = \varphi_h(h_\infty(v) - h)/\tau_h(v),\ r' = \varphi(r_\infty(v) - r)/\tau_r,$$

$$[Ca]' = \varepsilon(-I_{Ca} - I_T - k_{Ca}[Ca]),\ s'_{Gi} = \alpha(1 - s_{Gi})S_\infty(v) - \beta_{Gi} s_{Gi},$$

where $S_\infty(v)$ is given in section "Model Equations for STN, GPe and GPi Neurons".

$$X_\infty(v) = 1/(1 + e^{-(v-\theta_X)/\sigma_X}),$$

where $X = m, n, h, r, a, s$,
and

$$\tau_X(v) = \tau_X^0 + \tau_X^1/(1 + e^{-(v-\theta_X^\tau)/\sigma_X^\tau}),$$

where $X = n, h$.

GPi parameters:

$g_L = 0.1,\ g_{Na} = 120,\ g_K = 30,\ g_T = 0.5,\ g_{Ca} = 0.1,\ g_{AHP} = 30,\ g_{Sn \to Gi} = 0.5,$
$g_{Ge \to Gi} = 1,$

$v_L = -55,\ v_{Na} = 55,\ v_K = -80,\ v_{Ca} = 120,\ v_{Ge \to Gi} = -100,\ v_{Sn \to Gi} = 0,$

$\tau_n^0 = 0.05,\ \tau_n^1 = 0.27,\ \tau_h^0 = 0.05,\ \tau_h^1 = 0.27,\ \tau_r = 30,$

$\phi_r = 1,\ \phi_n = 0.1,\ \phi_h = 0.05$

$k_1 = 30,\ k_{Ca} = 15,\ \varepsilon = 0.0001\text{ms}^{-1},$

$\theta_r = -70,\ \theta_m = -37,\ \theta_n = -50,\ \theta_h = -58,\ \theta_a = -57,\ \theta_s = -35,\ \alpha = 2,\ \theta_n^\tau = -40,$
$\theta_h^\tau = -40,$

$\sigma_m = 10,\ \sigma_n = 14,\ \sigma_h = -12,\ \sigma_r = -2,\ \sigma_a = 2,\ \sigma_s = 2,\ \sigma_n^\tau = -12,\ \sigma_h^\tau = -12,$

$\beta_{Gi} = .08,\ k_{Ca} = 15.$

STN currents:

$I_L,\ I_{Na},\ I_K,\ I_{Ca},\ I_{AHP}$ are as given above for the GPi neuron, and $I_T = g_T a_\infty^3(v) b_\infty(r)(v - v_{Ca})$. The synaptic currents from GPe to STN are the following:

$I_{Ge \to Sn} = g_{Ge \to Sn} \sum_{j \in \Lambda} s_{Ge \to Sn}^j (v - v_{Sn \to Ge})$, where Λ is a subgroup of GPe cells and $s_{Ge \to Sn}$ is given in GPe equations.

For the stimulation current I^{stim}, see details in sections "Generation of Stimulation Current from Feedback" and "Comparison Between Open–Loop and Closed–Loop Stimulations".

STN equations and functions:

4 Computational Models of Closed–Loop Deep Brain Stimulation

$n, h, r, [Ca]$ equations and functions $X_\infty(v), \tau_X(v)$ are the same as given above for GPi neuron, except there is no $r_\infty(v)$ used and we introduce $b_\infty(r) = 1/(1+e^{(r-\theta_b)/\sigma_b}) - 1/(1+e^{-\theta_b/\sigma_b})$.

The synaptic input from STN to GPe and GPi is described as:

$$s_{Sn \to Ge}' = \alpha_{Sn \to Ge}(1 - s_{Sn \to Ge})s_\infty(v-30) - \beta_{Sn \to Ge}s_{Sn \to Ge},$$

$$s_{Sn \to Gi}' = \alpha_{Sn \to Gi}(1 - s_{Sn \to Gi})s_\infty(v-30) - \beta_{Sn \to Gi}s_{Sn \to Gi},$$

STN Parameters:

$$g_L = 2.25, g_{Na} = 37.5, g_K = 45, g_T = 0.5, g_{Ca} = 0.5, g_{AHP} = 9, g_{Ge \to Sn} = 0.9,$$

$$v_L = -60, v_{Na} = 55, v_K = -80, v_{Ca} = 140, v_{Ge \to Sn} = -100,$$

$$\tau_n^0 = 1, \tau_n^1 = 100, \tau_h^0 = 1, \tau_h^1 = 500, \tau_r^0 = 7.1, \tau_r^1 = 17.5,$$

$$\phi_r = 0.5, \phi_n = 0.75, \phi_h = 0.75,$$

$$k_1 = 15, k_{Ca} = 22.5, \varepsilon = 5 \times 10^{-5},$$

$$\theta_r = -67, \theta_m = -30, \theta_n = -32, \theta_h = -39, \theta_a = -63, \theta_s = -39, \theta_b = 0.25,$$

$$\theta_n^\tau = -80, \theta_h^\tau = -57, \theta_r^\tau = 68$$

$$\sigma_m = 15, \sigma_n = 8, \sigma_h = -3.1, \sigma_r = -2, \sigma_a = 7.8, \sigma_s = 8, \sigma_b = 0.07$$

$$\sigma_n^\tau = -26, \sigma_h^\tau = -3, \sigma_r^\tau = -2.2,$$

$$\alpha_{Sn \to Ge} = 5, \alpha_{Sn \to Gi} = 1, \beta_{Sn \to Ge} = 1, \beta_{Sn \to Gi} = 0.05, wk = 0.45.$$

GPe currents:

$I_L, I_{Na}, I_K, I_{Ca}, I_{AHP}$ are modeled as given above for the GPi neuron. The synaptic currents to GPe are:

$I_{Sn \to Ge} = g_{Sn \to Ge} \sum_{j \in \Lambda} s_{Sn \to Ge}(v - v_{Sn \to Ge})$, where Λ is a subgroup of STN neurons and $s_{Sn \to Ge}$ is given under STN equations.

$I_{Ge \to Ge} = g_{Ge \to Ge} \sum_{j \in \Lambda} s_{Ge \to Ge}(v - v_{Ge \to Ge})$ where Λ is a subgroup of STN neurons and $s_{Ge \to Ge}$ is the same as $s_{Ge \to Sn}$ given under GPe equations.

$I_{app} = -1.2$ is a constant applied current.

GPe equations and functions:

$n, h, r, [Ca]$ equations and function $X_\infty(v), \tau_X(v)$ are as given above for the GPi neuron.

The synaptic input from STN to GPe and GPi is described as:

$$s_{Ge \to Sn}' = \alpha_{Ge \to Sn}(1 - s_{Ge \to Sn})s_\infty(v-20) - \beta_{Ge \to Sn}s_{Ge \to Sn},$$

$$s_{Ge \to Gi}' = \alpha_{Ge \to Gi}(1 - s_{Ge \to Gi})s_\infty(v - 20) - \beta_{Ge \to Gi}s_{Ge \to Gi}.$$

GPe parameters:
Most parameters for GPe are the same as those for GPi. We only list those that have different values and the additional ones not present in the GPi model.

$$g_{Sn \to Ge} = 0.18,\ g_{Ge \to Ge} = 0.01,\ v_{Sn \to Ge} = 0,\ v_{Ge \to Ge} = -80,$$

$$\alpha_{Ge \to Sn} = 2,\ \alpha_{Ge \to Gi} = 1$$

$$\beta_{Ge \to Sn} = 0.04,\ \beta_{Ge \to Gi} = 0.1.$$

CRS parameters: $\rho = 0.7$, $a_1 = 0.9$, $a_0 = 58$, $\tau_0 = 30$, which is the phase shift in f_k, $k = 1,2,3,4$.

MDFS parameters: $a = 0.0025$, $b = 0.00136$, $\mu = 0.0002$, $\tau = 25$.

Acknowledgments Yixin Guo received support from NSF awards DMS 1226180.

References

1. Alexander GE, Crutcher MD, DeLong MR. Basal ganglia thalamo-cortical circuits: parallel substrates for motor, oculomotor, "prefrontal" and "limbic" functions. Prog Brain Res 1990;85:119–46.
2. Andrews RJ. Neuroprotection at the nanolevel—part ii nanodevices for neuromodulation—deep brain stimulation and spinal cord injury. Ann N Y Acad Sci. 2007;1122:185–96.
3. Andrews RJ. Neuromodulation: advances in the next five years. Ann N Y Acad Sci. 2010;1199:204–11.
4. Bergman H, Wichmann T, Karmon B, DeLong M. The primate subthalamic nucleus. ii. Neuronal activity in the mptp model of Parkinsonism. J Neurophysiol. 1994;72:507–20.
5. Best J, Park C, Terman D, Wilson C. Transitions between irregular and rhythmic firing patterns in excitatory-inhibitory neuronal networks. J Comput Neurosci. 2007;23:217–35.
6. Beuter A, Lefaucheur J-P, Modolo J. Closed-loop cortical neuromodulation in Parkinson's disease: an alternative to deep brain stimulation? Clin Neurophysiol. 2014;125:874–85.
7. Bevan MD, Jeremy AF, Jérôme B. Cellular principles underlying normal and pathological activity in the subthalamic nucleus. Curr Opin Neurobiol. 2006;16:621–8.
8. Boraud T, Bezard E, Guehl D, Bioulac B, Gross C. Effects of l-dopa on neuronal activity of the globus pallidus externalis (gpe) and globus pallidus internalis (gpi) in the mptp-treated monkey. Brain Res. 1998;787:157–60.
9. Brown P, Oliviero A, Mazzone P, Insola A, Tonali P, Lazzaro VD. Dopamine dependency of oscillations between subthalmaic nucleus and pallidum in Parkinson's disease. J Neurosci. 2001;21:1033–8.
10. Buzsáki G, Anastassiou CA, Koch C. The origin of extracellular fields and currents—eeg, ecog, lfp and spikes. Nat Rev Neurosci. 2012;13(6):407–20.
11. Cagnan H, Meijer HGE, van Gils SA, Krupa M, Heida T, Rudolph M, Wadman WJ, Martens HCF. Frequency-selectivity of a thalamocortical relay neuron during Parkinson's disease and deep brain stimulation: a computational study. Eur J Neurosci. 2009;30:1306–17.
12. Carron R, Chaillet A, Filipchuk A, Pasillas-Lépine W, Hammond C. Closing the loop of deep brain stimulation. Front Syst Neurosci. 2013;7(112):1–18.

13. Castro-Alamancos M, Calcagnotto M. High-pass filtering of corticothalamic activity by neuromodulators released in the thalamus during arousal: in vitro and in vivo. J Neurophysiol. 2001;85:1489–97.
14. Chen CC, Pogosyana A, Zrinzoa UL, Tischa S, Limousina P, Ashkana K, Yousry T, Hariza MI, Brown P. Intra-operative recordings of local field potentials can help localize the subthalamic nucleus in Parkinson's disease surgery. Exp Neurol. 2006;198:214–21.
15. Degos B, Deniau J, Thierry A, Glowinski J, Pezard L, Maurice N. Neuroleptic-induced catalepsy: electrophysiological mechanisms of functional recovery induced by high-frequency stimulation of the subthalamic nucleus. J. Neurosci. 2005;25(33):7687–96.
16. DeLong MR, Wichmann T. Circuits and circuit disorders of the basal ganglia. Arch Neurol. 2007;64(1):20–4.
17. DeVito JL, Anderson ME. An autoradiographic study of the efferent connections of the globus pallidus in macaca mullata. Brain Res. 1982;46:107–17.
18. Dovzhenok A, Park C, Worth RM, Rubchinsky LL. Failure of delayed feedback deep brain stimulation for intermittent pathological synchronization in Parkinson's disease. PLoS ONE. 2013;8(3):1–10.
19. Feng X, Shea-Brown E, Rabitz H, Greenwald B, Kosut R. Optimal deep brain stimulation of the subthalamic nucleus—a computational study. J Comput Neurosci. 2007a;23(3):265–82.
20. Gorzeli P, Schiff SJ, Sinha A. Model-based rational feedback controller design for closed-loop deep brain stimulation of Parkinson's disease. J Neural Eng. 2013;10(2):1–16.
21. Graupe D, Basu DI, Tuninetti D, Vannemreddy P, Slavin KV. Adaptively controlling deep brain stimulation in essential tremor patient via surface electromyography. Neurol Res. 2010;32:899–904.
22. Guillery R, Sherman SM. The role of thalamus in the flow of information to the cortex. Philos Trans R Soc Lond B Biol Sci. 2002a;357:1695–708.
23. Guillery R, Sherman SM. The thalamus as a monitor of motor outputs. Philos Trans R Soc Lond B Biol Sci. 2002b;357:1809–21.
24. Guillery R, Sherman SM. Thalamic relay functions and their role in corticocortical communication: generalizations from the visual system. Neuron. 2002c;33:163–75.
25. Guo Y, Rubin JE. Multi-site stimulation of subthalamic nucleus diminishes thalamocortical relay errors in a biophysical network model. Neural Netw. 2011;24(6):602–16.
26. Guo Y, Rubin JE, McIntyre CC, Vitek JL, Terman D. Thalamocortical relay fidelity varies across subthalamic nucleus deep brain stimulation protocols in a data-driven computational model. J Neurophysiol. 2008;99:1477–92.
27. Guo Y, Park C, Worth RM, Rubchinsky LL. Basal ganglia modulation of thalamocortical relay in Parkinson's disease and dystonia. Front Comput Neurosci. 2013;7(124):1–11.
28. Haber S. The primate basal ganglia: parallel and integrative networks. J Chem Neuroanat. 2003;2:317–30.
29. Hahn PJ, McIntyre CC. Modeling shifts in the rate and pattern of subthalamopallidal network activity during deep brain stimulation. J Comput Neurosci. 2010;28(3):425–41.
30. Halliday GM. Thalamic changes in Parkinson's disease. Parkinsonism Relat Disord. 2009;15(S3):S152–5.
31. Halliday GM, Macdonald V, Henderson JM. A comparison of degeneration in motor thalamus and cortex between progressive supranuclear palsy and Parkinson's disease. Brain. 2005 Oct;128(10):2272–80.
32. Hamani C, Saint-Cyr JA, Fraser J, Kaplitt M, Lozano AM. The subthalamic nucleus in the context of movement disorders. Brain 2004;127(1):4–20.
33. Hammond C, Ammari R, Bioulac B, Garcia L. Latest view on the mechanism of action of deep brain stimulation. Mov Disord. 2008;23(15):2111–21.
34. Hauptmann C, Popovych O, Tass PA. Effectively desynchronizing brain stimulation based on a coordinated delayed feedback stimulation via serveral sites: a computational study. Biol Cybern. 2005;93:463–70.
35. Hauptmann C, Omel'chenko OO, Popovych OV, Maistrenko Y, Tass PA. Control of spatially patterned synchrony with multisite delayed feedback. Phys Rev E. 2007 Dec;76(6):066209.

36. Hauptmann C, Omel'chenko O, Popovych OV, Maistrenko Y, Tass PA. Desynchronizing the abnormally synchronized neural activity in the subthalamic nucleus: a modeling study. Expert Rev Med Devices. 2008 Sept;4(5):633–50.
37. Holt GR, Koch C. Electrical interactions via the extracellular potential near cell bodies. J Comput Neurosci. 1998;6(2):169–84.
38. Hurtado J, Rubchinsky L, Sigvardt K, Wheelock V, Pappas C. Temporal evolution of oscillations and synchrony in gpi/muscle pairs in Parkinson's disease. J Neurophysiol. 2005;93:1569–84, 2005.
39. Jolliffe IT. Principal component analysis. New York: Springer; 1986.
40. Kelly RM, Strick PL. Macro-architecture of basal ganglia loops with the cerebral cortex: use of rabies virus to reveal multisynaptic circuits. Prog Brain Res. 2004;143:449–59.
41. Lee JYK, Deogaonkar M, Rezai A. Deep brain stimulation of globus pallidus internus for dystonia. Parkinsonism Relat Disord. 2007;13(5):261–5.
42. Leung LWS. Field potentials in the central nervous system: recording, analysis, and modeling. In neurophysiological techniques: applications to neural systems. Neuromethods. 1990;15:277–312.
43. Levy R, Hutchison W, Lozano A, Dostrovsky J. High-frequency synchronization of neuronal activity in the subthalamic nucleus of Parkinsonian patients with limb tremor. J Neurosci. 2003;20:7766–75.
44. Little S, Pogosyan A, Neal S, Zavala B, Zrinzo L, Hariz M, Foltynie T, Limousin P, Ashkan K, FitzGerald J, Green AL, Aziz TZ, Brown P. Adaptive deep brain stimulation in advanced Parkinson disease. Ann Neurol. 2013;74(3):449–57.
45. Lourens M. Neural network dynamics in Parkinson's disease. Ph. D. Thesis, University of Twente; 2013.
46. Lozano AM, Hamani C. The future of deep brain stimulation. J Clin Neurophysiol. 2004;21(1):68–9.
47. Magnin M, Morel A, Jeanmonod D. Single-unit analysis of the pallidum, thalamus, and subthalamic nucleus in parkinsonian patients. Neuroscience. 2000;96:549–64.
48. Marceglia S, Rossi L, Foffani G, Bianchi A, Cerutti S, Priori A. Basal ganglia local field potentials: applications in the development of new deep brain stimulation devices for movement disorder. Expert Rev Med Devices. 2007;4(5):610–4.
49. McFarland NR, Haber SN. Thalamic relay nuclei of the basal ganglia form both reciprocal and nonreciprocal cortical connections, linking multiple frontal cortical areas. J Neurosci. 2002;22(18):8117–32.
50. Middleton FA, Strick PL. Basal ganglia output and cognition: evidence from anatomical, behavioral, and clinical studies. Brain Cogn. 2000;42:183–200.
51. Mitzdorf U. Current source-density method and application in cat cerebral cortex: investigation of evoked potentials and EEG phenomena. Physiol Rev. 1985;65(1):37–100.
52. Modolo J, Beuter A. Linking brain dynamics, neural mechanisms, and deep brain stimulation in Parkinson's disease: an integrated perspective. Med Eng Phys. 2009 July;31(6):615–23
53. Modolo J, Beuter A, Thomas AW, Legros A. Using "smart stimulators" to treat Parkinson's disease: re-engineering neurostimulation devices. Front Comput Neurosci. 2012;6:69:1–3.
54. Nini A, Feingold A, Slovin H, Bergman H. Neurons in the globus pallidus do not show correlated activity in the normal monkey, but phase-locked oscillations appear in the mptp model of Parkinsonism. J Neurophysiol. 1995;74:1800–5.
55. Nunez PL. Electric fields of the brain. New York: Oxford University Press; 1981.
56. Obeso JA, Marin C, Rodriguez-Oroz C, Blesa J, Benitez-Temiño B, Mena-Segovia J, Rodríguez M, Olanow CW. The basal ganglia in Parkinson's disease: current concepts and unexplained observations. Ann Neurol. 2008;64(S2):S30–46.
57. O'Connor S, Angelo K, Jacob TJ. Burst firing versus synchrony in a gap junction connected olfactory bulb mitral cell network model. Front Comput Neurosc. 2012;6(75):1–18.
58. Pirini M, Rocchiand L, Sensi M, Chiari L. A computational approach to investigate different targets in deep brain stimulation for Parkinson's disease. J Comp Neurosci. 2009;26:91–107.
59. Plenz D, Kitai S. A basal ganglia pacemaker formed by the subthalamic nucleus and external globus pallidus. Nature. 1999;400:677–82.

60. Popovych OV, Hauptmann C, Tass PA. Control of neuronal sychrony by nonlinear delayed feedback. Biol Cybern. 2006;95:69–85.
61. Raz A, Vaadia E, Bergman H. Firing patterns and correlations of spontaneous discharge of pallidal neurons in the normal and tremulous 1-methyl-4-phenyl-1,2,3,6 tetrahydropyridine vervet model of parkinsonism. J Neurosci. 2000;20:8559–71.
62. Rinzel J. Bursting oscillations in an excitable membrane model. In: Sleeman B, Jarvis R, editors. Ordinary and partial differential equations. New York: Springer; 1985. pp. 304–16.
63. Rosenblum MG, Pikovsky AS. Controlling synchronization in an ensemble of globally coupled oscillators. Phys Rev Lett. 2004a March;92(11):114102.
64. Rosenblum MG, Pikovsky AS. Delayed feedback control of collective synchrony: An approach to suppression of pathological brain rhythms. Phys Rev E. 2004b;70(4):041904.
65. Rosin B, Slovik M, Mitelman R, Rivlin-Etzion M, Haber SN, Israel E, Vaadia Z, Bergman H. Closed-loop deep brain stimulation is superior in ameliorating Parkinsonism. Neuron. 2011;72(2):370–84.
66. Rubin JE, Terman D. High frequency stimulation of the subthalamic nucleus eliminaties pathological thalamic rythmicity in a computational model. J Comput Neurosci. 2004;16:211–35.
67. Rubin J, McIntyre CC, Turner RS, Wichmann T. Basal ganglia activity patterns in Parkinsonism and computational modeling of their downstream effects. Eur J Neurosci. 2012;36(2):2213–28.
68. Sabato S, Grill WM, Glielmo L, Fiengo G. Closed-loop control of deep brain stimulation: a simulation study. Trans Neurol Syst Rehabil Eng. 2011;19:15–25.
69. Schiff SJ. Towards model-based control of Parkinson's disease. Phil Trans R Soc A. 2010;368:2269–308.
70. Seebens H, Einsle U, Straile D. Devaiations from syn- chrony: spatio-temporal variability of zooplankton community dynamics in a large lake. J Plankt Res. 20120(0):1–11.
71. Silberstein P, Kühn AA, Kupsch A, Trottenberg T, Krauss JK, Wöhrle JC, Mazzone P, Insola A, Di Lazzaro V, Oliviero A, Aziz T, Brown P. Patterning of globus pallidus local field potentials differs between Parkinson's disease and dystonia. Brain. 2003;126:2597–608.
72. Sommer MA. The role of the thalamus in motor control. Curr Opin Neurobiol. 2003;13(6):663–70.
73. Starr PA, Rau GM, Davis V, Marks Jr WJ, Ostrem JL, Simmons D, Lindsey N, Turner RS. Spontaneous pallidal neuronal activity in human dystonia: comparison with Parkinson's disease and normal macaque. J Neurophysiol. 2005;93:3165–76.
74. Tass PA. A model of desynchronizing deep brain stimulation with a demand-controlled coordinated reset of neural subpopulations. Biol Cybern. 2003;89:81–8.
75. Tass PA, Qin L, Hauptmann C, Dovero S, Bezard E, Boraud T, Meissner WG. Coordinated reset has sustained aftereffects in Parkinsonian monkeys. Ann Neurol. 2012;72(5):816–20.
76. Terman D, Rubin JE, Yew AC, Wilson CJ. Activity patterns in a model for the subthalamopallidal network of the basal ganglia. J Nerurosci. 2002;2002(7):2963–76.
77. Tukhlina N, Rosenblum M, Pikovsky A, Kurths J. Feedback suppression of neural synchrony by vanishing stimulation. Phys RevE. 2007;75:011918.
78. Urbain N, Gervasoni D, Souliere F, Lobo L, Rentero N, Windels F, Astier B, Savasta M, Fort P, Renaud B. Unrelated course of subthalamic nucleus and globus pallidus neuronal activities across vigilance states in the rat. Eur J Neurosci. 2000;12:3361–74.
79. Urbain N, Rentero N, Gervasoni D, Renaud B, Chouvet G. The switch of subthalamic neurons from an irregular to a bursting pattern does not solely depend on their gabaergic inputs in the anesthetic-free rat. J Neurosci. 2002;22:8665–75.
80. Vitek JL. Pathophysiology of dystonia: a neuronal model. Mov Disord. 2002;17:S49–62.
81. Volkmann J. Deep brain stimulation for the treatment of Parkinson's disease. J Clin Neurophys. 2004;21:6–17.
82. Wichmann T, DeLong MR. Deep brain stimulation for neurologic disorders. Neuron. 2006;52:197–204.
83. Wichmann T, Soares J. Neuronal firing before and after burst discharges in the monkey basal ganglia is predictably patterned in the normal state and altered in parkinsonism. J Neurophysiol. 2006;95:2120–33.

84. Wichmann T, Bergman H, Starr P, Subramanian T, Watts R, DeLong M. Comparison of mptp-induced changes in spontaneous neuronal discharge in the internal pallidal segment and in the substantia nigra pars reticulata in primates. Exp Brain Res. 1999;125:397–409.
85. Yoshida M, Rabin A, Anderson ME. Monosynaptic inhibition of pallidal neurons by axon collaterals of caudatonigral fibers. Exp Brain Res. 1972;15:333–47.
86. Yoshida F, Martinez-Torres I, Pogosyan A, Holl E, Petersen E, Chen CC, Foltynie T, Limousin P, Zrinzo LU, Hariz MI, Brown P. Value of subthalamic nucleus local field potentials recordings in predicting stimulation parameters for deep brain stimulation in Parkinson's disease. J Neurol Neurosurg Psychiatry. 2010;81(8):885–9.

Chapter 5
A Multiscale "Working Brain" Model

P. A. Robinson, S. Postnova, R. G. Abeysuriya, J. W. Kim, J. A. Roberts,
L. McKenzie-Sell, A. Karanjai, C. C. Kerr, F. Fung, R. Anderson,
M. J. Breakspear, P. M. Drysdale, B. D. Fulcher, A. J. K. Phillips,
C. J. Rennie and G. Yin

Abstract By modeling salient features of the corticothalamic system over multiple spatial and temporal scales, physiologically based neural field theory has yielded numerous successful predictions that interrelate stimuli, neural activity, and measurements. Likewise, physiologically based neural mass theories of the brainstem-hypothalamus sleep-wake switch and associated systems have recently been developed and shown to quantitatively reproduce a wide variety of arousal-state phenomena. In both cases, model parameters have been independently constrained, and each model has integrated multiple phenomena and measurements into a single unified framework, thereby validating the modeling approach and enabling these features to be interrelated and interpreted in terms of underlying physiology and anatomy. Here, a first integration of the corticothalamic and arousal-state models is carried out by incorporating a simple model of their couplings: upward via the neuromodulatory effects of the ascending arousal system, and downward via the gating of light inputs by higher-level behavior. The resulting "working brain" system has a neural-mass-like limit, governed by delay differential equations that enable it to respond correctly to light-dark cycles, sleep deprivation, jetlag, and pharmacological inputs, while driving the corticothalamic system into parameter regions where it reproduces associated electroencephalograms, evoked response potentials, and other phenomena, whose properties are further elucidated by retaining the appropriate neural field equations. Overall, the combined model provides a simple, highly flexible framework for quantitatively modeling a variety of mesoscale to

P. A. Robinson (✉) · S. Postnova · R. G. Abeysuriya · J. W. Kim · L. McKenzie-Sell ·
A. Karanjai · C. C. Kerr · F. Fung · R. Anderson · P. M. Drysdale · B. D. Fulcher ·
C. J. Rennie · G. Yin
School of Physics, University of Sydney, Sydney, NSW 2006, Australia
e-mail: robinson@physics.usyd.edu.au

P. A. Robinson · S. Postnova · R. G. Abeysuriya · J. W. Kim
Center for Integrated Research and Understanding of Sleep, University of Sydney,
431 Glebe Point Rd, Glebe, NSW 2037, Australia

P. A. Robinson · S. Postnova · R. G. Abeysuriya · J. W. Kim
Cooperative Research Center for Alertness, Safety, and Productivity, University of Sydney,
NSW 2006, Australia

macroscale brain phenomena, ranging from normal behaviors to highly nonlinear dynamics such as found in seizures, and for examining interactions between such phenomena. these findings are illustrated with representative examples. Fitting of the model to data can be used to infer brain states and underlying parameters.

Keywords Neural field theory · Brain dynamics · Arousal state · Simulations · Modeling

Introduction

Brain dynamics involves interactions across many scales—spatially from microscopic to whole-brain, and temporally from the sub-millisecond range to years. Except under conditions that isolate a single scale, these multiscale aspects of the underlying physiology and anatomy must be included to model the behavior adequately at any scale, and to enable quantitative validation against experiment.

Neural field theories provide a natural basis for modeling and analyzing multiscale neural systems. Moreover, links to measurements can be included—an essential point because most measurement processes aggregate over many neurons and all modify signals in some way. In the class of models described here, averages are taken over microscopic neural structure to obtain mean-field descriptions on scales from tenths of a millimeter up to the whole brain, incorporating representations of the anatomy and physiology of separate excitatory and inhibitory neural populations, nonlinear neural responses, multiscale interconnections, synaptic, dendritic, cell-body, and axonal dynamics, and feedbacks between structures and neural

---

P. A. Robinson
Center for Integrative Brain Function, University of Sydney, Sydney, NSW 2006, Australia

P. A. Robinson · S. Postnova · R. G. Abeysuriya · J. W. Kim
NeuroSleep: The Center for Translational Sleep and Circadian Neurobiology,
University of Sydney, 431 Glebe Point Rd, 2037 Glebe, NSW, Australia

P. A. Robinson · R. G. Abeysuriya · J. W. Kim · C. C. Kerr · C. J. Rennie
Brain Dynamics Center, Westmead Millennium Institute, Darcy Rd,
Westmead, NSW 2145, Australia

J. A. Roberts · M. J. Breakspear
QIMR Berghofer Medical Research Institute, Herston, QLD 4006, Australia

C. C. Kerr
Department of Physiology and Pharmacology, Downstate Medical Center, State University of New York, 450 Clarkson Ave, Brooklyn, NY 11203, USA

A. J. K. Phillip
Division of Sleep Medicine, Brigham and Women's Hospital, 221 Longwood Ave, Suite 438, Boston, MA 02115, USA

populations [11, 20, 29, 43, 44, 63, 64, 67, 68, 74, 76, 79, 77, 72, 75, 73, 78, 83, 40, 93, 99, 100].

Essential features of any realistic neurodynamic model are that it: (i) be based on physiology and anatomy, including salient features at many spatial and temporal scales; (ii) be quantitative, with predictions that can be calculated analytically or numerically and verified against experiment; (iii) have parameters that directly relate to physiology and anatomy, and that can be measured or constrained via independent experiments (this does not exclude the theory itself enabling improved estimates of parameters); (iv) be applicable to multiple phenomena and data types; and (v) be invertible, if possible, allowing parameters to be deduced by fitting model predictions to data (the parameters obtained must be consistent with independent measurements). These criteria rule out (among others) highly idealized models of abstract neurons, as are sometimes used in computer science, theories of single phenomena, and models with completely free parameters.

We have developed a physiologically based neural-field model of brain dynamics that satisfies the above criteria and which has been extensively validated against a wide range of experiments. When applied to the corticothalamic system, it reproduces and unifies many features of EEGs, including background spectra and the spectral peaks seen in waking and sleeping states [79, 72, 75], evoked response potentials [64], measures of coherence and spatiotemporal structure [47, 46, 67, 68], and generalized epilepsies and low-dimensional seizure dynamics [11, 76]. Similarly, we have developed, calibrated, and applied a neural mass model of sleep-wake dynamics, based on nuclei in the brainstem and hypothalamus, plus the suprachiasmatic nucleus (SCN) [51, 52, 53, 22, 23, 57, 59, 60].

In the present work, we unite our corticothalamic and arousal-state models into the single "working brain" model seen in Fig. 5.1. This model has the correct arousal dynamics, shows appropriate spontaneous and evoked responses, and reproduces nonlinear dynamics such as those found in epilepsy, among other outcomes. This provides a workbench for understanding a wide variety of mesoscale to macroscale brain phenomena and their interrelationships. The structure of the chapter is as follows: in section "Modeling" we explain neural field theory and its applications to the corticothalamic system, sleep-wake system, and their interactions. In section

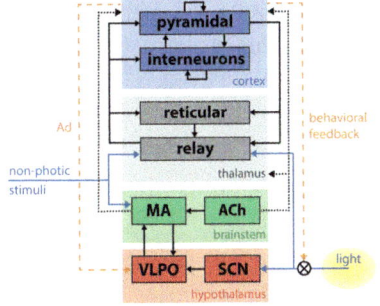

Fig. 5.1 Schematic of "working brain" model, showing structures involved in corticothalamic generation of EEGs and the brainstem-hypothalamus ascending arousal system (see section "Arousal State Modeling" for abbreviations), with some of their main inputs, connections, and feedbacks shown by arrows. Adapted from Ref. [82]

"Brain Dynamics" we outline some of the key areas in which the model is validated against experiment, including a number of new comparisons. Section "Summary" provides a brief summary.

Modeling

In this section we briefly review the key corticothalamic and arousal-system sectors of our model, parameter calibration, and links between the two sectors. More detailed discussion and further generalizations can be found elsewhere [63, 64, 68, 70, 72]. We note that other variants of most individual aspects of the model exist in the literature, but that detailed discussion of these variants is beyond the scope of the present chapter, where our aim is to unite the sectors as shown in Fig. 5.1.

General Neural-Field Modeling

The brain contains multiple populations of neurons, which we distinguish by a subscript a that designates both the structure in which a given population lies (e.g., a particular nucleus) and the type of neuron (e.g., interneuron, pyramidal cell). We average their properties over scales of ~0.1 mm and seek equations for the resulting mean-field quantities.

The mean soma potential $V_a(r,t)$ (measured relative to rest here) is approximated as the sum of contributions $V_{ab}(r,t)$ arriving as a result of activity at each type of (mainly) dendritic synapse b, where b denotes both the population and neurotransmitter type, r is the spatial coordinate, and t the time. This gives

$$V_a(\mathbf{r},t) = \sum_b V_{ab}(\mathbf{r},t). \tag{5.1}$$

The potential V_{ab} is generated when synaptic inputs from afferent neurons b are temporally low-pass filtered and smeared out in time as a result of receptor dynamics and passage through the dendrites of neurons a (i.e., by dynamics of ion channels, membranes, etc.). It approximately obeys the differential equation [69, 79, 72, 75]

$$D_{ab}V_{ab}(\mathbf{r},t) = N_{ab}s_{ab}\varphi_b(\mathbf{r},t-\tau_{ab}), \tag{5.2}$$

$$D_{ab} = \frac{1}{\alpha_{ab}\beta_{ab}}\frac{d^2}{dt^2} + \left(\frac{1}{\alpha_{ab}}+\frac{1}{\beta_{ab}}\right)\frac{d}{dt}+1, \tag{5.3}$$

where $1/\beta_{ab}$ and $1/\alpha_{ab}$ are the characteristic rise and decay times of the potential change due to an impulse at a dendritic synapse. The right of Eq. (5.2) describes the influence of the firing rates φ_b from neuronal populations b, in general delayed by a time τ_{ab} due to discrete anatomical separations between different structures. The quantity N_{ab} is the mean number of synapses to neurons of type a from type b, and s_{ab} is the time-integrated response in neurons of type a to a unit signal from neurons of type b, implicitly weighted by the neurotransmitter release probability. In the present chapter we treat the s_{ab} as constants and ignore their dynamics, which can be driven by neuromodulators, firing rate, and other effects; however, such dynamics can be incorporated straightforwardly [e.g., [13]].

Action potentials are produced at the axonal hillock when the soma potential V_a exceeds a threshold, at a rate that rises steeply with V_a before leveling off. In a population, this dependence is smeared out by differences in individual neurons and their environments to yield the population-average response function

$$Q_a(\mathbf{r},t) = S[V_a(\mathbf{r},t)], \qquad (5.4)$$

where S is a sigmoid function that increases from 0 to Q_{max} as V_a increases from $-\infty$ to $+\infty$ [20, 69, 72]. We use

$$S[V_a(\mathbf{r},t)] = \frac{Q_{max}}{1+\exp\{-[V_a(\mathbf{r},t)-\theta_a]/\sigma'\}}, \qquad (5.5)$$

where we assume a common mean neural firing threshold θ relative to resting, with $\sigma'\pi/\sqrt{3}$ being its standard deviation (here, θ, σ', and Q_{max} are assumed to be the same in all populations for simplicity). When treating linear perturbations relative to a given state, we make the approximation

$$Q_a(\mathbf{r},t) = \rho_a V_a(\mathbf{r},t), \qquad (5.6)$$

where Q_a and V_a are now perturbations and ρ_a is the derivative of the sigmoid at an assumed steady state of the system in the absence of perturbations (we discuss the existence and stability of such states below).

Each neuronal population a in the corticothalamic system produces a field φ_a of pulses, that travels to other neuronal populations at a velocity v_a through axons with a characteristic range r_a. This activity spreads out and dissipates if not regenerated. To a good approximation, this type of propagation obeys a damped wave equation [29, 44, 72]:

$$D_a \varphi_a(\mathbf{r},t) = S[V_a(\mathbf{r},t)], \qquad (5.7)$$

$$D_a = \frac{1}{\gamma_a^2}\frac{\partial^2}{\partial t^2} + \frac{2}{\gamma_a}\frac{\partial}{\partial t} + 1 - r_a^2 \nabla^2, \qquad (5.8)$$

where the damping coefficient is $\gamma_a = v_a/r_a$. Eqs. (5.7) and (5.8) yield propagation ranges in good agreement with anatomical results [10]. It is sometimes erroneously claimed that this propagation is an approximation to propagation with delta-function delays of the form $\delta(t-|r|/v_a)$; in reality, both mathematical approaches approximate the same *physical* system. More relevant is the fact that there is actually a range of velocities of propagation in axons [7, 30], which leads to a range of transit times for signals between any two points; however, this effect can be approximated via a corresponding increase in the effective synaptodendritic time constants (thereby spreading the response over a longer interval), which should then be interpreted as axo-synapto-dendritic quantities. More generally, a velocity spread can be introduced into the underlying propagator [8], or an appropriate generalized propagator could be derived.

Equations (5.1)–(5.5), (5.7), and (5.8) form a closed nonlinear set, which can be solved numerically, or examined analytically in various limits. Once a set of specific neural populations has been chosen, and physiologically realistic values have been assigned to their parameters, these equations can be used to make predictions of neural activity. These equations govern spatiotemporal dynamics of firing rates; oscillations predicted from them are emergent changes of the *average* rate of spiking, whose frequencies do not usually equal the spiking frequency itself.

Corticothalamic System

Much work has been done on applications of our neural field theory to the corticothalamic system, which is the first sector of the combined model of Fig. 5.1 to be discussed. This work has also included extensive verification and validation of the predictions against experiment, as discussed in the next section.

Figure 5.2 shows the large-scale structures and connectivities incorporated, including the thalamic reticular nucleus r, which inhibits relay (or *specific*) nuclei s, and is lumped here with the perigeniculate nucleus, which has an analogous role [90, 94]. Relay nuclei convey external stimuli φ_n and pass on corticothalamic

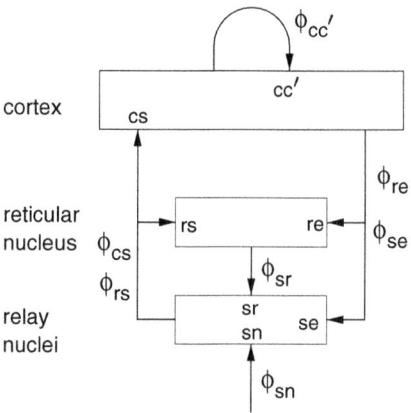

Fig. 5.2 Schematic of corticothalamic interactions, showing the locations at which the ν_{ab} of Eq. (5.9) and linear gains G_{ab} act, where c and c' can both denote the cortical populations e and i. Figure from Ref. [69]

feedback. In this section we consider long-range excitatory cortical neurons ($a = e$), short-range mainly inhibitory cortical neurons ($a = i$), neurons in the reticular nucleus ($a = r$), neurons of thalamic relay nuclei ($a = s$), and external inputs ($a = n$) from non-corticothalamic neurons. These populations are discussed further below.

If intracortical connectivities are proportional to the numbers of neurons involved—the *random connectivity approximation*—then $s_{ib} = s_{eb}$ and $L_{ib} = L_{eb}$ for each b, whence $V_i = V_e$ and $Q_i = Q_e$ [73, 100]. This lets us concentrate on excitatory quantities, with inhibitory ones derivable from them (we stress that inhibitory effects are still included). The short range of i neurons and the small size of the thalamic nuclei enables us to assume $r_a \approx 0$ and, hence, $\gamma_a \approx \infty$ for $a = i, r, s$ for many purposes. The only nonzero discrete delays are $\tau_{es} = \tau_{se} = \tau_{re} = t_0/2$, where t_0 is the time for signals to pass from cortex to thalamus and back. We also assume that all the synaptodendritic time constants are equal, for simplicity, and set $\alpha_{ab} = \alpha$ and $\beta_{ab} = \beta$ for all a and b in what follows; this allows us to drop the subscripts ab in Eqs. (5.2) and (5.3) and write D_α in place of D_{ab}.

Including only the connections shown in Fig. 5.2 and making the approximations mentioned above, our nonlinear model has 16 parameters, not all of which appear separately in the linear limit. By defining

$$\nu_{ab} = N_{ab} s_{ab}, \tag{5.9}$$

these are $Q_{max}, \theta, \sigma', \alpha, \beta, \gamma_e, r_e, t_0, \nu_{ee}, \nu_{ei}, \nu_{es}, \nu_{se}, \nu_{sr}, \nu_{sn}, \nu_{re},$ and ν_{rs}. These are sufficient to allow adequate representation of the most important anatomy and physiology, but few enough to yield useful interpretations and to enable reliable calibration of their values by fitting theoretical predictions to data. The parameters are approximately known from experiment [69, 70, 79, 75, 83] leading to the indicative values in Table 5.1. We use only values compatible with physiology. Sensitivities of the model to parameter variations have been explored in general [75] and in connection with variations between arousal states [76], as discussed below.

An important implication of the parameters above is that the corticothalamic loop delay t_0 places any oscillations that involve this loop at frequencies of order 10 Hz. This means that inclusion of the thalamus and the dynamics of these loops is essential to understand phenomena at frequencies below ~20 Hz. At frequencies $\ll 10$ Hz it is sufficient to include a static corticothalamic feedback strength, and at frequencies $\gg 10$ Hz the corticothalamic feedback is too slow and too attenuated by low-pass effects to influence the dynamics strongly.

The above connectivities and parameters imply, using Eqs. (5.1)–(5.3),

$$D_\alpha V_e(t) = \nu_{ee}\varphi_e(t) + \nu_{ei}\varphi_i(t) + \nu_{es}\varphi_s(t - t_0/2), \tag{5.10}$$

$$D_\alpha V_i(t) = \nu_{ee}\varphi_e(t) + \nu_{ei}\varphi_i(t) + \nu_{es}\varphi_s(t - t_0/2), \tag{5.11}$$

$$D_\alpha V_r(t) = \nu_{re}\varphi_e(t - t_0/2) + \nu_{rs}\varphi_s(t), \tag{5.12}$$

$$D_\alpha V_s(t) = \nu_{se}\varphi_e(t - t_0/2) + \nu_{sr}\varphi_r(t) + \nu_{sn}\varphi_n(t), \tag{5.13}$$

Table 5.1 Indicative parameters for the eyes-open state and sleep-spindle state in normal adults, based on previous work [79, 2]. Parameters used in the figures in this chapter are similar, but not always identical because some are from previous publications where different values were employed

Parameter	Description	Eyes-open	Spindle	Unit
Q_{max}	Maximum firing rate	340	340	s^{-1}
v_e	Axonal velocity	10	10	ms^{-1}
r_e	Axonal range	86	86	mm
θ	Firing threshold	12.9	12.9	mV
σ'	Threshold spread	3.8	3.8	mV
γ_e	Cortical damping rate	116	116	s^{-1}
t_0	Corticothalamic loop delay	85	85	ms
$1/\alpha$	Synaptodendritic decay time	12	22	ms
$1/\beta$	Synaptodendritic rise time	1.3	5.4	ms
ν_{ee}	Connection strength (e from e)	7.85	3.06	mV s
$-\nu_{ei}$	Connection strength (e from i)	9.88	3.24	mV s
ν_{es}	Connection strength (e from s)	0.90	0.92	mV s
ν_{se}	Connection strength (s from e)	2.68	4.73	mV s
$-\nu_{sr}$	Connection strength (s from r)	1.31	1.95	mV s
ν_{sn}	Connection strength (s from n)	6.60	2.70	mV s
ν_{re}	Connection strength (r from e)	0.21	0.26	mV s
ν_{rs}	Connection strength (r from s)	0.06	2.88	mV s
$\varphi_e^{(0)}$	Steady state firing rate (e)	5.2	8.5	s^{-1}
$\varphi_r^{(0)}$	Steady state firing rate (r)	16.3	27.8	s^{-1}
$\varphi_s^{(0)}$	Steady state firing rate (s)	8.4	0.5	s^{-1}
$\varphi_n^{(0)}$	Steady state firing rate (n)	1	1	s^{-1}

whence $V_i = V_e$ and $Q_i = Q_e$, as asserted above. The right side of each of Eqs (5.10)–(5.13) describes the spatial summation of all afferent activity (including via self-connections) for one neural population, and D_α on the left describes temporal dynamics. The short ranges of the axons i, r, and s imply that the corresponding damping rates are large and that $D_a \approx 1$ for these populations, further implying

$$\varphi_a = Q_a = S(V_a), \tag{5.14}$$

for $a = i, r, s$. For the e population, Eqs. (5.7) and (5.8) yield

$$\left(\frac{1}{\gamma_e^2}\frac{\partial^2}{\partial t^2} + \frac{2}{\gamma_e}\frac{\partial}{\partial t} + 1 - r_e^2 \nabla^2\right)\varphi_e(\mathbf{r},t) = S[V_e(\mathbf{r},t)], \tag{5.15}$$

with $\gamma_e = v_e/r_e$. Equations (5.10)–(5.15) describe our corticothalamic model. The *neural mass* limit correponds to $\gamma_e \to \infty$, when all delays within populations are negligible compared to the timescales of the phenomena of interest.

Once neural activity has been predicted from stimuli, one must relate it to measurements to interpret experimental results. The limited spatiotemporal resolution of such measurements often provides an additional justification for the use of mean-field modeling, since finer structure is not resolvable. The quantity φ_e dominates in determining both EEG and fMRI signals—the former because of the dominance and highly aligned dipoles of pyramidal cells, the latter because of pyramidal cells' predominance in cortical metabolic load. In both cases, it is the synaptic dynamics that predominate in determining the measured signals, so φ_e (not Q_e) is the relevant variable. These points have been discussed in detail elsewhere [78, 71, 45, 74, 30].

Arousal State Modeling

Transitions between wake and sleep states are primarily governed by the nuclei of the ascending arousal system (AAS) of the brainstem and hypothalamus, that project diffusely to the corticothalamic system, receive inputs from the suprachiasmatic nucleus (SCN) and elsewhere, and interact with sleep-promoting nuclei [86, 85]. A full description of sleep–wake transitions and their EEG correlates requires an integrated model of both the ascending arousal system and the corticothalamic system, including their mutual interactions. This section briefly describes how the nuclei of the AAS are modeled using the same methods as above [82].

The most important nuclei to model in the AAS are well established from physiology, and are shown in Fig. 5.1. These include the wake-promoting monoaminergic (MA) group and the sleep-promoting ventrolateral preoptic nucleus (VLPO), which inhibit one another, resulting in flip-flop dynamics if they interact strongly—only one is active at a time, and suppresses the other [84, 86, 85] to form the *sleep-wake switch*. In wake the MA is dominant, and in sleep the VLPO is dominant. State transitions are driven by inputs to the sleep-wake switch, including the circadian drive C from the SCN and the homeostatic sleep drive H from buildup of metabolites (likely including adenosine, Ad, related compounds, or their byproducts) in wake and their clearance in sleep [55, 38, 101]. Cholinergic (ACh) and orexinergic (Orx, not shown in Fig. 5.1) inputs to the MA group are also present [48, 86, 85].

Most models of human sleep have been either nonmathematical (e.g., based on sleep diaries) or abstract (mathematical, but not derived directly from physiology). The widely known *two-process model* is of the latter form, and includes circadian and homeostatic influences [15, 3]. Recent advances in sleep neurophysiology have enabled development of physiologically based models [51, 97, 16, 56, 62, 9]; here we use neural mass theory (NMT) to model the dynamics of the AAS nuclei.

Phillips and Robinson [51] argued that (i) since the system spends little time in transitions, the generation rate of H can be approximated as having two values, one for wake and one for sleep, (ii) the clearance rate of H is assumed to be proportional to H with a characteristic time scale χ, and (iii) the production rate of H is μQ_m, where μ is a constant and Q_m serves as a proxy for arousal state. These steps yield equations for H and the mean soma voltages V_a in the MA group ($a=m$) and VLPO ($a=v$):

$$\tau \frac{dV_v}{dt} + V_v = \nu_{vm} Q_m + D, \qquad (5.16)$$

$$\tau \frac{dV_m}{dt} + V_m = \nu_{mv} Q_v + A, \qquad (5.17)$$

$$\chi \frac{dH}{dt} + H = \mu Q_m, \qquad (5.18)$$

$$Q_a = S(V_a), \qquad (5.19)$$

$$D = \nu_{vc} C + \nu_{vh} H + D_0, \qquad (5.20)$$

where the time constants τ_m and τ_v of the responses have been assumed equal to a common value τ [these replace $1/\alpha$ in Eq. (5.3), with $\beta \to \infty$ formally], χ is the somnogen clearance time, v denotes VLPO, m denotes MA, the ν_{ab}, V_a, and Q_a have the same meanings as in previous sections, $\propto$ gives the proportionality between monoaminergic activity and somnogen generation rate.

The total sleep drive D to the VLPO comprises C and H, where C can be interpreted as the SCN firing rate and H is a firing rate change due to somnogenic effects; in both cases only the terms $\nu_{vc} C$ and $\nu_{vh} H$ influence D. The D_0 is a constant baseline level for the total sleep drive. When circadian entrainment to the natural light-dark cycle can be assumed, one has

$$C \approx \cos(\Omega t), \qquad (5.21)$$

where $\Omega = 2\pi/(24\mathrm{h})$ is the angular rate of Earth's rotation, and the amplitude of C is absorbed into ν_{vc}.

In other cases, such as jetlag and shiftwork, C is modeled using the human circadian pacemaker model of St Hilaire et al. [96], which is the most recent version of the well-known model of the human circadian oscillator by Kronauer et al. [34]. The model includes (i) a light processing component, (ii) a component for the effects of non-photic stimuli, and (iii) a van der Pol oscillator component.

In light processing, retinal photoreceptors are converted from ready to activated state by photons at a rate α', dependent on light intensity I:

$$\alpha' = \alpha'_0 \left(\frac{I}{I_0} \right)^p \frac{I}{I + I_1}, \qquad (5.22)$$

where p, I_0, and I_1 are constants used to adjust the model dynamics to experimental data. Activated photoreceptors are converted back to ready at a constant rate β'. Thus the fraction n of activated photoreceptors follows

$$\frac{dn}{dt} = \alpha'(1-n) - \beta'n. \tag{5.23}$$

The resultant photic drive B is proportional to the rate α' and the fraction of photoreceptors that can be activated $(1-n)$:

$$B = G\alpha'(1-n)(1-\varepsilon x)(1-\varepsilon x_c), \tag{5.24}$$

where G and ε are constants adjusted to fit experimental data [96], and x and x_c are the circadian variables of the oscillator. The term $(1-\varepsilon x)(1-\varepsilon x_c)$ accounts for the experimentally observed phase-dependent sensitivity of the circadian pacemaker to light.

The non-photic effects, such as meals and locomotion, on the circadian oscillator are modeled as increased stimulation during wake hours:

$$N_s = \rho\left(\frac{1}{3} - s\right)[1 - \tanh(10x)], \tag{5.25}$$

where ρ is a constant reflecting strength of the non-photic stimulation, and s equals 1 during wakefulness and 0 during sleep, providing for state-dependency. The factor in the square brackets accounts for stronger nonphotic effects near the minimum of core body temperature [96].

The circadian oscillator thus follows

$$\frac{dx}{dt} = \Omega\left[x_c + \gamma\left(\frac{1}{3}x + \frac{4}{3}x^3 - \frac{256}{105}x^7\right) + B + N_s\right], \tag{5.26}$$

$$\frac{dx_c}{dt} = \Omega\left[qBx_c - x\left\{\left(\frac{\delta}{\tau_c}\right)^2 + kB\right\}\right], \tag{5.27}$$

where x is the pacemaker activity, x_c is a complementary variable, τ_c is an intrinsic circadian period, γ is the stiffness of the oscillator, k and q determine strength of the photic drive B, δ is a correction factor that ensures the correct intrinsic circadian period, and k is positive for diurnal animals [34, 28]. The numerical coefficients in Eqs. (5.26) and (5.27) were chosen to achieve unit amplitude of the limit cycle.

Core body temperature (CBT) demonstrates circadian fluctuations and timing of CBT minimum is often used as a marker of circadian phase. In entrained individuals it typically appears about 2 h before awakening, and is calculated as:

$$t_{CBT} = t_{\varphi_{crit}} + t_0, \tag{5.28}$$

$$\varphi_{crit} = \tan^{-1}(x/x_c) \approx -2.98, \tag{5.29}$$

where $t_0 = 0.97$ h is a constant and φ_{crit} is a phase difference between the circadian variables x and x_c given in radians.

Connections between the AAS and the circadian oscillator systems are reciprocal: (i) the circadian variable $C = (1+x)/2$ has an effect on the VLPO, as described in Eq. (5.20), and (ii) the AAS system has an effect on C by reducing (or setting to zero lux) the light input $I(t)$ during sleep.

The parameters in the above model have the nominal values in Table 5.2, determined by physiological constraints from the literature and comparison with a restricted set of experiments on normal sleep, sleep deprivation, and the effects of light on circadian phase [51, 52, 96, 34, 28]. The theory then predicts phenomena in regimes outside those of the calibration experiments, with only slight adjustments to account for individual state and trait differences—these adjustments can be compared with what is expected from independent physiological analyses; e.g., a long circadian period τ_c should be associated with the evening ("night owl") chronotype, while short τ_c should be associated with morning type [54].

Table 5.2 Parameter values of the AAS model and their units. The sigmoid parameters carry an extra prime to indicate that they can differ from the cortical values in Table 5.1

AAS			Circadian pacemaker		
Quantity	Nominal	Unit	Quantity	Nominal	Unit
Q'_{max}	100	s^{-1}	Ω	$\dfrac{2\pi}{24 \times 3600}$	s^{-1}
θ'	10	mV	q	1/3	–
σ''	3	mV	k	0.55	–
ν_{vm}	−2.1	mV s	δ	$\dfrac{24 \times 3600}{0.99729}$	s
ν_{mv}	−1.8	mV s	τ_c	24.2×3600	s
τ	10	s	β'	0.007/60	s^{-1}
ν_{vc}	−5.1	mV s	p	0.5	–
ν_{vh}	1.0	mV s	I_0	9500	lux
μ	4.5	–	I_1	100	lux
χ	60×3600	s	G	37	–
D_0	−11.6	mV	ρ	0.032	–
A	1.3	mV	ε	0.4	–
			α'_0	1.7×10^{-3}	s^{-1}

Feedbacks and the Combined Model

To study the combined dynamics of how the ascending arousal system drives the brain between wake and sleep states we make a first-approximation integration of the sleep-wake and corticothalamic sectors of the model by including the gating of light via eye-opening/closure as a function of arousal state, plus changes in synaptic strengths [and hence the ν_{ab} in Eqs. (5.10)–(5.13)] as arousal changes. In general, we write

$$\nu_{ab} = \nu_{ab}(\text{sleep}) + [\nu_{ab}(\text{wake}) - \nu_{ab}(\text{sleep})] \times \frac{Q_m - Q_m(\text{sleep})}{Q_m(\text{wake}) - Q_m(\text{sleep})}, \quad (5.30)$$

to approximate the change between wake at high Q_m and sleep at low Q_m (different sleep stages can be included in a similar way). More detail of this aspect is provided below.

Numerics

Equations (5.1)–(5.27) can be solved numerically. Since state evolution via the change in synaptic strengths occurs on a timescale that is vastly slower than that of the neural firing rates, we exploit this separation of timescales to simplify the numerical integration, so that we separately solve for the fast activity dynamics in Eqs. (5.1)–(5.9), and the slow sleep-wake cycle dynamics in Eqs. (5.16)–(5.20), (5.23), (5.26), and (5.27).

To solve the delay-partial differential equations that determine the fast dynamics, we have written a C++ code *Neurofield*. This code allows arbitrary numbers of neural populations and uses a typical timestep of 0.1 ms, and typical spatial grid spacing of ~ 1 cm (other combinations are allowed, consistent with the Courant condition) to solve on scales up to the whole brain. Equations (5.2) and (5.3) are solved via explicit direct integration, and Eqs. (5.7) and (5.8) are solved via explicit finite differences using a nine-point spatial stencil, with delayed signals retrieved from stored history to cover the time interval t_0 in corticothalamic loops. Postprocessing of time series then enables quantities like power spectra and evoked potentials to be calculated for suitably specified external inputs to the brain.

The ordinary differential equations governing arousal state dynamics (5.16)–(5.27) are solved with a standard fourth-order Runge-Kutta algorithm in our code *SleepCode*. These determine the arousal state and, via Eq. (5.30), the synaptic strengths for the fast activity dynamics.

Brain Dynamics

In this section we show how the working brain model yields realistic outcomes for both fast (EEG) and slow (sleep-wake cycle) dynamics, including the correct relationships between these sets of phenomena, as validated by comparison with many different types of experiment.

Steady States

We find spatially uniform steady states of our corticothalamic system by setting all the spatial and temporal derivatives to zero in Eqs. (5.10)–(5.15); since these points have been discussed elsewhere [72, 79], we only summarize them here. The resulting equations can be rearranged to yield a single equation for the steady state value of φ_e [79]:

$$\begin{aligned}
0 = & S^{-1}(\varphi_e^{(0)}) - (\nu_{ee} + \nu_{ei})\varphi_e^{(0)} \\
& - \nu_{es} S\left(\nu_{se}\varphi_e^{(0)} + \nu_{sn}\varphi_n^{(0)} + \nu_{sr} S\left[\nu_{re}\varphi_e^{(0)}\right.\right. \\
& \left.\left. + \frac{\nu_{rs}}{\nu_{es}}\left\{S^{-1}(\varphi_e^{(0)}) - (\nu_{ee} + \nu_{ei})\varphi_e^{(0)}\right\}\right]\right),
\end{aligned} \quad (5.31)$$

where S^{-1} denotes the inverse of the sigmoid function S and $\varphi_a^{(0)}$ is the steady state value of φ_a. The function on the right of Eq. (5.31) is continuous and asymptotes to $-\infty$ as $\varphi_e^{(0)} \to 0$ and to $+\infty$ as $\varphi_e^{(0)} \to Q_{max}$. Normally there are two stable zeros separated by an unstable one. As in previous work [72, 79], our current analysis shows that one stable zero occurs at low $\varphi_e^{(0)}$, as seen in Fig. 5.3, and we identify this as the baseline activity level of normal brain function. This state ceases to exist via a saddle-node bifurcation as the mean external input $\varphi_n^{(0)}$ is increased [72]. The other stable zero is at high $\varphi_e^{(0)}$ with all neurons firing near their physiological maximum. This would thus represent some kind of seizure state, but would require

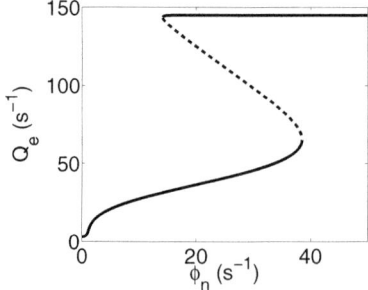

Fig. 5.3 Steady states of the corticothalamic model, showing Q_e vs φ_n. The stable states are shown as two *solid curves*, one with low firing rates (<60 Hz^{-1}) and the other with firing rates near saturation (>145 Hz^{-1}), with parameters other than φ_n as for eyes-open in Table 5.1. These two branches are linked by an unstable branch (*dotted*)

further physiology (e.g., of hemodynamics and hypoxia) to be treated adequately. The states are shown in Fig. 5.3, where they are linked by the unstable branch.

Transfer Functions, Linear Waves, and Spectra

Small perturbations relative to steady states can be treated using linear analysis. A stimulus $\varphi_n(\mathbf{k},\omega)$ of angular frequency $\omega = 2\pi f$, where f is the usual frequency in Hz, and wave vector k, which has magnitude $k = 2\pi/\lambda$, where λ is the wavelength, has a transfer function that relates $\varphi_e(\mathbf{k},\omega)$ to $\varphi_n(\mathbf{k},\omega)$, given by

$$\frac{\varphi_e(\mathbf{k},\omega)}{\varphi_n(\mathbf{k},\omega)} = \frac{G_{es}L}{1-G_{ei}L} \frac{G_{sn}Le^{i\omega t_0/2}}{1-G_{srs}L^2} \frac{1}{q^2(\omega)r_e^2 + k^2 r_e^2}, \qquad (5.32)$$

$$q^2(\omega)r_e^2 = (1-i\omega/\gamma_e)^2 - \frac{L}{1-G_{ei}L}$$
$$\times \left[G_{ee} + \frac{(G_{ese}+G_{esre}L)L}{1-G_{srs}L^2} e^{i\omega t_0} \right], \qquad (5.33)$$

$$G_{ab} = \frac{\varphi_a^{(0)}}{\sigma'}\left(1 - \frac{\varphi_a^{(0)}}{Q_{max}}\right)\nu_{ab}, \qquad (5.34)$$

where $L = (1-i\omega/\alpha)^{-1}(1-i\omega/\beta)^{-1}$ embodies the lowpass filter characteristics of synaptodendritic dynamics [64, 69, 75]. The ratio (5.32) is the cortical excitatory response per unit external stimulus [64, 69, 75]. The gain G_{ab} is the differential output produced by neurons a per unit change in input from neurons b, and the static gains for loops in Fig. 5.2 are $G_{ese} = G_{es}G_{se}$ for feedback via relay nuclei only, $G_{esre} = G_{es}G_{sr}G_{re}$ for the loop through reticular and relay nuclei, and $G_{srs} = G_{sr}G_{rs}$ for the intrathalamic loop.

Corticothalamic waves obey the dispersion relation [72]

$$q^2(\omega) + k^2 = 0, \qquad (5.35)$$

which corresponds to singularity of the transfer function (5.32). Solutions of this equation satisfy $\omega = kv_e - i\gamma_e$ at high frequencies [72], while their general dispersion has been investigated in detail previously [46, 63, 72].

The EEG frequency spectrum is obtained by squaring the modulus of $\varphi_e(\mathbf{k},\omega)$ and integrating over $\mathbf{k}$. It can be written in terms of the transfer function (5.32) as

$$P_e(\omega) = \int \left|\frac{\varphi_e(\mathbf{k},\omega)}{\varphi_n(\mathbf{k},\omega)}\right|^2 |\varphi_n(\mathbf{k},\omega)|^2 d^2\mathbf{k}. \qquad (5.36)$$

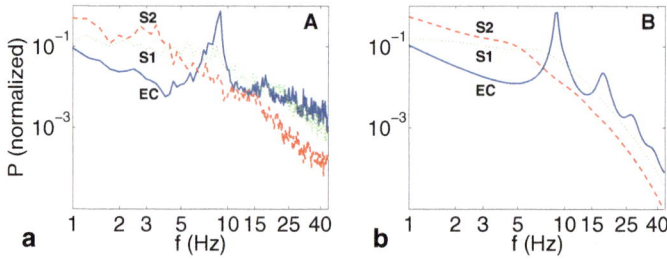

Fig. 5.4 Spectra in eyes-closed state, sleep stage 1, and sleep stage 2 (*non-spindle*). **a** Clinical, with some muscle artifact at high frequencies in S1 and EC. **b** Model without inclusion of muscle artifact terms

If we assume that under conditions for spontaneous EEG the external stimuli $\varphi_n(\mathbf{k},\omega)$ can be approximated by spatiotemporal white noise, $|\varphi_n(\mathbf{k},\omega)|^2 = $ constant, then [71]

$$P_e(\omega) = \frac{\langle \varphi_n^2 \rangle}{4\pi r_e^4} \left| \frac{G_{esn}L^2}{(1-G_{ei}L)(1-G_{srs}L^2)} \right|^2 \frac{\text{Arg } q^2}{\text{Im } q^2}, \qquad (5.37)$$

where $\langle \varphi_n^2 \rangle$ is the mean-square noise level.

Figure 5.4 shows that the power spectrum obtained from Eq. (5.37) has a form that closely matches what is seen in experimental recordings from normal wake and sleep states, with trajectories between these states approximated via Eq. (5.30). The features reproduced include the alpha and beta peaks at frequencies $f \approx 1/t_0, 2/t_0$, and the asymptotic low- and high-frequency behaviors; key differences between waking and sleep spectra can also be reproduced, including the strong increase in low-frequency activity in sleep, where our model predicts a steepening of the spectrum from $1/f$ to $1/f^3$ [75]. Notably, each of the features can be related to underlying anatomy and physiology. The low-frequency $1/f$ behavior is a signature of marginally stable, near-critical dynamics, which allow complex behavior [76, 72, 75], while the steep high-frequency fall-off results from synaptodendritic low-pass filtering. As discussed elsewhere [76, 79, 80], corticothalamic loop resonances account for the alpha and beta peaks, the correlated changes in spectral peaks between sleep and waking, and splitting of the alpha peak in a substantial fraction of normal subjects, for example [76, 75, 78], unlike suggested alternative mechanisms, including "pacemakers" and purely cortical resonances, each of which only accounts for a subset of the above features of the data [75]. Overall, these points imply that the thalamus must be incorporated to account for most salient EEG features at frequencies below about 20 Hz.

Evoked Responses

The transfer function (5.32) can be used to calculate the response to an impulsive stimulus, which transforms to $\varphi_n(\mathbf{k},\omega) = $ constant, simply by inverse-transforming

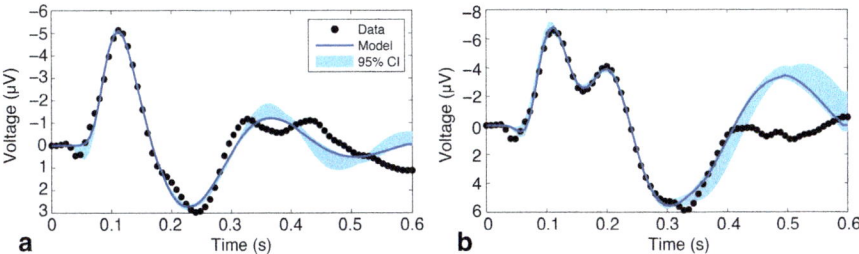

Fig. 5.5 Example auditory evoked potentials. The *dots* are from experiment and the *blue curves* with associated uncertainty bands are model fits. Fits were obtained by starting from parameters similar to those in Table 5.1, but Monte-Carlo initialized for multiple realizations spread over a range of neighboring values, then performing gradient descent to obtain a good fit starting from each set of initial conditions. **a** Response to a standard stimulus. **b** Response to a target stimulus

Eq. (5.32). More generally, one can include spatiotemporal profiles of the stimulus to find the appropriate form of $\varphi_n(\mathbf{k},\omega)$, multiply Eq. (5.32) by this, then inverse transform [64, 31].

Figure 5.5 shows the evoked potentials measured and calculated for an auditory oddball paradigm in which 1000 Hz tones are randomly interspersed amid a train of 500 Hz standard tones, with a probability of 20%, and the subject is instructed to respond to target tones with a button press. The standard response is well matched by model calculations for parameters similar to those in Table 5.1. The target response is more complex, corresponding to stronger corticothalamic feedback that is likely associated with increased attention. The agreement is good in both cases, up to about 400 ms. Beyond this point the agreement decreases, most likely due to parameters changing in time (e.g., due to cortical feedback) and the involvement of other brain structures.

Moving beyond responses to individual impulsive stimuli, periodic stimuli are known to evoke various EEG dynamics. For example, periodic flashing light (termed "flicker") yields steady-state visual evoked potentials (SSVEPs), which markedly alter EEG spectra [26]. To model these, we add a periodic drive to the background white-noise external input to the thalamus [76, 66], with total drive

$$\varphi_n(t) = \varphi_n^{\text{noise}}(t) + \varphi_n^D(t), \tag{5.38}$$

where φ_n^{noise} is Gaussian white noise with mean $\varphi_n^{(0)}$ and standard deviation σ_n, and φ_n^D is a periodic drive with zero mean. Commonly-used forms for φ_n^D include pure sinusoids $\Phi_n \cos(2\pi f_D t)$ and square waves $\text{sgn}[\Phi_n \cos(2\pi f_D t)]$, where Φ_n is the drive amplitude, f_D is the drive frequency, and sgn is the sign function.

Figure 5.6(a) shows the spectral response as a function of drive frequency. There are several notable features of these SSVEPs. First, there is a strong response at the fundamental frequency $f = f_D$. Second, there are nonlinear responses at frequencies harmonically related to the fundamental—these are spectral peaks due to frequency-locking at ratios of 2:1, 3:1, 4:1, ... Third, the background alpha peak at f_α is diminished over a wide range around $f_D = f_\alpha$—this is entrainment of the alpha rhythm. Fourth, for a range of f_D around $f_D = 2f_\alpha$ there is a strong response at $f = f_D/2$—this is 1:2 frequency-locking—with the alpha peak diminished here,

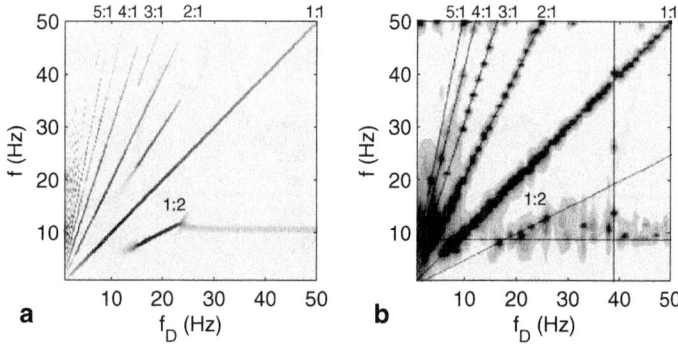

Fig. 5.6 Steady state visual evoked potential (SSVEP) dynamics. Spectral responses for a square-wave drive with $f_D = 1$–50 Hz are shown. Responses to the N th harmonic of f_D are labeled $N:1$; subharmonic entrainment is labeled 1:2, and the alpha frequency corresponds to the horizontal feature near 10 Hz. Darker shading denotes higher power, and spectra have been multiplied by f to enhance power at high frequencies, as in the original Fig. 5.5 of [26]. **a** Model. Spectra are smoothed with a 0.4 Hz moving window to display sharp peaks more clearly. **b** Adult human SSVEP data, adapted from Fig. 5.5 of [26]. Figure adapted from Ref. [66]

thus exhibiting subharmonic entrainment. All of these features are directly observable in the human EEG spectra in Fig. 5.6b.

Stability Zone, Instabilities, Seizures, and Phase Transitions

Stability

Linear waves obey the dispersion relation (5.35), with instability boundaries occurring where this equation is satisfied for real ω [76, 72, 75]. In most circumstances, waves with $k = 0$ (i.e., spatially uniform) are the most unstable [72], and it is found that only the lowest frequency spectral resonances can become unstable. Analysis of stability of perturbations relative to the steady state that represents normal activity for realistic parameter ranges finds just four $k = 0$ instabilities, leading to global nonlinear dynamics [11, 76, 77]: (a) Slow-wave instability ($f \approx 0$) via a saddle–node bifurcation that leads to a low frequency spike-wave limit cycle; (b) theta instability, via a supercritical Hopf bifurcation that saturates in a nonlinear limit cycle near 3 Hz, with a spike-wave form unless its parameters are close to the instability boundary; (c) alpha instability, via a subcritical Hopf bifurcation, giving a limit cycle near 10 Hz; and (d) spindle instability at $\omega \approx (\alpha\beta)^{1/2}$, leading to a limit cycle at 10–15 Hz (the nature of this bifurcation has not yet been investigated). The boundaries defined by these instabilities are interpreted as corresponding to onsets of generalized seizures, as discussed in more detail below [11, 76, 77].

The occurrence of only a few instabilities, at low frequencies, enables the state and physical stability of the brain to be approximately represented in a 3D space with axes

5 A Multiscale "Working Brain" Model

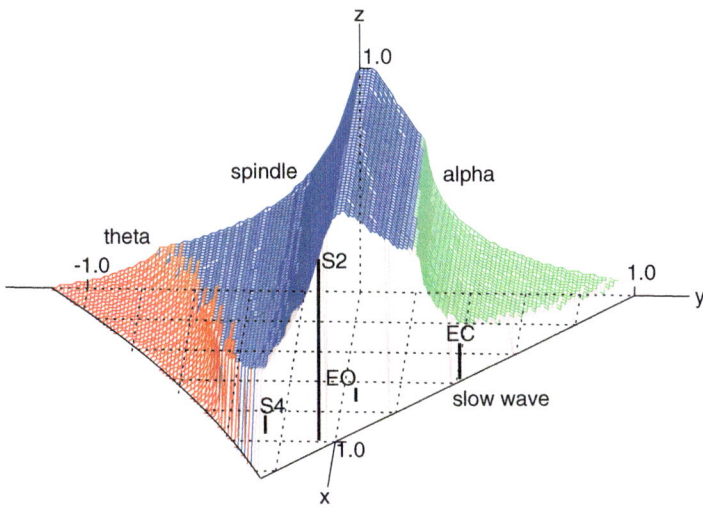

Fig. 5.7 Brain stability zone. The surface is shaded according to instability, as labeled (*blue*=spindle, *green*=alpha, *red*=theta), with the transparent front right-hand face corresponding to a slow-wave instability. Approximate locations are shown of alert eyes-open (EO), relaxed, eyes-closed (EC), sleep stage 2 (S2; the spindle substate is at the top of the bar, the non-spindle state near the base), and sleep stage 4 (S4) states, with each state located at the top of its bar, whose (x, y) coordinates can be read from the grid. Figure adapted from Ref. [76]

$$x = G_{ee} / (1 - G_{ei}), \tag{5.39}$$

$$y = (G_{ese} + G_{esre}) / [(1 - G_{srs})(1 - G_{ei})], \tag{5.40}$$

$$z = -G_{srs} \alpha \beta / (\alpha + \beta)^2, \tag{5.41}$$

which parameterize cortical, corticothalamic, and thalamic stability, respectively [11, 76]. In terms of these quantities, parameters corresponding to linearly stable brain states lie in a *stability zone* illustrated in Fig. 5.7. The back is at $x = 0$ and the base at $z = 0$. A pure spindle instability occurs at $z = 1$, which couples to the alpha instability, with spindle dominating at top and left, and alpha at right. At small z, the left surface is defined by a theta instability [11, 76]. The vertical front right surface corresponds to slow-wave instability at $x + y = 1$. Equation (5.30) parametrizes trajectories followed between the states shown.

Non-seizure states lie within the stability zone in Fig. 5.7, which is a three-dimensional (3D) projection of an 8D space parameterized by the corticothalamic ν_{ab}, with state evolution driven by the AAS and approximated here via Eq. (5.30). Detailed arguments regarding the sign of feedback via the thalamus, proximity between neighboring behavioral states, and the results of explicit fitting to data (which is enabled by using the present model), place the arousal sequence, from alert eyes-open (EO) to deep sleep, including relaxed eyes-closed (EC) and sleep stages 1–4 (S1–S4), some of which are shown in Fig. 5.7 [76].

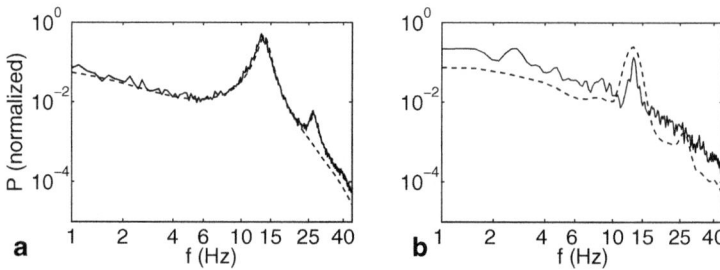

Fig. 5.8 Spectra in the sleep spindle state. **a** Numerical spindle spectrum (*solid*), showing fundamental peak near 14 Hz and second harmonic near 28 Hz. The linear spectrum is shown *dashed*. **b** Experimental spectra averaged over 30 s epochs (*solid*) that include both spindle and non-spindle periods, and averaged over just those 2 s intervals in which spindles are present (*dashed*). Frames adapted from Ref. [1]

Weak Nonlinearity

In section "Evoked Responses" we saw that nonlinearities can occur when the brain receives sufficiently strong oscillating inputs. In most normal brain states nonlinear effects are small and the linear power spectrum accurately represents the spectrum obtained by numerical integration of the nonlinear model equations. However, nonlinear effects can be significant in sleep spindles, which are strong 12–15 Hz oscillations mostly seen in sleep stage 2 [2]. Figure 5.8 compares the power spectrum obtained by integrating the model equations with the linear approximation to the power spectrum for typical model parameters (see Table 5.1). There is a harmonic peak clearly visible at twice the frequency of the spindle oscillation, which arises due to nonlinear effects in our full model.

Analysis of the voltage transfer functions for each population revealed that the voltage fluctuations in the relay nuclei are over 6 times larger than fluctuations in any of the other populations, and nonlinearity in the relay nuclei is largely responsible for the sleep spindle harmonic. Nonlinearity in the relay nuclei is included by retaining the second order term in the Taylor expansion of the sigmoid in Eq. (5.5). The Fourier transform of this nonlinear response results in a convolution, making the nonlinear component of the voltage at any particular **k** and ω depend on V_s at all **k** and ω, thus transferring energy between spatial modes and frequencies. In particular, self-convolution of the strong spindle peak seen in Fig. 5.8 gives rise to the harmonic, an example of wave-wave coalescence. We also predict a quadratic relationship of the power in the spindle harmonic peak to that in the fundamental spindle oscillation. Analysis of experimental data from 9 healthy subjects has confirmed both of these predictions [1].

Seizures

Two of the most common generalized epilepsies are absence and tonic-clonic seizures. In absence epilepsy, seizures last 5–20 s, cause loss of consciousness, show

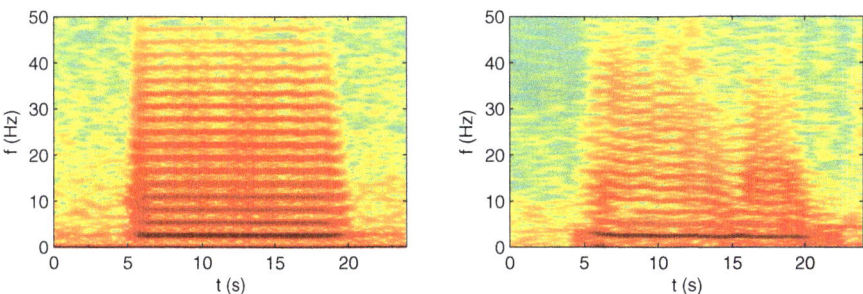

Fig. 5.9 Dynamic spectra during an absence seizure. **a** Simulated from the model. **b** Clinical. Adapted from Ref. [11], which contains further details

a spike-wave cycle which starts and stops abruptly across the whole scalp, and the subject reaches a post-seizure state similar to the pre-seizure one. Tonic-clonic seizures display a tonic phase of roughly 10 Hz oscillations lasting about 10 s, followed by a clonic phase of similar duration dominated by polyspike-wave complexes, with an unresponsive post-seizure state very different from the pre-seizure one [11, 42, 91]. Fig. 5.9(a) and (b) show results from our model under conditions for theta instability, and for a clinical absence seizure, respectively. The dynamic spectrum in Fig. 5.9(a) shows onset of an approximately 3-Hz spike-wave cycle as the system is forced across the instability boundary by ramping one of its parameters, in this case ν_{se}, followed by termination when it is ramped back. This closely resembles observed absence seizure dynamic spectrum in Fig. 5.9(b) [11, 18, 76, 77]. When the destabilizing parameter is ramped back, the system returns smoothly to very nearly its initial state, also consistent with Fig. 5.9(b).

Similar agreement has been found with generalized tonic-clonic seizure dynamics near 10 Hz. However, in this case, the oscillations set in with nonzero amplitude and, when the control parameter is ramped back, hysteresis is observed, with the seizure terminating in a quiescent state, with return to normal only after further parameter evolution, all consistent with clinical observations [11, 42].

Normal Sleep Dynamics, Deprivation, and Recovery

The first key result from the AAS sector of our model is that the steady states of Eqs. (5.16)–(5.20) exhibit bistability when plotted against the sleep drive D. The upper and lower branches represent wake and sleep, respectively, with an unstable branch between. Cycles of D cause the system to move around the hysteresis loop shown in Fig. 5.10, with saddle-node bifurcations from wake to sleep and back again. Near-stable *ghost* states [95] are located just beyond these bifurcations and play an important role in sleep dynamics, as we see below [22].

The existence of different sleep-drive thresholds for sleep-to-wake and wake-to-sleep transitions is the result of mutual inhibition between the wake-active MA and sleep-active VLPO populations, and is a characteristic shared with the

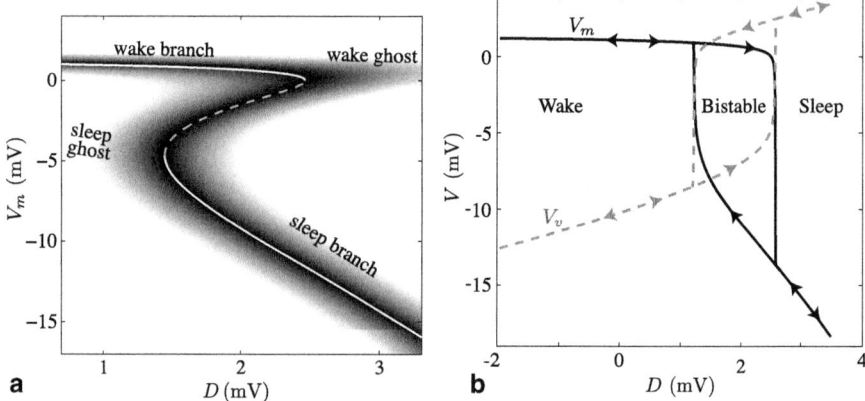

Fig. 5.10 Sleep dynamics. **a** Stable wake and sleep branches are linked by an unstable branch (*dashed*). Wake and sleep ghosts are shown. Regions of small $|\dot{V}_m|$ are shaded. **b** V_m versus sleep drive D over 24 h. As D oscillates, arrows show how V_m and V_v cycle around hysteresis loops between wake and sleep. Adapted from Ref. [22]

"two-process" model [15], which was not constructed directly from the physiology, but incorporated key features of the observed dynamics. The phenomenological two-process model is thus seen to approximate the physiologically derived dynamics. An important distinction from the two-process model is that the present model's state passes beyond both transition points during normal cycling rather than staying strictly between these thresholds. Furthermore, because our model's states are defined in terms of physiological quantities such as firing rates and voltages, state transitions are not instantaneous, but have their own characteristic timescale related to τ [52].

The present model has been used to simulate the effects of total sleep deprivation and recovery by fixing V_m and V_v to their wake values during deprivation, then releasing them to undergo their usual dynamics during recovery. The results reproduce observed time courses of recovery sleep (about 13 h the first night, 11 h the second night, and 9 h the third night after: 100 h deprivation) and recovery of sleep latency to baseline, as in Fig. 5.11, and predicts that initiating sleep near normal bedtime minimizes required recovery sleep [52].

Stimuli

External stimuli to the MA or VLPO are modeled as perturbations to the drives A and D [22]. Here, auditory stimuli excite the MA [5] via an added contribution to A on the right of Eq. (5.17), causing an excursion from equilibrium. To simulate sleep fragmentation by auditory stimuli [92], stimuli via A perturb the model from the sleep branch to higher V_m. For small A, the system returns quickly to the sleep branch, but if A is sufficiently large, the return is via the wake ghost, where the

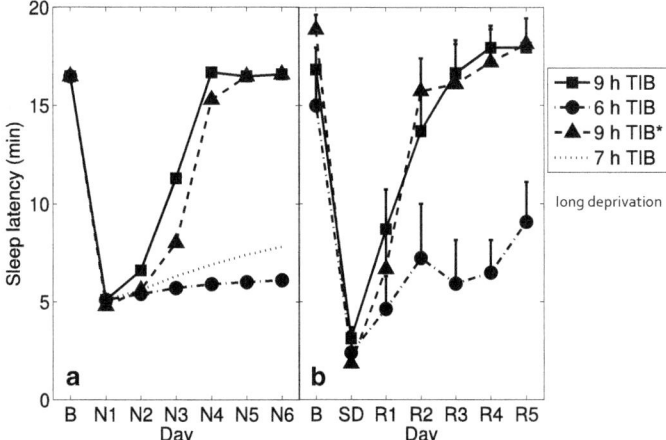

Fig. 5.11 Sleep latencies after deprivation for **a** model and **b** experiment (mean±SEM) of [37]. Triangles are for 9 h time in bed (TIB) after 63 h deprivation. Other curves are after 39 h deprivation, with 9 h TIB (*squares*), 6 h TIB (*circles*), plus a simulated 7 h TIB case (*dotted*), for which there are no experimental data. In **a** sleep latencies are measured at the end of each day, for baseline (B) and recovery nights N1, N2, ... In **b** latencies are measured for B, sleep deprivation (SD), and recovery nights R1 ... R5. Adapted from Ref. [52]

system lingers in a brief awakening [22]. Hence we define the arousal threshold to be the magnitude of the drive $|A|$ required to perturb the system to the wake ghost and cause momentary awakening (in the bistable region a large perturbation will shift the system to the wake branch, causing full awakening). A linear fit to a clinical auditory decibel scale [as expected given that sensory responses are roughly logarithmic in stimulus intensity [30]] allows us to predict the arousal threshold vs. time since sleep onset [22], as seen in Fig. 5.12. The approach of adding drives to represent stimuli can also be used to include pharmacological agents whose actions can be represented as drives to the MA and VLPO [cf., section "Caffeine"].

Wake Effort and Fatigue

Ordinarily, as D increases past its wake-sleep bifurcation value (≈ 2.5 mV), the system enters sleep. By applying an external drive, it is possible to keep the system in the wake ghost to maintain wakefulness. We term this additional drive 'wake effort' W and add it to the MA drive A on the right of Eq. (5.17). This drive tends to keep the MA active and thus opposes the effects of D on the system. The effort required to remain awake is zero on the wake branch, and increases with D in the wake ghost, consistent with common experience that it is easy to stay awake during the day if not sleep deprived and that subjective wake effort increases with length of wake extension.

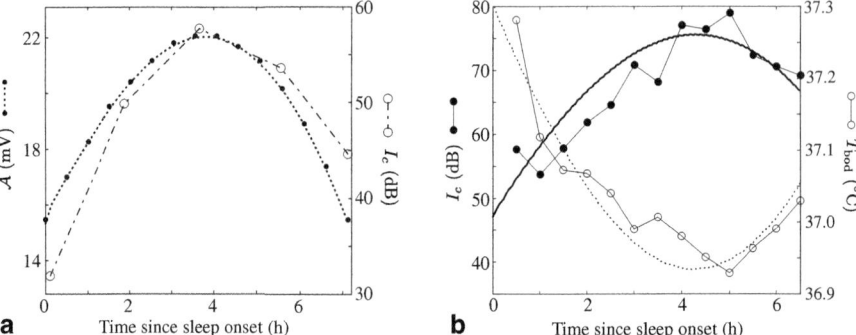

Fig. 5.12 Model predictions of arousal threshold. **a** The model arousal threshold A (*in mV*) agrees well with experimental auditory arousal thresholds I_c (dB) measured over a normal night of sleep [8]. **b** The model also predicts the arousal threshold and body temperature in a sleep fragmentation study [36]. Data are *circles*, model predictions are *curves*. Adapted from Ref. [22]

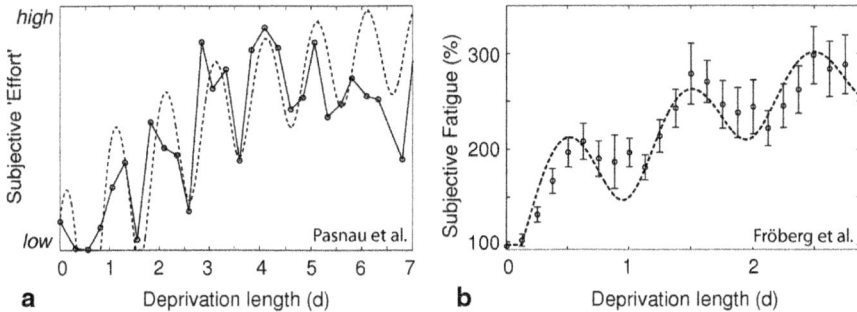

Fig. 5.13 Model fits to clinical subjective fatigue data. Data are shown with *circles* and model fits with a *dashed line*. **a** Subjective 'effort' data from a 7-day sleep deprivation study of [49]. Simulated wake-effort W over the same period is rescaled by a constant factor to reproduce the observations. **b** Subjective fatigue from linearly scaled to the model's W drive. Adapted from Ref. [23]

Simulating sleep deprivation this way produces a W time series that can be compared to experiment. We have confirmed our hypothesized correlation between W and subjective fatigue levels through comparisons with data [23], as illustrated in Fig. 5.13. Performance levels are also expected to correlate with W, although the relationship may be nonlinear and/or task-dependent, e.g., due to motivational input from the limbic system that decreases that required from the cortex and improves performance [87, 23], factors that have yet to be modeled.

Caffeine

Caffeine is a competitive antagonist of adenosine (Ad), which competes for Ad receptor sites in the brain, thereby partially masking the effects of Ad [19]. It also reduces inhibition of basal forebrain acetylcholine (ACh) by Ad, increasing the firing rate of ACh nuclei [12], and the value of A. These effects are modeled by the replacements

$$\nu_{vh} \to \nu_{vh}[1 - \zeta_H Z_C(t)], \qquad (5.42)$$

$$A \to A + \zeta_A Z_C(t), \qquad (5.43)$$

where Z_C is the concentration of caffeine in mg/kg and ζ_A and ζ_H are constants determined from comparison with experiment [60]. Puckeridge et al. [60] modeled caffeine pharmacokinetics by assuming it is absorbed and eliminated at rates proportional to dose and concentration, respectively, giving

$$Z_C(t) = \gamma_C [e^{-k_e(t-t_0)} - e^{-k_a(t-t_0)}], \qquad (5.44)$$

for $t \geq t_0$, where $k_a \approx 10^{-3}\,\mathrm{s}^{-1}$ and $k_e \approx 4.5 \times 10^{-5}\,\mathrm{s}^{-1}$ are rates of elimination and absorption, respectively, and γ_C and t_0 are the size and time of dose.

Figure 5.14 shows model output for a well-entrained subject with habitual bedtime of 23:00 who takes 200 mg of caffeine at 22:00. Caffeine delays sleep onset as shown in Fig. 5.14(a), shortens sleep duration, and raises homeostatic sleep drive in Fig. 5.14b because: (i) D decreases, shifting the system away from sleep as demonstrated in Fig. 5.14c, and (ii) the increase in A stabilizes wake at larger D. There is a only a small delay in waking because the circadian drive C moves to its wake phase at its usual time. As a result of reduced sleep time, the total sleep drive D is

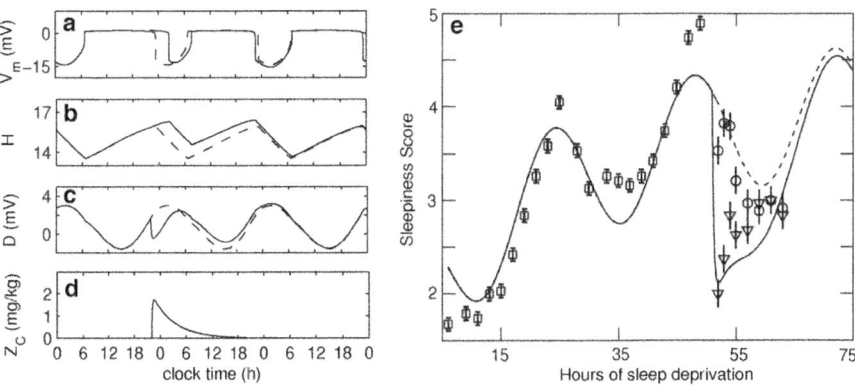

Fig. 5.14 Model dynamics with and without a caffeine. Panels (**a–d**) show effects of 200 mg caffeine intake at 22:00 in a fully entrained individual with habitual sleep times from 23:00 to 07:00. The caffeine case is shown solid, while a control case (no caffeine) is shown dashed. Panel (**e**) compares model simulation for the effects of caffeine on subjective fatigue during sleep deprivation with experimental results from [50]. Subjects took caffeine or placebo at 54 h (49th h since start of total deprivation). Experimental data are shown as squares, triangles, and circles, for before either dose, after caffeine, and after placebo, respectively. Simulated values are shown solid and dashed for subjects with low caffeine sensitivity who took or did not take caffeine, respectively. Adapted from Ref. [60]

higher the next day than it would have been without caffeine taken the night before. Caffeine concentration is shown in Fig. 5.14d.

The model successfully matches clinical data of [50], who followed subjective sleepiness over 60 h of total sleep deprivation. They compared the effectiveness of 600 mg of caffeine against a placebo given on the 49th h of the study. Sleepiness was assessed via the Stanford Sleepiness Scale [27]. From Fig. 5.14e, the sleepiness score S is found to be linearly related to D, with

$$S = c_1 D + c_2, \qquad (5.45)$$

where $c_1 \approx 0.23 (\text{mV})^{-1}$ and $c_2 \approx 2.33$ [60]. Comparison with experiment in Fig. 5.14e shows that both exhibit a secular increase in S, with superposed circadian oscillations. Theory and experiment agree that 600 mg of caffeine reduces S by approximately 2.5 and that its effects are lost after about 15 h (≈ 3 decay times).

Circadian Misalignment

For an entrained individual light-dark cycle, the homeostatic process and circadian oscillator are synchronized, so sleep occurs at a similar circadian phase every cycle. However, in conditions like jetlag and shiftwork this synchrony is disturbed, leading to insufficient sleep, fatigue, and risk of accidents [61, 4, 88]. The light-dark cycle has the strongest synchronizing effect on the circadian system [34, 65, 24]. Generally, light exposure after the time t_{CBT} of core body temperature minimum causes phase advance, while light before t_{CBT} causes phase delay according to the circadian phase response curve [41]. Higher light intensity results in larger phase shifts [6]. The effects of light on the circadian phase in the model have been extensively validated against experiment [96, 34, 28]. The model thus allows study of dynamics during circadian misalignment and the development of interventions to improve adaptation and decrease sleepiness.

Jetlag

A change of time zone is implemented in the model by shifting the daylight profile by a corresponding number of hours and incorporating a flight. Flight is implemented by specifying the flight time, lighting intensity, and whether wakefulness occurs during the flight. Forced wakefulness is implemented in the same way as in sleep deprivation, discussed in sections "Normal Sleep Dynamics, Deprivation, and Recovery" and "Wake Effort and Fatigue".

A raster plot of the model dynamics for +6 h time zone change is shown in Fig. 5.15a. This time difference corresponds to a flight from Sydney, Australia to the west coast of the United States. In this example a 14 h flight starts at 08:00 on the 8th day of the simulation, and wakefulness is forced during the flight (red) and at the airport (yellow).

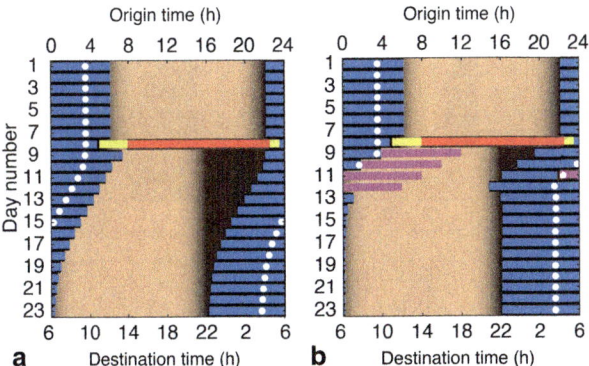

Fig. 5.15 Circadian adaptation to a time difference of +6 h. Panel (**a**) shows a case of natural adaptation, and (**b**) intervention with specified waking up times and light exposure. Background *shades of brown* indicate lighting intensity, *blue stripes* show sleep times, *white circles* show t_{CBT}, *red line* is the flight time, and *yellow* is the transfer and time spent at the airport. In (**b**) *magenta* show the times of forced wakefulness and scheduled exposure to natural lighting

To adjust to the new time zone the circadian oscillator needs to advance by 6 h. In natural conditions re-entrainment takes about 1.5 days per hour of circadian change [35], which is close to what is seen in our simulations in Fig. 5.15a. In this case for the first week after arrival natural sleep (blue lines) is initiated late in the destination night time and continues well into the daytime. Adaptation is achieved only after about 12 days since arrival. However, it is possible to improve adaptation by scheduling wake-up times and light exposure, as demonstrated in Fig. 5.15b. In this case forced wakefulness and light exposure are scheduled for 8 h directly after t_{CBT} for three days since arrival in order to facilitate phase advance. This achieves adaptation in 3 days instead of 12.

Shiftwork

Shiftwork is implemented in the model by specifying lighting and introducing forced wakefulness during the scheduled shift times [57, 58, 59]. Examples of night shiftwork simulations with and without light interventions are shown in Fig. 5.16. Normally, night workers are exposed to low lighting during shifts and bright daylight during their commute home. This prevents adaptation to the work schedule and leads to disturbed sleep and increased sleepiness, as is also predicted by the model and seen in Fig. 5.16a and 5.16c. In quantitative agreement with experimental observations [14, 58] the model predicts that bright light intervention during the night shifts and reduced light exposure at home facilitate circadian adaptation thereby improving sleep and reducing total sleep drive as shown in Fig. 5.16b and 5.16c.

The model has been used to study the effects of shift start times, rotation speed, and days off on circadian adaptation and dynamics of the total sleep drive D. The model predicts that, for permanent schedules, shifts starting around 2300 have the longest adaptation times and increased sleep drive [57]. For forward rotating shift

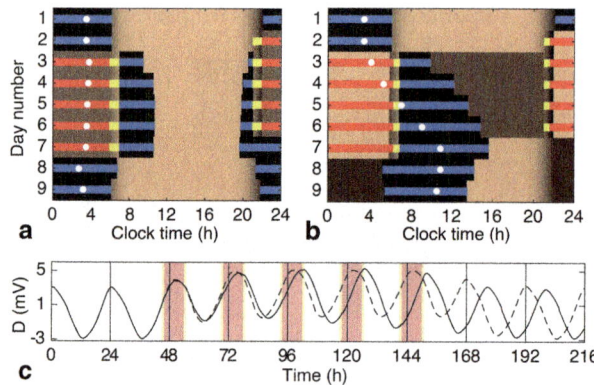

Fig. 5.16 Simulations of night shiftwork. **a** A raster plot from simulations with low shift lighting of 350 lx. **b** The effects of light intervention are shown with shift lighting of 5000 lx. **c** Comparison of total sleep drives, with a *dashed line* showing D in case (**a**) and a solid line for case (**b**). *Red* and *yellow* shading show shift and commute times. In (**a**) and (**b**) background *shades of brown* illustrate levels of light intensity with up to 5000 lx of natural daylight. Daytime lighting during the shift days in (**b**) is 150 lx. Sleep times are shown in *blue*, and *white circles* show t_{CBT}

schedules we demonstrated that intermediate rotation speeds of 3–6 days lead to irregular circadian dynamics and increased mean sleep drive, whereas the effects of shift start times is weak [59]. These results are in good agreement with real-world data [32, 33, 25].

Microsleeps and Critical Fluctuations

The sleep sector of our model allows more detailed study of the transition from wake to sleep by monitoring the MA voltage signal V_m for sleep drive D near its wake-sleep bifurcation value D_c. Similar catastrophic bifurcations have been investigated in a number of other complex systems, such as seismology [17] and climate science [98, 39]. There is a focus on characterization of precursors to transitions, particularly increased sensitivity to noise and critical slowing down [89].

When small amplitude white noise was added to the sleep drive D to the VLPO in Eq. (5.16), the variance σ_m^2 of the corresponding noise induced in V_m was found to scale as

$$\sigma_m^2 \propto (D_c - D)^{-\alpha}, \tag{5.46}$$

with the exponent $\alpha = 0.51 \pm 0.015$ determined from our numerical results, in agreement with a preliminary finding [81]. This scaling is illustrated in Fig. 5.16a, and is in excellent agreement with a theoretical prediction obtained by perturbing our dynamical equations with white noise, and integrating the resulting spectral power of the system response over frequency, which gives

$$\sigma^2 \propto (D_c - D)^{-1/2}. \tag{5.47}$$

Critical slowing down can be investigated via the power spectrum of the noise induced in the V_m signal. Power spectra calculated for sleep drive D close to D_c are shown in Fig. 5.16(b), where they have been smoothed and normalized to coincide at high frequencies. At high frequencies the spectra all fall off like $1/\omega^2$, while at lower frequencies the relative power grows as $D \to D_c$, consistent with critical slowing down. These features agree with the predicted response power spectral density $P(\omega)$,

$$P(\omega) = \frac{1}{m^2 \left(\omega_1^2 - \omega^2\right)^2 + \omega^2 \kappa^2}, \tag{5.48}$$

where $\omega_1 \to 0$ is a resonant frequency that decreases toward zero as $D \to D_c$. The constants m and κ are analogous to mass and friction terms in the dynamical equations, and are related to characteristic time constants τ_m and τ_v [53]. This form predicts that the spectrum should scale like $1/\omega^2$ provided the second term in the denominator dominates the first, which is the case in our numerical calculations.

The behavior of the variance and spectrum have potential to signal the approach to the critical point, particularly if their amplitudes are precalibrated for an individual subject. Such precursors of the wake-to-sleep transition could be of great value in safety-related contexts to prevent traffic accidents and other lapses (Fig. 5.17).

Summary

Physiologically based neural-field theories of the brain are able to incorporate essential physiology and anatomy across the many spatial and temporal scales necessary to treat a host of phenomena involving neural activity. They achieve this for physiologically realistic parameters, and yield numerous predictions that accord with observations using a variety of experimental methods in both the linear and nonlinear regimes. This agreement validates the theoretical approach and verifies its outcomes. Moreover, NFT achieves these outcomes in a way that unifies disparate subfields and measurement modalities in a single framework, and which permits parameter determination via fits of model predictions to experimental data.

In the present work, we have reviewed specific neural-field theories of the corticothalamic system and the ascending arousal system, and key predictions and corresponding experimental tests for a wide range of phenomena, ranging from normal brain activity to highly nonlinear dynamics of seizures, plus the arousal-

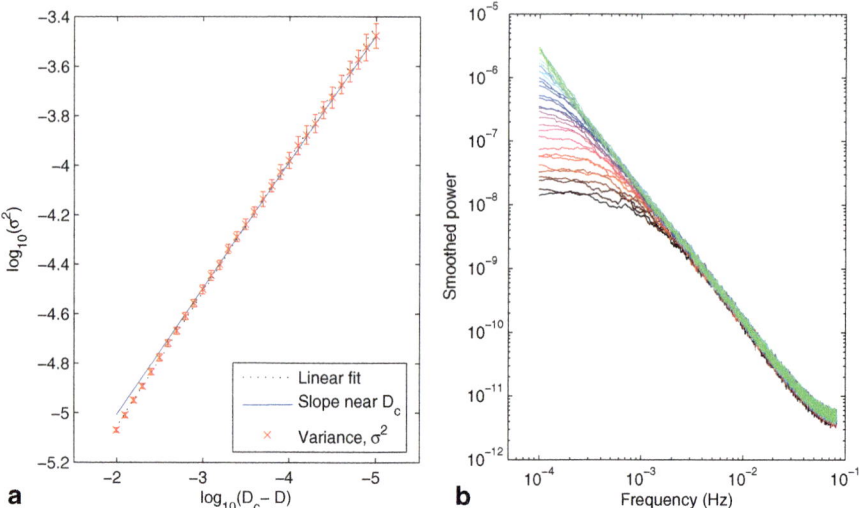

Fig. 5.17 a Log-log plot of variance σ_m^2 against the difference between the critical drive D_c and the sleep drive D. Error bars were calculated as the standard deviation of 50 realizations. The *dotted line* is an equally weighted linear fit, with $m = 0.54$, while the *solid line*, has slope $\alpha = 0.51 \pm 0.015$ and illustrates the asymptotic scaling closest to the critical point. **b** Numerically obtained power spectra on logarithmic axes, with frequency ω in Hz and power in arbitrary units, smoothed with boxcar averaging and normalized to overlap at right to facilitate comparison of shape. The spectra were calculated for various sleep drives D near the critical point D_c, with curves corresponding to $(D_c - D)$ ranging from 10^{-2} for the *lowest curve at left* to 10^{-5} for the *top curve*

state dynamics that drives the corticothalamic system between states. These two main sectors are coupled to yield an overall "working brain" model that exhibits appropriate dynamics over a wide range of arousal states. This model can thus serve as a tractable tool for predicting and analyzing numerous phenomena, including by extending it to include additional structures, measurement modalities, physiology, and pharmacology, for example.

Acknowledgements This work was supported by the Australian Research Council, the National Health and Medical Research Council, the Westmead Millennium Foundation, the National Institutes of Health through 1K99HL119618, and Brain Resource Ltd.

References

1. Abeysuriya RG, Rennie CJ, Robinson PA, Kim JW. Experimental observation of a theoretically predicted nonlinear sleep spindle harmonic in human EEG. Clin Neurophysiol. 2014;125:2016–2023.
2. Abeysuriya RG, Rennie CJ, Robinson PA. Prediction and verification of nonlinear sleep spindle harmonic oscillations. J Theor Biol. 2014;344:70–77.

3. Achermann P, Borbély AA. Mathematical models of sleep regulation. Front Biosci. 2003;8:s683–3.
4. Barger LK, Lockley SW, Rajaratnam SM, Landrigan CP. Neurobehavioral, health, and safety consequences associated with shift work in safety-sensitive professions. Curr Neurol Neurosci Rep. 2009 March;9(2):155–64.
5. Berridge CW, Warehouse BD. The locus coeruleus-noradrenergic system: modulation of behavioral state and state-dependent cognitive processes. 2003;42(1):33–84.
6. Boivin DB, Duffy JF, Kronauer RE, Czeisler CA. Dose-response relationships for resetting of human circadian clock by light. Nature. 1996 Feb;379(6565):540–2.
7. Bojak I, Liley DTJ. Axonal velocity distributions in neural field equations. PLoS Comput Biol. 2010;6(1):e1000653.
8. Bonnet MH, Webb WB, Barnard G. Effect of flurazepam, pentobarbital, and caffeine on arousal threshold. Sleep. 1979;1:271–9.
9. Booth V, Diniz Behn CG. Physiologically-based modeling of sleep-wake regulatory networks. Math Biosci. 2014 April;250:54–68.
10. Braitenberg V, Shüz A. Anatomy of the cortex: statistics and geometry. Berlin: Springer; 1991.
11. Breakspear M, Roberts JA, Terry JR, Rodrigues S, Mahant N, Robinson PA. A unifying explanation of primary generalized seizures through nonlinear brain modeling and bifurcation analysis. Cereb Cortex. 2006;16:1296–313.
12. Carter AJ, O'Connor WT, Carter MJ, Ungerstedt U. Caffeine enhances acetylcholine release in the hippocampus in vivo by a selective interaction with adenosine A1 receptors. J Pharmacol Exp Ther. 1995;273:637–42.
13. Clearwater JM, Rennie CJ, Robinson PA. Mean field model of acetylcholine mediated dynamics in the cerebral cortex. Biol Cybern. 2007;97:449–60.
14. Czeisler CA, Johnson MP, Duffy JF, Brown EN, Ronda JM, Kronauer RE. Exposure to bright light and darkness to treat physiologic maladaptation to night work. N Engl J Med. 1990 May;322(18):1253–9.
15. Daan S, Beersma DG, Borbély AA. Timing of human sleep: recovery process gated by a circadian pacemaker. Am J Physiol. 1984 Feb; 246(2 Pt 2):R161–83.
16. Diniz Behn CG, Brown EN, Scammell TE, Kopell NJ. Mathematical model of network dynamics governing mouse sleep-wake behavior. J Neurophysiol. 2007 June;97(6):3828–40.
17. Dubrovskii VA, Sergeev VN. Universal precursor of geomechanical catastrophes. Doklady Phys. 2004;49(4):231–3.
18. Feucht M, Möller U, Witte H, Schmidt K, Arnold M, Benninger F, Steinberger K, Friedrich MH. Nonlinear dynamics of 3 Hz spike-and-wave discharges recorded during typical absence seizures in children. Cereb Cortex. 1998;8:524–33.
19. Fredholm BB. Astra award lecture. Adenosine, adenosine receptors and the actions of caffeine. Pharmacol Toxicol. 1995 Feb;76(2):93–101.
20. Freeman WJ. Mass action in the nervous system. New York: Academic Press; 1975.
21. Fröberg JE, Karisson CG, Levi L, Lidberg L. Circadian rhythms of catecholamine excretion, shooting range performance and self-ratings of fatigue during sleep deprivation. Biol Psychol. 1975;2:175–88.
22. Fulcher BD, Phillips AJK, Robinson PA. Modeling of the impact of impulsive stimuli on sleep-wake dynamics. Phys Rev E. 2008;78:051920.
23. Fulcher BD, Phillips AJK, Robinson PA. Quantitative physiologically based modeling of subjective fatigue during sleep deprivation. J Theor Biol. 2010;264:407–19.
24. Gooley JJ, Rajaratnam SM, Brainard GC, Kronauer RE, Czeisler CA, Lockley SW. Spectral responses of the human circadian system depend on the irradiance and duration of exposure to light. Sci Transl Med. 2010 May;2(31):31–3.
25. Härmä M, Tarja H, Irja K, Mikael S, Jussi V, Anne B, Pertti M. A controlled intervention study on the effects of a very rapidly forward rotating shift system on sleep-wakefulness and well-being among young and elderly shift workers. Int J Psychophysiol. 2006 Jan;59(1):70–9.

26. Herrmann CS. Human EEG responses to 1–100 Hz flicker: resonance phenomena in visual cortex and their potential correlation to cognitive phenomena. Exp Brain Res. 2001;137:346–53.
27. Hoddes E, Zarcone V, Smythe H, Phillips R, Dement WC. Quantification of sleepiness: a new approach. Psychophysiology. 1973;10:431–6.
28. Jewett ME, Forger DB, Kronauer RE. Revised limit cycle oscillator model of human circadian pacemaker. J Biol Rhythms. 1999;14:493–9.
29. Jirsa VK, Haken H. Field theory of electromagnetic brain activity. Phys Rev Lett. 1996;77:960–3.
30. Kandel ER, Schwartz JH, Jessell TM. Principles of neural science, 4th edition. New York: MacGraw-Hill; 2000.
31. Kerr CC, Rennie CJ, Robinson PA. Model-based analysis and quantification of age trends in auditory evoked potentials. Clin Neurophysiol. 2011;122:134–47.
32. Knauth P. The design of shift systems. Ergonomics. 1993;36(1–3):15–28.
33. Knauth P. Speed and direction of shift rotation. J Sleep Res. 1995 Dec;4(S2):41–6.
34. Kronauer RE, Forger DB, Jewett ME. Quantifying human circadian pacemaker response to brief, extended, and repeated light stimuli over the phototopic range. J Biol Rhythms. 1999;14:500–15.
35. Kunz D, Herrmann WM. Sleep-wake cycle, sleep-related disturbances, and sleep disorders: a chronobiological approach. Compr Psychiatry. 2000; 41(2 Suppl 1):104–15.
36. Lammers WJ, Badia P, Hughes R, Harsh J. Temperature, time-of-night testing, and responsiveness to stimuli presented while sleeping. Psychophysiology. 1991;28:463–7.
37. Lamond N, Jay S, Dorrian J, Ferguson S, Jones C, Dawson D. The dynamics of neurobehavioral recovery following sleep loss. J Sleep Res. 2007;16:33–41.
38. Landolt HP. Sleep homeostasis: a role for adenosine in humans? Biochem Pharmacol. 2008 June;75(11):2070–9.
39. Lenton TM, Livina VN, Dakos V, van Nes EH, Scheffer M. Early warning of climate tipping points from critical slowing down: comparing methods to improve robustness. Philos Trans R Soc A—Math Phys Eng Sci. 2012;370(1962):1185–204.
40. Lopes da Silva FH, Hoeks A, Smits H, Zetterberg LH. Model of brain rhythmic activity: the alpha-rhythm of the thalamus. Kybernetik. 1974;15:27–37.
41. Minors DS, Waterhouse JM, Wirz-Justice A. A human phase-response curve to light. Neurosci Lett. 1991 Nov;133(1):36–40.
42. Niedermeyer E. The normal EEG of the waking adult. In Niedermeyer E, Lopes da Silva FH, editors. Electroencephalography: basic principles, clinical applications, and related fields. Baltimore; Williams & Wilkins; 1999. p 149–173.
43. Nunez PL. The brain wave equation: a model for the EEG. Math Biosci. 1974;21:279–97.
44. Nunez PL. Neocortical dynamics and Human EEG Rhythms. New York: Oxford University Press; 1995.
45. Nunez PL, Srinivasan R. Electric fields of the brain: The neurophysics of EEG, 2nd edition. New York: Oxford University Press; 2006.
46. O'Connor SC, Robinson PA. Wave-number spectrum of electrocorticographic signals. Phys Rev E. 2003;67:051912.1–051912.13.
47. O'Connor SC, Robinson PA, Chiang AKI. Wave-number spectrum of electroencephalographic signals. Phys Rev E. 2002;66:061905.1–061905.12.
48. Pace-Schott EF, Hobson JA. The neurobiology of sleep: genetics, cellular physiology and subcortical networks. Nat Rev Neurosci. 2002;3:591–605.
49. Pasnau RO, Naitoh P, Stier S, Kollar EJ. The psychological effects of 205 hours of sleep deprivation. Arch Gen Psychiatry. 1968;18:495–505.
50. Penetar D, McCann U, Thorne D, Kamimori G, Galinski C, Sing H, Thomas M, Belenky G. Caffeine reversal of sleep deprivation effects on alertness and mood. Psychopharmacology. 1993;112:359–65.
51. Phillips AJK, Robinson PA. A quantitative model of sleep-wake dynamics based on the physiology of the brainstem ascending arousal system. J Biol Rhythms. 2007;22:167–79.

52. Phillips AJK, Robinson PA. Sleep deprivation in a quantitative physiologically based model of the ascending arousal system. J Theor Biol. 2008;255:413–23.
53. Phillips AJK, Robinson PA. Potential formulation of sleep dynamics. Phys Rev E. 2009;79(2):021913.
54. Phillips AJK, Chen PY, Robinson PA. Probing the mechanisms of chronotype using quantitative modeling. J Biol Rhythms. 2010;25:217–27.
55. Porkka-Heiskanen T, Strecker RE, McCarley RW. Brain site-specificity of extracellular adenosine concentration changes during sleep deprivation and spontaneous sleep: an in vivo microdialysis study. Neuroscience. 2000;99:507–17.
56. Postnova S, Voigt K, Braun HA. A mathematical model of homeostatic regulation of sleep-wake cycles by hypocretin/orexin. J Biol Rhythms. 2009 Dec;24(6):523–35.
57. Postnova S, Layden A, Robinson PA, Phillips AJ, Abeysuriya RG. Exploring sleepiness and entrainment on permanent shift schedules in a physiologically based model. J Biol Rhythms. 2012 Feb;27(1):91–102.
58. Postnova S, Robinson PA, Postnov DD. Adaptation to shift work: physiologically based modeling of the effects of lighting and shifts' start time. PLoS ONE. 2013;8(1):e53379.
59. Postnova S, Postnov DD, Seneviratne M, Robinson PA. Effects of rotation interval on sleepiness and circadian dynamics on forward rotating 3-shift systems. J Biol Rhythms. 2014 Feb;29(1):60–70.
60. Puckeridge M, Fulcher BD, Phillips AJK, Robinson PA. Incorporation of caffeine into a quantitative model of fatigue and sleep. J Theor Biol. 2011;273:44–54.
61. Rajaratnam SM, Howard ME, Grunstein RR. Sleep loss and circadian disruption in shift work: health burden and management. Med J Aust. 2013 Oct;199(8):S11–5.
62. Rempe MJ, Best J, Terman D. A mathematical model of the sleep/wake cycle. J Math Biol. 2010;60:615–44.
63. Rennie CJ, Robinson PA, Wright JJ. Effects of local feedback on dispersion of electrical waves in the cerebral cortex. Phys Rev E. 1999;59:3320–9.
64. Rennie CJ, Robinson PA, Wright JJ. Unified neurophysical model of EEG spectra and evoked potentials. Biol Cybern. 2002;86:457–71.
65. Reppert SM, Weaver DR. Coordination of circadian timing in mammals. Nature. 2002 Aug;418(6901):935–41.
66. Roberts JA, Robinson PA. Quantitative theory of driven nonlinear brain dynamics. NeuroImage. 2012;62:1947–55.
67. Robinson PA. Interpretation of scaling properties of electroencephalographic fluctuations via spectral analysis and underlying physiology. Phys Rev E. 2003;67:032902.
68. Robinson PA. Neurophysical theory of coherence and correlations of electroencephalographic and electrocorticographic signals. J Theor Biol. 2003;222:163–75.
69. Robinson PA. Propagator theory of brain dynamics. Phys Rev E. 2005;72:011904.
70. Robinson PA. Patchy propagators, brain dynamics, and the generation of spatially structured gamma oscillations. Phys Rev E. 2006;73:041904.
71. Robinson PA. Interrelating anatomical, effective, and functional brain connectivity using propagators and neural field theory. Phys Rev E. 2012;85:011912.
72. Robinson PA, Rennie CJ, Wright JJ. Propagation and stability of waves of electrical activity in the cerebral cortex. Phys Rev E. 1997;56:826–40.
73. Robinson PA, Rennie CJ, Wright JJ, Bourke P. Steady states and global dynamics of electrical activity in the cerebral cortex. Phys Rev E. 1998;58:3557–71.
74. Robinson PA, Loxley PN, O'Connor SC, Rennie CJ. Modal analysis of corticothalamic dynamics, electroencephalographic spectra, and evoked potentials. Phys Rev E. 2001;63:041909.
75. Robinson PA, Rennie CJ, Wright JJ, Bahramali H, Gordon E, Rowe DL. Prediction of electroencephalographic spectra from neurophysiology. Phys Rev E. 2001;63:021903.
76. Robinson PA, Rennie CJ, Rowe DL. Dynamics of large-scale brain activity in normal arousal states and epileptic seizures. Phys Rev E. 2002;65:041924.
77. Robinson PA, Rennie CJ, Rowe DL, O'Connor SC, Wright JJ, Gordon E, Whitehouse RW. Neurophysical modeling of brain dynamics. Neuropsychopharmocology. 2003;28:s74–9.

78. Robinson PA, Whitehouse RW, Rennie CJ. Nonuniform corticothalamic continuum model of electroencephalographic spectra with application to split-alpha peaks. Phys Rev E. 2003;68:021922.
79. Robinson PA, Rennie CJ, Rowe DL, O'Connor SC. Estimation of multiscale neurophysiologic parameters by electroencephalographic means. Hum Brain Mapp. 2004;23:53–72.
80. Robinson PA, Rennie CJ, Rowe DL, O'Connor SC, Gordon E. Multiscale brain modelling. Philos Trans R Soc B. 2005;360:1043–50.
81. Robinson PA, Rennie CJ, Phillips AJK, Kim JW, Roberts JA. Phase transitions in physiologically based multiscale mean-field models. In Steyn-Ross DA, Steyn-Ross M, editors. Modeling phase transitions in the brain. New York: Springer; 2010. pp. 179–201.
82. Robinson PA, Phillips AJK, Fulcher BD, Puckeridge M, Roberts JA, Rennie CJ. Quantitative modeling of sleep dynamics. In Hutt A, editor. Sleep and anesthesia: neural correlates in theory and experiment. New York: Springer; 2011. pp. 45–68.
83. Rowe DL, Robinson PA, Rennie CJ. Estimation of neurophysiological parameters from the waking EEG using a biophysical model of brain dynamics. J Theor Biol. 2004;231:413–33.
84. Saper CB, Chou TC, Scammell TE. The sleep switch: hypothalamic control of sleep and wakefulness. Trends Neurosci. 2001;24:726–31.
85. Saper CB, Scammell TE, Lu J. Hypothalamic regulation of sleep and circadian rhythms. Nature. 2005 Oct;437(7063):1257–63.
86. Saper CB, Fuller PM, Pedersen NP, Lu J, Scammell TE. Sleep state switching. Neuron. 2010 Dec;68(6):1023–42.
87. Scammell TE. The neurobiology, diagnosis, and treatment of narcolepsy. Ann Neurol. 2003;53:154–66.
88. Scheer FA, Hilton MF, Mantzoros CS, Shea SA. Adverse metabolic and cardiovascular consequences of circadian misalignment. Proc Natl Acad Sci U S A. 2009 March;106(11):4453–8.
89. Scheffer M, Carpenter SR, Lenton TM, Bascompte J, Brock W, Dakos V, van de Koppel J, van de Leemput IA, Levin SA, van Nes EH, Pascual M, Vandermeer J. Anticipating critical transitions. Science. 2012;338(6105):344–8.
90. Sherman SM, Guillery RW. Exploring the thalamus. New York: Academic Press; 2001.
91. Srinivasan R, Nunez PL, Silberstein RB. Spatial filtering and neocortical dynamics: estimates of EEG coherence. IEEE Trans Biomed Eng. 1998;45:814–26.
92. Stepanski EJ. The effect of sleep fragmentation on daytime function. Sleep. 2002;25:268–76.
93. Steriade M, Gloor P, Llinás RR, Lopes da Silva FH, Mesulam MM. Basic mechanisms of cerebral rhythmic activities. Electroencephalogr Clin Neurophysiol. 1990;76:481–508.
94. Steriade M, Jones EG, McCormic DA. Thalamus. Amsterdam: Elsevier; 1997.
95. Strogatz SH. Nonlinear dynamics and chaos: with applications to physics, biology, chemistry, and engineering. Reading: Addison-Wesley; 1994.
96. St Hilaire MA, Klerman EB, Khalsa SB, Wright KP, Czeisler CA, Kronauer RE. Addition of a non-photic component to a light-based mathematical model of the human circadian pacemaker. J Theor Biol. 2007 Aug;247(4):583–99.
97. Tamakawa Y, Karashima A, Koyama Y, Katayama N, Nakao M. A quartet neural system model orchestrating sleep and wakefulness mechanisms. J Neurophysiol. 2006;95:2055–69.
98. Thompson JMT, Sieber J. Climate tipping as a noisy bifurcation: a predictive technique. IMA J Appl Math. 2011;76(1):27–46.
99. Wilson HR, Cowan JD. A mathematical theory of the functional dynamics of cortical and thalamic nervous tissue. Kybernetik. 1973;13:55–80.
100. Wright JJ, Liley DTJ. Dynamics of the brain at global and microscopic scales: neural networks and the EEG. Behav Brain Sci. 1996;19:285–309.
101. Xie L, Kang H, Xu Q, Chen MJ, Liao Y, Thiyagarajan M, O'Donnell J, Christensen DJ, Nicholson C, Iliff JJ, Takano T, Deane R, Nedergaard M. Sleep drives metabolite clearance from the adult brain. Science. 2013;342(6156):373–7.

Chapter 6
How to Render Neural Fields More Realistic

Axel Hutt, Meysam Hashemi and Peter beim Graben

Abstract Conventional neural field models describe well some experimental data, such as Local Field Potentials or electroencephalographic data. The work reviews recent extensions of neural field models and describes the activation and attenuation of spectral power in certain frequency bands subjected to the statistical properties of an external input and subjected to the properties of synaptic receptor efficacy and the heteroclinic transitions between meta-stable state as a model for event-related potentials.

Keywords Systems · Kernels · Heteroclinic orbits · Input · Models

Introduction

Complex systems are omnipresent in nature. They exhibit multiple spatial and temporal scales whose understanding and control is one of the most challenging problems of research for the last centuries. The notion of *complexity* is not well-defined [30], but typically a system is said to be complex if, in addition to the multiple scales, there exist self-organised sub-units in the system which are generated by smaller units. In many biological systems, climbing this hierarchy from smaller units to larger units implies increasing the spatial and temporal scale.

A. Hutt (✉) · M. Hashemi
INRIA Grand Est—Nancy, Team NEUROSYS, 615 rue du Jardin Botaniqu,
54600 Villers-lès-Nancy, France

CNRS, Loria, UMR no. 7503, 54500 Vandoeuvre-lès-Nancy, France

Universitè de Lorraine, Loria, UMR no. 7503, 54500 Vandoeuvre-lès-Nancy, France
e-mail: axel.hutt@inria.fr

M. Hashemi
e-mail: meysam.hashemi@inria.fr

P. beim Graben
Department of German Studies and Linguistics, Humboldt-Universität zu Berlin
and Bernstein Center for Computational Neuroscience, Berlin, Germany
e-mail: peter.beim.graben@hu-berlin.de

© Springer International Publishing Switzerland 2015
B. S. Bhattacharya, F. N. Chowdhury (eds.), *Validating Neuro-Computational Models of Neurological and Psychiatric Disorders,* Springer Series in Computational Neuroscience 14, DOI 10.1007/978-3-319-20037-8_6

Before discussing neural systems, we would like to clarify a major scientific problem in this context: the choice of the description level. Let us assume the task to describe a water wave mathematically, i.e., derive a mathematical model whose solution describes the spatiotemporal phenomenon of a wave. Some researchers may ask: *we want to understand how water works, and especially, how a wave is generated.* Hence, a straightforward simple and fundamental approach is to take a closer look at the building blocks of water (remark: this is very similar to the way how biologists work in neuroscience today). The researchers find H_2O-molecules, study the properties of the atoms and the inter-molecule interactions like hydrogen bridges and van-der-Waals bounds. Now, to describe a water wave, one has to study millions and millions of these molecules and their interactions. To this end, the researchers take the reasonable approach to start simulating two, three, then even hundreds of these molecules. The numerical analysis is demanding, but still there is no description of the water waves, since the system appears to be so complex taking into account all the different interactions of molecules and the research in this hard task got stuck in some way.

Of course it is well-known that it is not necessary to study single molecules to describe water waves, today we take the famous Navier-Stokes equation (NSE) which includes the solution for this phenomenon. This equation involves mean fluid properties, e.g., inner friction and viscosity [52]. Hence the NSE considers average properties of interacting single molecules. In other words, it does not know single molecules, but provides a powerful description of a large ensemble or mass of molecules. It allows to describe several different rather complex fluid phenomena, but of course *no phenomena related to single molecules*. Hence the NSE equation provides a very good mathematical description of the system at a macroscopic description level only. Going back to the intended description of water waves by a single molecule study, this approach is not reasonable, probably it will not lead to a good description level of macroscopic phenomena and hence it is not constructive.

In todays' neuroscience, the approach of linking single neuron activity to macroscopic phenomena is attractive, e.g. in the context of cognition [7, 31, 57], sleep [67] or anaesthesia [4, 13, 39]. These studies state a relationship between the single neuron activity (microscopic scale) and behavioral phenomenon (macroscopic scale). However, to our best knowledge it is not understood how the different experimental findings on different scales are linked to each other, i.e., the link between the two scales is not understood and no model linking the scales has been developed yet. This situation resembles closely the water wave task described above: it is clear that there is a relation between small sub-units (molecules or neurons) to large complex systems units (water wave or cognition), since the dynamics of the large units is generated by the sub-units, but a link appears to be too complex. Consequently, learning from physics and the NSE, it is necessary to consider more abstract, intermediate models whose elements are based on small sub-units properties but which allow to model large unit phenomena. In other words, it is much more effective to consider mesoscopic population models involving average properties of interacting neurons and which allow to describe macroscopic experimental phenomena, such as Local Field Potentials (LFP), encephalographic activity (EEG/MEG) or even behaviour. Promising candidates for such models are neural mass or neural field models which have been validated by LFPs and EEG in many previous studies, cf. [18, 27, 69, 71]. The present book chapter discusses recent advances in these models rendering the standard neural population models more realistic.

The subsequent sections do not give a complete overview over the recent advances in the field, a recent excellent review article already provides this information [17]. The present chapter first introduces briefly two types of neural field models. Then it gives some details of few selected extensions, always introducing the neuroscientific problem by experimental data before presenting a mathematical description of the phenomena described. In each example, the experimental data validates the neural population model and elucidates how to model experimental data.

Two Classes of Neural Field Models

Amari Model

The first neural field models developed by Wilson and Cowan [68] and Amari [3] are continuum limits of large-scale neural networks. Typically, their dynamic variables describe either mean voltage [3] or mean firing rate [46, 68] of a population element of neural tissue, see also the excellent review article of Bressloff [17]. In some subsequent sections we consider the paradigmatic Amari equation [3] describing the spatiotemporal dynamics of mean potential $V(x,t)$ over a cortical d-dimensional manifold $\Omega \subset \mathbf{R}^d$:

$$\tau \frac{\partial V(x,t)}{\partial t} = -V(x,t) + \int_\Omega K(x,y) S[V(y,t)] y + I(x,t). \qquad (6.1)$$

with the spatial synaptic kernel $K(x,y)$, which defines the connectivity between site $y \in \Omega$ and site $x \in \Omega$. The transfer function S is nonlinear and typically of sigmoidal shape. This model considers external inputs $I(x,t)$, e.g., originating from other extra-cortical populations and from external stimulation. The model (6.1) takes into account a single synaptic time scale τ assuming an exponential synaptic response function. However, we point out that re-scaling of time allows to set $\tau = 1$.

In general, the connectivity kernel $K(x,y)$ fully depends on both sites x and y reflecting *spatial heterogeneity*. If the connectivity solely depends on the difference between x and y, i.e. $K(x,y) = K(x-y)$, then the neural field activity does not depend on specific spatial locations and hence is translational invariant. This case is called *spatially homogeneity* [3]. If the connectivity even depends on the distance between x and y only, i.e. $K(x,y) = K(\|x-y\|)$, with $\|x\|$ as some norm in Ω, then the neural field is *spatially homogeneous and isotropic* [24].

Several extensions of the Amari model (6.1) are possible, such as the consideration of finite axonal transmission speeds [41, 42], constant feedback delays [36, 61] (see also section "Delayed Nonlocal Feedback Between Populations"), heterogeneity [9, 10] (see also section "Heterogeneous Neural Fields"), spike-frequency adaption [22], statistical properties of single neurons [28], the combination of several brain areas [60], electromagnetic fields [11, 12] and many more [17].

Mathematically, Eq. (6.1) is an integro-differential equation. Spatially homogeneous (respectively isotropic) neural fields have been intensively studied in the literature due to their nice analytical properties [23, 29, 41, 42]. Moreover, these models may be transformed to derive partial differential wave equations [23, 37, 47] for certain classes of synaptic kernels.

Robinson Model

As mentioned at the end of the previous paragraph, under certain conditions integro-differential equations may be transformed to partial differential equations. In the 1990s James Wright and David Liley started developing partial differential equation models for neural population activity [70]. Their work has inspired other teams, e.g., Peter Robinson and colleagues, who developed a similar neural model which has been proven to be very successful. This type of neural field model [34, 59, 66] is based on a population-level model of a single thalamo-cortical module consisting of excitatory (E) and inhibitory (I) cortical population, thalamic relay neurons (S), and thalamic reticular neurons (R). The average soma membrane potential is modeled by

$$V_a(t) = \sum_{b=E,I,R,S} h(t) \otimes K_{a,b} \phi_b(t - \tau_{a,b}), a = E, I, S, R \quad (6.2)$$

where $\otimes$ denotes temporal convolution and ϕ_b is the pulse firing rate of the population b. The constants $K_{a,b}$ are the strengths of the connections from population of type b to population of type a. The delay term, $\tau_{a,b}$ is zero for intra-cortical and intra-thalamic connections and non-zero for thalamocortical or corticothalamic connections [66].

The model assumes that only axons of excitatory cortical neurons are long enough to emit axonal propagating pulses. Moreover, ϕ_E obeys the damped oscillator equation

$$D\phi_E = S(V_E), \quad (6.3)$$

with the operator D

$$D = \left(\frac{1}{\gamma}\frac{\partial}{\partial t} + 1\right)^2. \quad (6.4)$$

6 How to Render Neural Fields More Realistic

In Eq. (6.2) $h(t) = H\bar{h}(t)$, where $\bar{h}(t)$ denotes the mean synaptic response function

$$\bar{h}(t) = \frac{\alpha\beta}{\beta - \alpha}(e^{-\alpha t} - e^{-\beta t}), \tag{6.5}$$

where α and β are the synaptic decay and rise rate of synaptic response function, respectively.

In addition, the pre-factor H defines the response function amplitude subject to the aneasthetic concentration. For more details of the model and the nominal parameter values, see [34].

Delayed Nonlocal Feedback Between Populations

In neural fields, one might include delays in several ways. The finite axonal transmission delay is proportional to the fraction of distance between two spatial locations and transmission speed and takes into account the finite propagation speed of action potentials along axonal branches [43], or in more general terms, it considers the finite-time interaction between two elements in a spatially extended system [37]. In addition, one could argue that delayed interactions happen on a single-neuron scale between neurons and it is more reasonable to treat these inter-neuron delays as a kind of effective delay [61, 62]. This latter type of delay is constant. In addition to these two delay types, the nonlocal feedback delay takes into account the finite axonal transmission speed along axonal pathways between two brain areas. Since this axonal pathway has a finite defined length, the transmission delay is fixed and hence also constant [36]. Variations of all these delay types may be considered by distributed transmission speeds and/or distributed delays [6, 42]. The subsequent paragraphs consider constant delays reflecting the finite transmission speed along axonal branches between brain areas, experimental data validate quantitatively the models proposed.

The Primary Sensory Area in Weakly-Electric Fish

To examine the neural decoding in weakly electric fish, Doiron et al. [26] had performed an experimental *in vivo* stimulation study. A dipole was placed near the skin of a fish to stimulate only a part of the receptive field. Figure 6.1 sketches the experimental setup, the spike autocorrelation $A(t)$ and the interspike interval (ISI) histograms of a typical pyramidal cell response to a local and global stimulus. The stimulus was temporal noise evoking spatially weakly correlated sensory receptor activity for local stimulation and strong spatial correlations in the receptor dynamics in case of the experimentally global stimulation. The global random stimulus evokes bursting in the neuron activity whereas the local stimulus evokes a single

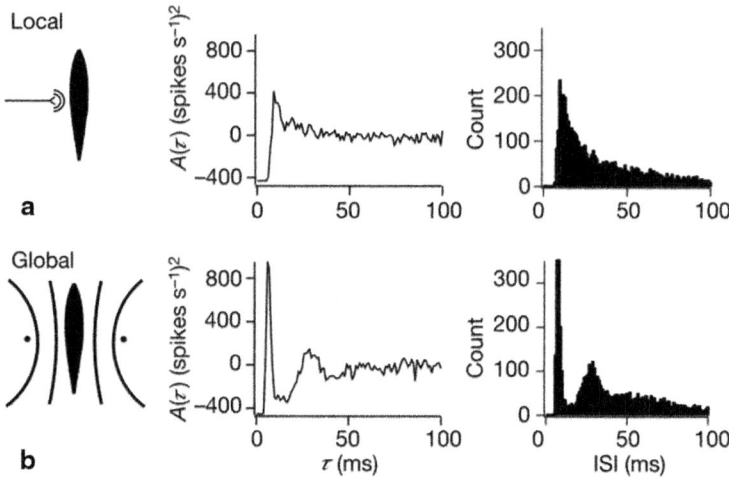

Fig. 6.1 Experimental setup and firing statistics of stimulation experiment [26]. In (**a**), the electric skin stimulation is local (*left panel*) and induces a single main oscillation as seen in the spike autocorrelation function $A(\tau)$ (*center panel*) and the corresponding histogram (*right panel*). In (**b**) the stimulus was global inducing an additional oscillation mode. Taken from [26] by permission

principal firing mode. This experiment raises the question how the spatial correlations in the input stimulus interact with those imposed by the physiological system.

A rather simple population model considers the primary sensory areas in the electro-sensory system of weakly electric fish [14], but similar configurations can also be found in parts of the vertebrate brain [2]. The model sketched in Fig. 6.2 [44] is made up of the ELL, a layer of pyramidal cells driven by the primary receptors that receive an external stimulus, and the higher area Np. These areas are spatially coupled via a delayed topographic feedback with connectivity kernels $K_{en}(x)$ and $K_{ne}(x)$ which reflect connections from the $Np(n)$ to the $ELL(e)$ and vice versa, respectively, see Fig. 6.2. The neurons in both populations have insignificant direct

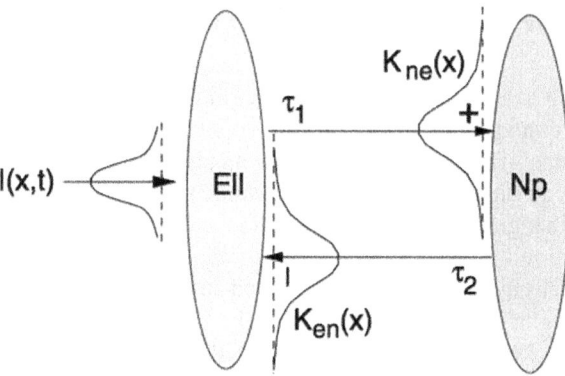

Fig. 6.2 Topography of the delayed feedback model. The plus and minus signs indicate the excitatory and inhibitory connections, respectively

couplings and the coupling from the ELL to Np is excitatory and delayed in time by τ_1, the coupling back to the ELL is inhibitory with delay τ_2. Moreover, according to the experimental setup in [26], the model considers excitatory spatiotemporal stimuli $I(x,t)$ to the ELL.

The aim of the study is to learn more about the mechanism how to change the principal oscillation frequency by the properties of an external stimulus. The model considers spatiotemporal noise with a well-defined and adjustable spatial correlation length. The population model [44] describes two coupled neural fields whose activities are strongly related to experimentally observable local field potentials [55].

Then the theoretical power spectrum in the ELL reads

$$P(\nu) = \int_{-\infty}^{\infty} R(\nu,l)\tilde{C}(l)dl \qquad (6.6)$$

where $R(\nu,l)$ is the spectral response function and $\tilde{C}(l)$ is the scaled Fourier transform of the input correlation function. It turns out, that the power spectrum Eq. (6.6) does not depend on the spatial scale of the feedback loop σ_f and the input correlation scale σ_i independently, but just depends on their ratio, called η. This finding reflects the coupling of the spatial scale of the external input to the intrinsic spatial scale of the system.

Figure 6.3 shows the resulting power spectra for two values of the spatial scale ratio η. We observe that a small ratio $\eta = \sigma_f/\sigma_i = 1$ generates a spectral peak at about 20 Hz, whereas the large ratio $\eta ?1$ generates a power peak at about 0 Hz. Hence retaining the topographic feedback but decreasing the input correlation function from large values of $\sigma_i = \sigma_f/\eta$ (global noise) to small values of σ_i (local

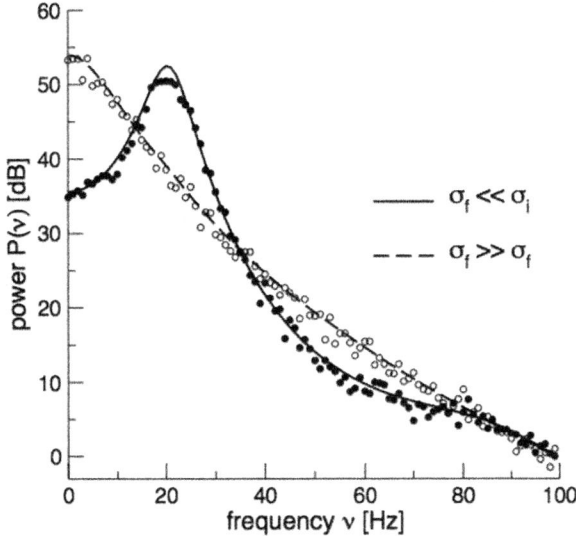

Fig. 6.3 Theoretical power spectrum computed for $\eta = 40$ (*solid line*) and $\eta = 1/40$ (*dashed line*)

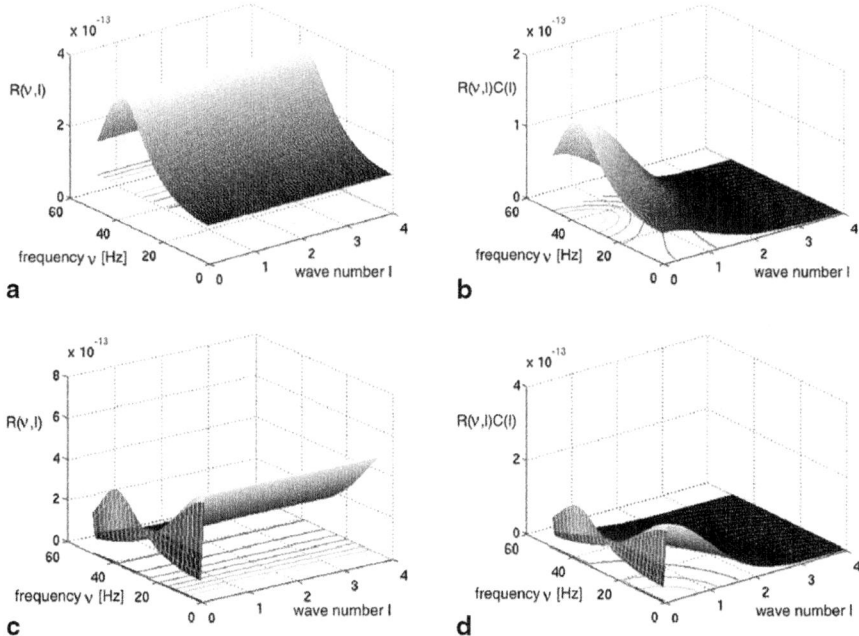

Fig. 6.4 The response function $R(\nu, l)$ and the integrand of the power spectrum integral $R(\nu, l)\tilde{C}(l)$ for (**a, b**) $\sigma_i = 40\sigma_f$ (*global noise*), (**c, d**) $\sigma_i = \sigma_f / 40$ (*local noise*). Taken from Ref. [44] by permission

noise) switches the spectral peak, similar as observed experimentally by Doiron et al. [26].

To understand this, Fig. 6.4 shows the response function $R(\nu, l)$ and the integrand $R(\nu_l)\tilde{C}(l)$ in the definition of the power spectrum Eq. (6.6). The response function R and $R\tilde{C}$ have a single maximum at about 20 Hz for global noise (Fig. 6.4a and 6.4b), whereas R and $R\tilde{C}$ have two local maxima at 20 Hz and 0 Hz (Fig. 6.4c and 6.4d). Since the peak of $R\tilde{C}$ at 0 Hz is broader than the peak about 20 Hz and the power spectrum is the integral over $R(\tilde{l})$, cf. Eq. (6.6), the contribution of $R\tilde{C}$ to the power at 0 Hz exceeds the contribution at 20 Hz yielding a strong peak at 0 Hz.

This result reveals that the spectral peak in Fig. 6.3 at 0 Hz results from the selection of one mode out of two possible modes at frequencies of 0 Hz and 20 Hz, whereas the spectral peak at 20 Hz is the only oscillation mode present in the system. The switch between these two configurations depends on the spatial correlation of the stimulus noise. These modes reflect activity subnetworks from which only one is engaged. Hence the rather simple population model reveals the underlying mechanism of the switch-on of oscillations at a certain frequency induced by an external stimulus. This study may explain the occurrence of certain oscillations observed in EEG by an underlying change of stimulation, as found in the mammalian olfactory bulb during breathing [19] or in anaesthesia at increased levels of anaesthetic concentrations [56, 64].

General Anaesthesia

General anaesthesia is an important medical application in today's hospital surgery, but its underlying neural interactions is still a mystery. In the last decades, general anaesthesia has attracted theoretical researchers [15, 20, 45, 53, 65]. Most theoretical studies aim to explain signal features of electroencephalographic data (EEG) observed during anaesthesia, such as the attenuation or enhancement of α-activity accompanied by a subsequent enhancement of δ-activity while increasing anaesthetic concentration [21, 54], cf. Fig. 6.5. The subsequent paragraphs show how neural field models may explain on the power enhancement and the frequency shift of maximum power while increasing the anaesthetic concentration.

To this end, we consider a derivative of the Robinson model introduced in Sect. "Two Classes of Neural Field Models" and introduce a new sigmoid function derived from properties of type-I neurons [39]

$$S(V_a) = F(V_a, 0) - F(V_a, \gamma), \qquad (6.7)$$

with

$$F(V_a, \gamma) = \frac{S_{max}}{2}\left[1 + \mathrm{erf}\left(\frac{V_a - \theta - \gamma\sigma^2}{\sqrt{2}\sigma}\right)\right]e^{-\gamma(V_a-\theta)+\gamma^2\sigma^2/2}, \qquad (6.8)$$

in which the parameter $\gamma < \infty$ takes into account the properties of type I-neurons, S_{max} is the maximum population firing rate, θ is the mean firing threshold, and σ is related to the standard deviation of firing thresholds in the populations.

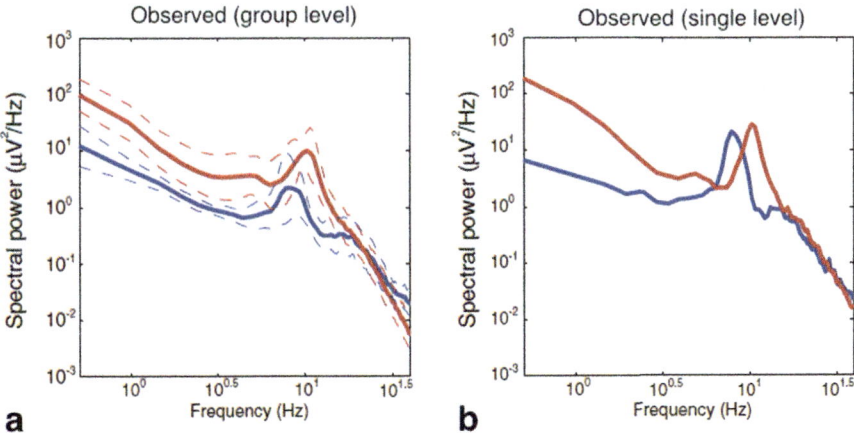

Fig. 6.5 The power spectra measured in frontal EEG electrodes in the absence of anaesthesia (*blue line*) and during propofol anaesthesia (*red line*) in a group of subjects (**a**) and for a single subject (**b**). Taken from [34] by permission

External input to the system can be considered as a non-specific input to relay neurons as

$$\phi_N = \langle \phi_N \rangle + \sqrt{2\kappa}\xi(t), \tag{6.9}$$

where $\langle \phi_N \rangle$ is its mean value and $\xi(t)$ is Gaussian white noise with the intensity κ and zero mean.

It is well-known that general anaesthetics (GA) bind directly to sensitive target receptors. For instance, a large number of studies link the effect of many GAs to altered function of the $GABA_A$ receptors. Recent clinical findings have revealed several sites of elicited anaesthetic action of the anaesthetic propofol in the human brain, e.g., propofol suppresses field potentials in the rat thalamus and cortex [5].

In detail, the anaesthetic propofol increases the decay time constant of synaptic $GABA_A$ receptors, and hence increases the total charge transfer in these synapses but not that of excitatory synapses [50]. To integrate physiological observations into neural models such as the Robinson model detailed in Sect. "Two Classes of Neural Field Models", the anaesthetic action on synaptic receptors is modelled by $\alpha \to \alpha/p$ with $p \geq 1$ leading to a decrease of the decay rate constant α, or equivalently, an increase in decay time constant of $GABA_A$ receptors [32]. The factor $p=1$ reflects absent anaesthetic action, i.e, the baseline condition. The model considers inhibitory synapses at excitatory cortical neurons (factor p_1), and at inhibitory cortical neurons (factor p_2) and at thalamic relay neurons (factor p_3).

To compute the power spectrum of the system, we consider the stationary state of Eq. (6.2), which obeys $dV_a(t)/dt = 0$ where V_a is taken from Eq. (6.2). Increasing the anaesthetic concentration, i.e. increasing the three factors p_1, p_2 and p_3 changes the stationary states dependent on the relation of these three factors, cf. Fig. 6.6. In (a) the two lower stationary states collide to a single state whereas in (c) the two upper states collide. This difference indicates two fundamentally different mechanisms which may yield different power spectra.

The power spectrum characterises small fluctuations about this stationary state. Assuming that excitatory activity generates the EEG, and by virtue of the specific choice of external input to relay neurons, the power spectrum of the EEG depends just on one matrix component of the Greens function [32] by

$$P_E(\omega) = 2\kappa\sqrt{2\pi}\left|\tilde{G}_{1,3}(\omega)\right|^2. \tag{6.10}$$

in which $\tilde{G}_{1,3}(\omega)$ is a matrix element of the 4×4 matrix

$$\tilde{\mathbf{G}}(\omega) = \frac{1}{\sqrt{2\pi}}\begin{bmatrix} \tilde{L}\dfrac{\tilde{D}}{K_{11}} - K_1 & -K_2 & -K_3 e^{-i\omega\tau} & 0 \\ -K_4 & \tilde{L} - K_5 & -K_6 & 0 \\ -K_7 e^{-\omega\tau} & 0 & \tilde{L} & -K_8 \\ -K_9 e^{-\omega\tau} & 0 & -K_{10} & \tilde{L} \end{bmatrix}^{-1} \tag{6.11}$$

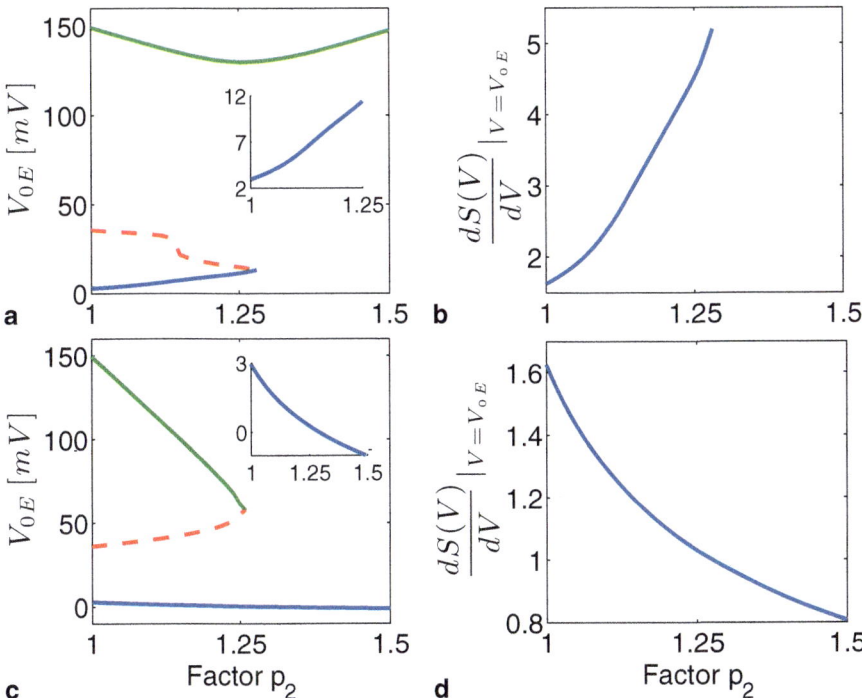

Fig. 6.6 The stationary states and the nonlinear gain dS/dV computed at the lowest stationary state of pyramidal neurons V_E subjected to the factor p_2. (**a, b**) $p_1 = p_3 = 1 + 0.3(p_2 - 1)$, (**c, d**) $p_1 = p_2 = p_3$. We observe three states in (**a**) and (**c**) for $p_2 = 1$ where (**a**) the two lower states collide and (**c**) the two upper states collide. The center branch (*dashed red*) is linearly unstable, whereas the other branches are linearly stable. For clarification, the lower branches are shown in the insets. Parameters are $S_{max} = 250$Hz, $V_{th} = 15$mV, $\gamma = 0.08$mV and $\sigma = 10$mV

with $\tilde{L} = (1 + \omega/\alpha)(1 + i\omega/\beta)$, $\tilde{D} = (1 + \omega/\gamma)^2$, and the constants $K_i, i = 1, 2, \ldots, 11$ are proportional to the synaptic strengths and the nonlinear gains $\partial S/\partial V$ computed at the stationary state of the system.

Figure 6.7 shows the theoretical power spectrum P_E in the baseline condition and after the administration of propofol for two different relations of the anaesthetic factors p_1, p_2 and p_3. At first we note that the spectra resemble well the spectrum obtained from experimental observations: increasing p_2 in a specific relation to the p_1 and p_3 yields increases in delta and theta power as well as more pronounced alpha oscillation with increased peak-frequency. The dynamical analysis of the model [34] reveals three resonances in the baseline condition, including an oscillatory resonance corresponding to the peak in the alpha-band and a pair of zero-frequency resonances. Increasing the anaesthetic concentration diminishes the damping rate of alpha resonances (and hence increases its magnitude) while its frequency increases. Hence increases in alpha power and its peak-frequency results from the approach of the system of an oscillatory instability [38]. Moreover, the two zero-frequency

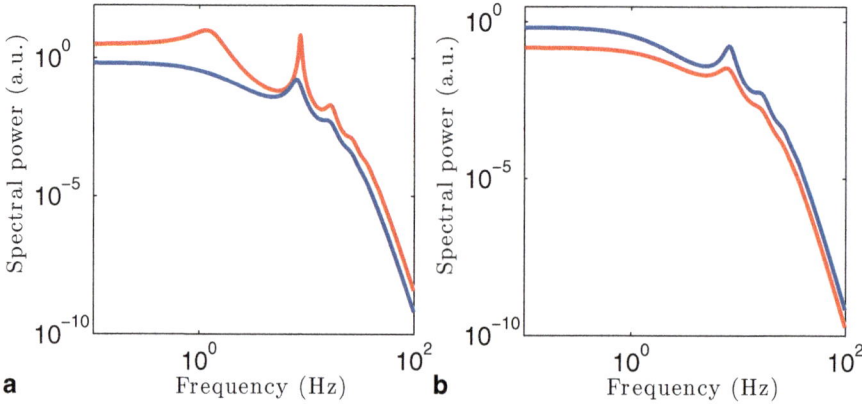

Fig. 6.7 Theoretical power spectrum in the baseline condition ($p_1 = p_2 = p_3 = 1.0$ encoded in *blue*) and in the anaesthesia condition (*red line*). (**a**) $p_1 = p_3 = 1 + 0.3(p_2 - 1)$ and $p_2 = 1.15$ (*red line*) and (**b**) $p_1 = p_2 = p_3 = 1.15$ (*red line*)

resonances collide and gradually increases in frequency leading to a magnitude increases in delta and theta power (solid red line in Fig. 6.7.)

Summarizing, the shown results of previous studies [32, 34] reveal the differential role of synaptic inhibition at GABAergic receptors located on the dendrites of different neural types. We conclude that synaptic potentiation within local cortical inhibitory neurons suffices to reproduce experimental observation in EEG during anaesthesia [34, 38]. The model approach presented permits to describe further EEG signal features what may be achieved by adding further sub-cortical areas, such as the brain stem or other thalamic areas.

Heterogeneous Neural Fields

Previous experimental studies on the structure of biological neural networks and their connections, e.g., as the exciting work of Hellwig [33], reveals a large rather simple mesoscopic structure at the spatial scale of several hundreds of micrometers with an underlying more complex structure at smaller scales. For instance, [33] reveals a rather homogeneuous Gaussian or exponential connectivity distribution of single neurons in layers II and III in rat visual cortex on a scale of: 500 μm with an overlying complex connectivity structure at much shorter scales reflecting heterogeneous connections. Hence, a homogeneous model serves as a first approximation at larger spatial scales, whereas heterogeneous connections need to be considered in more realistic models.

Several previous theoretical studies have investigated the effect of heterogeneous connections in neural fields, primarily in the context of pattern formation [16, 25, 48, 49, 63]. Most of the previous studies consider the heterogeneous fields as small heterogeneous perturbations of homogeneous states. The subsequent paragraphs show a new approach taking into account the heterogeneous structure not as a limit of the homogeneous case.

Starting from the Amari Eq. (6.1), we expand the integral into a power series [10]

$$\frac{\partial V(x,t)}{\partial t} + V(x,t) \\ = \int_\Omega K_1(x,y) V(y,t) \, \mathrm{d}y - \int_\Omega \int_\Omega K_2(x,y,z) V(y,t) V(z,t) \, \mathrm{d}y \, \mathrm{d}z, \tag{6.12}$$

where the kernels K_1 and K_2 have to be reconstructed from experimental data, e.g. by fitting the dynamics to event-related potentials.

Event-Related Potentials

In order to illustrate this, we combine a recently developed technique for nonlinear data analysis [9] with our new method for training heterogeneous neural fields [10]. Figure 6.8 shows the grand average event-related potentials (ERP) $U_n(t), n = 1\ldots,N$ of $N = 25$ recording electrodes from a language processing experiment of subject-object ambiguities in German [8]. The first column displays the control condition with a canonical subject-verb-object order: "*Die Rednerin hat **den** Berater beim Kongress gesucht*" ("The speaker has sought the advisor at the congress"), while the second column (Fig. 6.8e, Fig. 6.8f, Fig. 6.8g, Fig. 6.8h)) shows ERPs for a non-canonical, yet in German grammatical, object-verb-subject order: "*Die Rednerin hat **der** Berater beim Kongress gesucht*" ("The speaker has been sought by the advisor at the congress"). The panels (a) and (b) in the first row of Fig. 6.8 present the ERPs averaged over 14 subjects elicited by the disambiguating article "***den***" vs. "***der***" at time zero. Each trace shows the voltage of one of $N = 25$ EEG electrodes. Comparing panels (a) and (b) in Fig. 6.8, one recognises a difference between conditions at about 600 ms after onset of the critical stimulus, known as the P600 ERP component.

Heteroclinic Orbits

To analyze these spatiotemporal ERP patterns we apply our recently developed recurrence domain segmentation technique [9] whose results are shown in Fig. 6.8c and 6.8d. This method computes the recurrence plots from the N-dimensional trajectories of each condition using a ball-size $\varepsilon > 0$ as parameter. Here, we optimized ε according to a reasonable heuristics for obtaining a relatively low number of recurrence domains with relatively long dwell times and good (visual) discrimination of conditions which worked quite well for $\varepsilon = 1.9 \mu V$. Following [9], the recurrence plots are interpreted as rewriting grammars to replace large time indices by lower,

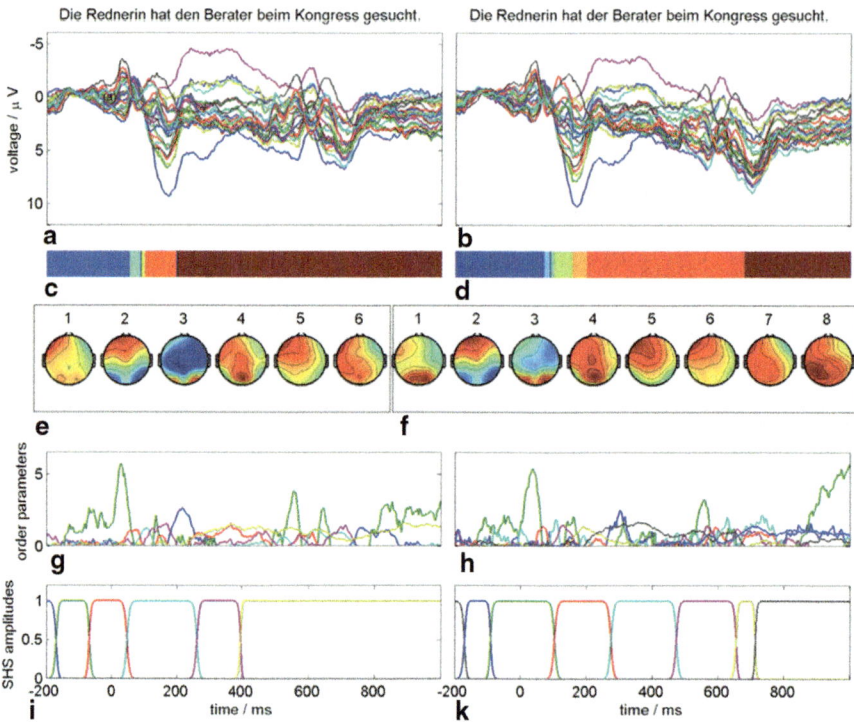

Fig. 6.8 Experimental event-related potentials (*ERP*) [8] in two experimental conditions (*left and right column*) and their interpretation as heteroclinic orbits. (**a, b**) ERPs of $N = 25$ single EEG time series, (**c, d**) temporal sequence of extracted recurrence domains, (**e, f**) the spatial EEG activity distributions on the scalp (ERP components, seen from above, nose on top, back of head on bottom of the maps) averaged over the time windows of recurrence domains shown in panels (**c, d**), projections of the multivariate EEG signal (shown in panels (**a, b**)) on the corresponding ERP component maps (shown in panels (**e, f**), (**g, h**) sequence of heteroclinic saddles (*SHS*) modeled by the Lotka-Volterra model

recurrent, ones. These transformed time indices are plotted in Fig. 6.8c and 6.8d as different colors. The processing differences between conditions are clearly seen as different segments, indicating that ERP components, such as the P600, can be regarded as recurrence domains, or, in first approximation, as saddle nodes [40, 35].

These saddles are estimated in Fig. 6.8c, 6.8d, 6.8e, 6.8f, 6.8g as follows: For each segment k of Fig. 6.8c and 6.8d we compute its center of gravity, i.e. its temporal average, to obtain a spatial EEG activity distribution on the scalp, namely a spatial voltage distribution $U_k(z), k = 1 \ldots, P$ with number of segments P and where z is the spatial location on the scalp. The patterns $U_k(z)$ are activity patterns extrapolated in space on the basis of the average ERP components which are discrete in space. They are shown in Fig. 6.8e and 6.8f.

These distributions are assumed to originate from underlying neural patterns embedded in neural populations, e.g. sub-networks in neural populations. These underlying corresponding neural patterns may be called $V_k(x)$. Let us call $V(x,t)$

the neural population activity generating the EEG signal. Then the Moore-Penrose pseudoinverses $V_k(x)^+$ of $V_k(x)$ form a bi-orthogonal basis with the patterns $V_k(x)$, such that the projections

$$\xi_k(t) = \int_\Omega V_k(x)^+ V(x,t) x$$

serve as order parameters in a decomposition

$$V(x,t) = \sum_k \xi_k(t) V_k(x).$$

The $\xi_k(t)$ are plotted in Figs. 6.8g and 6.8h where each trace depicts the projection onto one segment from the recurrence domain analysis in Figs. 6.8c and 6.8d.

The dynamics of the order parameters $\{\xi_k\}$ is of high temporal complexity, indicating that a large number of time scales is involved in the ERP language processing dynamics. However, in a first approximation we assume that the recurrence domains are in fact isolated saddles that are connected along stable heteroclinic sequences (SHS). This can be modeled through winnerless competition in a network of Lotka-Volterra populations [1, 58]. This assumption leads to the order parameter dynamics shown in Fig. 6.8g and 6.8h. From the growth rates $\sigma_k > 0$, and interaction weights $\rho_{kj} > 0$, $\rho_{kk} = 1$, of the Lotka-Volterra dynamics and from the bi-orthogonal pattern systems $V_j(x), V_k(x)^+$, one can construct the kernels of the neural field Eq. (6.12) as

$$K_1(x,y) = \sum_k (\sigma_k + 1) V_k^+(y) V_k(x)$$

$$K_2(x,y,z) = \sum_{kj} \rho_{kj} \sigma_j V_k^+(y) V_j^+(z) V_k(x),$$

We observe that the kernel $K_1(x,y)$ describes Hebbian synapses between sites y and x trained with pattern sequence V_k [10]. Interestingly, this memory storage mechanism resembles well the storage of patterns in a Bidirectional Associative Memory (BAM) [51].

Moreover, to our best knowledge the three-point kernel $K_2(x,y,z)$ has not been studied yet in the context of neural fields. It further generalizes Hebbian learning to interactions between three sites $x, y, z \in \Omega$. Interestingly, one can write

$$K_1(x,y) = \sum_k (\sigma_k + 1) K_{1,k}(x,y)$$

$$K_2(x,y,z) = \sum_j \left[\sum_k \rho_{kj} K_{1,k}(x,y) \right] \sigma_j V_j^+(z). \qquad (6.13)$$

with $K_{1,k}(x,y) = V_k^+(y) V_k(x)$. According to Eq. (6.13) the three-point kernel can be see as a linear superposition of two-point kernels weighted by the interaction weights ρ_{kj} and the underlying stored patterns $V_j^+(z)$ and its corresponding growth rates σ_k. The additional dependence of the spatial location z is reasonable in heterogeneous fields: the action at spatial location x does not only depend on the activity in spatial location y (as assumed in homogeneous systems), but may also depend on the path between x and y which in turn depends on the underlying memory structure. This argument motivates the presence of an additional spatial variable z, but does neither explain the terms in $K_2(x, y, z)$ nor its dependence on the specific choice of the heteroclinic dynamics (here a Lotka-Volterra system). Future work will attempt to elaborate more details on the presence of three-point kernels by fitting a Lotka-Volterra dynamics to the sequence of ERP-data and hence validate the two-point and three-point kernels K_2 and K_3, respectively.

Conclusion

This chapter has presented some theoretical neural field studies describing the power spectra in homogeneous neural fields and the nonlinear dynamics in heterogeneous neural fields. The models have been validated quantitatively and qualitatively by experimental data. We conclude that effective, i.e. low-dimensional, neural field models are able to capture the major neural mechanisms observed in experimental data. These effective models indicate basic neural mechanisms. The work elucidates the importance to validate the neural field models by experimental data to reveal an effective model.

Acknowledgements The authors thank Stefan Frisch and Heiner Drenhaus for conducting the ERP experiment. AH and MH acknowledge funding from the European Research Council for support under the European Union's Seventh Framework Programme (FP7/2007-2013) ERC grant agreement No.257253. PbG acknowledges financial support through a Heisenberg Fellowship of the German Research Foundation DFG (GR 3711/1-2).

References

1. Afraimovich VS, Zhigulin VP, Rabinovich MI. On the origin of reproducible sequential activity in neural circuits. Chaos. 2004;14(4):1123–29.
2. Ahissar E, Kleinfeld D. Closed-loop neuronal computations: focus on vibrissa somatosensation in rat. Cereb Cortex. 2003;13:53–62.
3. Amari SI. Dynamics of pattern formation in lateral-inhibition type neural fields. Biol Cybern. 1977;27:77–87.
4. Antkowiak B. Different actions of general anesthetics on the firing patterns of neocortical neurons mediated by the GABAA-receptor. Anesthesiology. 1999;91:500–11.
5. Antkowiak B. In vitro networks: cortical mechanisms of anaesthetic action. Brit J Anaesth. 2002;89(1):102–11.
6. Atay FM, Hutt A. Neural fields with distributed transmission speeds and constant feedback delays. SIAM J Appl Dyn Syst. 2006;5(4):670–98.

7. Barraclough NE, Perrett DI. From single cells to social perception. Phil Trans R Soc B. 2011;366(1571):1739–52.
8. Beim Graben P, Gerth S, Vasishth S. Towards dynamical system models of language-related brain potentials. Cogn Neurodyn. 2008;2(3):229–55.
9. Beim Graben P, Hutt A. Detecting recurrence domains of dynamical systems by symbolic dynamics. Phys Rev Lett. 2013;110(15):154101.
10. Beim Graben P, Hutt A. Attractor and saddle node dynamics in heterogeneous neural fields. EPJ Nonlin Biomed Phys. 2014;2:4.
11. Beim Graben P, Rodrigues S. A biophysical observation model for field potentials of networks of leaky integrate-and-fire neurons. Front Comput Neurosci. 2013;6(100). doi: 10.3389/fncom.2012.00100.
12. Beim Graben P, Rodrigues S. On the electrodynamics of neural networks. In: Coombes S, Beim Graben P, Potthast R, Wright JJ, editors. Neural fields: theory and applications. Berlin: Springer; 2014.
13. Belelli D, Harrison NL, Maguire J, Macdonald RL, Walker MC, Cope DW. Extra-synaptic gabaa receptors: form, pharmacology, and function. J Neurosc;2009;29(41):12757–63.
14. Berman NJ, Maler L. Neural architecture of the electrosensory lateral line lobe: adaption for coincidence detection, a sensory searchlight and frequency-dependent adaptive filtering. J Exp Biol. 1999;202:1243–53.
15. Bojak I, Liley DTJ. Modeling the effects of anesthesia on the electroencephalogram. Phys Rev E. 2005;71:041902.
16. Bressloff PC. Traveling fronts and wave propagation failure in an inhomogeneous neural network. Physica D. 2001;155:83–100.
17. Bressloff PC. Spatiotemporal dynamics of continuum neural fields. J Phys A. 2012;45(3):033001.
18. Bressloff PC, Cowan JD, Golubitsky M, Thomas PJ, Wiener MC. What geometric visual hallucinations tell us about the visual cortex. Neural Comput. 2002;14:473–91.
19. Buonviso N, Amat C, Litaudon P, Roux S, Royet JP, Farget V, Sicard G. Rhythm sequence through the olfactory bulb layers during the time window of a respiratory cycle. Eur J Neurosci. 2003;17:1811–19.
20. Ching S, Cimenser A, Purdon PL, Brown EN, Kopell NJ. Thalamocortical model for a propofol-induced-rhythm associated with loss of consciousness. Proc Natl Acad Sci U S A. 2010;107(52):22665–70.
21. Cimenser A, Purdon PL, Pierce ET, Walsh JL, Salazar-Gomez AF, Harrell PG, Tavares-Stoeckel C, Habeeb K, Brown EN. Tracking brain states under general anesthesia by using global coherence analysis. Proc Natl Acad Sci U S A. 2011;108(21):8832–7.
22. Coombes S, Owen MR. Bumps, breathers, and waves in a neural network with spike frequency adaptation. Phys Rev Lett. 2005;94:148102.
23. Coombes S, Lord GJ, Owen MR. Waves and bumps in neuronal networks with axo-dendritic synaptic interactions. Physica D. 2003;178:219–41.
24. Coombes S, Venkov NA, Shiau L, Bojak I, Liley DTJ, Laing CR. Modeling electrocortical activity through improved local approximations of integral neural field equations. Phys Rev E. 2007;76(5):051901.
25. Coombes S, Laing CR, Schmidt H, Svanstedt N, Wyller JA. Waves in random neural media. Discrete Contin Dyn Syst A. 2012;32:2951–70.
26. Doiron B, Chacron MJ, Maler L, Longtin A, Bastian J. Inhibitory feedback required for network burst responses to communication but not to prey stimuli. Nature. 2003;421:539–43.
27. Ermentrout GB, Cowan JD. A mathematical theory of visual hallucination patterns. Biol Cybern. 1979;34:137–50.
28. Faugeras OD, Touboul JD, Cessac B. A constructive mean-field analysis of multi population neural networks with random synaptic weights and stochastic inputs. Front Comput Neurosci. 2008;3:1.
29. Folias SE, Bressloff PC. Stimulus-locked waves and breathers in an excitatory neural network. SIAM J Appl Math. 2005. 65:2067–92.

30. Grassberger P. Toward a quantitative theory of self-generated complexity. Int J Theor Phys. 1986;25(9):907–38.
31. Gross CG. Genealogy of the grandmother cell. Neuroscientist. 2002;8(5):512–8.
32. Hashemi M, Hutt A. A thalamocortical model to explain EEG during anaesthesia. submitted, 2013.
33. Hellwig B. A quantitative analysis of the local connectivity between pyramidal neurons in layers 2/3 of the rat visual cortex. Biol Cybern. 2000;82:11–121.
34. Hindriks R, van Putten MJAM. Meanfield modeling of propofol-induced changes in spontaneous EEG rhythms. Neuroimage. 2012;60:2323–44.
35. Hutt A. An analytical framework for modeling evoked and event-related potentials. Int J Bifurcat Chaos. 2004;14(2):653–66.
36. Hutt A. Effects of nonlocal feedback on traveling fronts in neural fields subject to transmission delay. Phys Rev E. 2004;70:052902.
37. Hutt A. Generalization of the reaction-diffusion, Swift-Hohenberg, and Kuramoto-Sivashinsky equations and effects of finite propagation speeds. Phys Rev E. 2007;75:026214.
38. Hutt A. The anaesthetic propofol shifts the frequency of maximum spectral power in EEG during general anaesthesia: analytical insights from a linear model. Front Comp Neurosci. 2013;7:2.
39. Hutt A, Buhry L. Study of gabaergic extra-synaptic tonic inhibition in single neurons and neural populations by traversing neural scales: application to propofol-induced anaesthesia. J Comput Neurosci, in press, 2014.
40. Hutt A, Riedel H. Analysis and modeling of quasi-stationary multivariate time series and their application to middle latency auditory evoked potentials. Physica D. 2003;177(1–4):203–32.
41. Hutt A, Rougier N. Activity spread and breathers induced by finite transmission speeds in two-dimensional neural fields. Phys Rev E. 2010;82:R055701.
42. Hutt A, Zhang L. Distributed nonlocal feedback delays may destabilize fronts in neural fields, distributed transmission delays do not. J Math Neurosci. 2013;3:9.
43. Hutt A, Bestehorn M, Wennekers T. Pattern formation in intracortical neuronal fields. Netw Comput Neural Syst. 2003;14:351–68.
44. Hutt A, Sutherland C, Longtin A. Driving neural oscillations with correlated spatial input and topographic feedback. Phys Rev E. 2008;78:021911.
45. Hutt A, Sleigh J, Steyn-Ross A, Steyn-Ross ML. General anaesthesia. Scholarpedia. 2013;8(8):30485.
46. Jancke D, Erlhagen W, Dinse HR, Akhavan AC, Giese M, Steinhage A, Schöner G. Parametric population representation of retinal location: neuronal interaction dynamics in cat primary visual cortex. J Neurosci. 1999;19(20):9016–28.
47. Jirsa VK, Haken H. Field theory of electromagnetic brain activity. Phys Rev Lett. 1996;77(5):960–3.
48. Jirsa VK, Kelso JAS. Spatiotemporal pattern formation in neural systems with heterogeneous connection topologies. Phys Rev E. 2000;62(6):8462–5.
49. Kilpatrick ZP, Folias SE, Bressloff PC. Traveling pulses and wave propagation failure in inhomogeneous neural media. SIAM J Appl Dyn Syst. 2008;7(1):161–85.
50. Kitamura A, Marszalec W, Yeh JZ, Narahashi T. Effects of halothane and propofol on excitatory and inhibitory synaptic transmission in rat cortical neurons. J Pharmacol. 2002;304(1):162–71.
51. Kosko B. Bidirectional associated memories. IEEE Trans Syst Man Cybern. 1988;18(1):49–60.
52. Landau LD, Lifshitz EM. Fluid mechanics. Boston: Butterworth-Heinemann; 1987.
53. McCarthy MM, Brown EN, Kopell N. Potential network mechanisms mediating electroencephalographic beta rhythm changes during propofol-induced paradoxical excitation. J Neurosci. 2008;28(50):13488–504.
54. Murphy M, Bruno M-A, Riedner BA, Boveroux P, Noirhomme Q, Landsness EC, Brichant J-F, Phillips C, Massimini M, Laureys S, Tononi G, Boly M. Propofol anesthesia and sleep: a high-density EEG study. Sleep. 2011;34(3):283–91.

55. Nunez PL. Global contributions to EEG dynamics. In: Nunez PL, editor. Neocortical dynamics and human EEG rhythms. New York: Oxford University Press; 1995. p. 475–533.
56. Purdon PL, Pierce ET, Mukamel EA, Prerau MJ, Walsh JL, Wong KF, Salazar-Gomez AF, Harrell PG, Sampson AL, Cimenser A, Ching S, Kopell NJ, Tavares-Stoeckel C, Habeeb K, Merhar R, Brown EN. Electroencephalogram signatures of loss and recovery of consciousness from propofol. Proc Natl Acad Sci U S A. 2012;110:E1142–50.
57. Quiroga RQ, Reddy L, Kreiman G, Koch C, Fried I. Invariant visual representation by single neurons in the human brain. Nature. 2005;435(7045):1102–7.
58. Rabinovich MI, Huerta R, Varona P, Afraimovichs VS. Transient cognitive dynamics, metastability, and decision making. PLoS Comput Biol. 2008;4(5):e1000072.
59. Robinson PA, Rennie CJ, Rowe DL, O'Connor SC. Estimation of multiscale neurophysiologic parameters by electroencephalographic means. Hum Brain Mapp. 2004;23:53–72.
60. Rougier N, Vitay J. Emergence of attention within neural population. Neural Netw. 2006;19(5):573–81.
61. Roxin A, Brunel N, Hansel D. The role of delays in shaping the spatio-temporal dynamics of neuronal activity in large networks. Phys Rev Lett. 2005;94:238103.
62. Roxin A, Brunel N, Hansel D. Rate models with delays and the dynamics of large networks of spiking models. Prog Theor Phys. 2006;161:68–85.
63. Schmidt H, Hutt A, Schimansky-Geier L. Wave fronts in inhomogeneous neural field models. Physica D. 2009;238(14):1101–12.
64. Sellers KK, Bennett DV, Hutt A, Frohlich F. Anesthesia differentially modulates spontaneous network dynamics by cortical area and layer. J Neurophysiol. 2013;110(12):2739–51.
65. Steyn-Ross ML, Steyn-Ross DA, Sleigh JW, Liley DTJ. Theoretical electroencephalogram stationary spectrum for a white-noise-driven cortex: evidence for a general anesthetic-induced phase transition. Phys Rev E. 1999;60(6):7299–311.
66. Victor JD, Drover JD, Conte MM, Schiff ND. Mean-field modeling of thalamocortical dynamics and a model-driven approach to EEG analysis. Proc Natl Acad Sci U S A. 2011;118:15631–8.
67. Vyazovskiy VV, Harris KD. Sleep and the single neuron: the role of global slow oscillations in individual cell rest. Na Rev Neurosci. 2013;14:443–51.
68. Wilson HR, Cowan JD. A mathematical theory of the functional dynamics of cortical and thalamic nervous tissue. Kybernetik. 1973;13:55–80.
69. Wolf F. Symmetry, multistability, and long-range interactions in brain development. Phys Rev Lett. 2005;95:208701.
70. Wright JJ, Liley DTJ. Dynamics of the brain at global and microscopic scales: Neural networks and the EEG. Behav Brain Sci. 1996;19:285–320.
71. Yildiz IB, Kiebel SJ. A hierarchical neuronal model for generation and online recognition of birdsongs. PLoS Comput Biol. 2011;7(12):e1002303.

Chapter 7
Multilevel Computational Modelling in Epilepsy: Classical Studies and Recent Advances

Wessel Woldman and John R. Terry

Abstract In this chapter we present a review of computational models for studying the dynamic mechanisms that describe the function of the human brain, with a specific focus on epilepsy. Epilepsy is a neurological disorder characterised by an increased likelihood of recurrent seizures, which in turn are characterised by transient, pathological episodes of hypersynchronised neural activity resulting in a variety of behavioural symptoms. Our chapter introduces some of the key concepts of epilepsy from a clinical perspective, before describing some of the classical approaches to modelling brain activity across multiple levels of description. We then focus on how these models have been used to explain and predict experimental and clinical phenomena within the field of epilepsy research. Here we focus on techniques that seek to integrate computational modelling with experimental and clinical measures, as we believe this "systems approach" to epilepsy research is from where the most significant new advances, particularly with regards model validation, will occur. We highlight some of the key studies, as well as emphasising more recent breakthroughs to provide a useful entry point into this rapidly expanding field of research.

Keywords Epilepsy · Networks · Seizures · Microscale · Neurons

Introduction to Epilepsy

Epilepsy is a serious neurological condition characterised by an increased predisposition to seizures. In this regard, epilepsy should not be understood as a single disorder but rather as a collection of conditions manifesting from underlying brain abnormalities, as seizures can affect sensory (i.e., somatosensory, auditory, visual,

---

W. Woldman (✉) · J. R. Terry
College of Engineering, Mathematics and Physical Sciences, University of Exeter,
Exeter EX4 4QF, UK
e-mail: www201@exeter.ac.uk

J. R. Terry
e-mail: J.Terry@exeter.ac.uk

olfactory), motor, and autonomic function; consciousness; emotional state; memory; cognition (i.e., problems with perception, attention, emotion, memory, speech) or behaviour [10]. This wide variety of behavioural manifestations is presumed to result from the involvement of a number of large-scale brain networks in the generation of seizures. As a result of the myriad of combinations that lead to seizures and ultimately result in a diagnosis of epilepsy, precise definitions for both "epilepsy" and "seizures" have proved challenging. In an attempt to provide consensus on the usage of the terms *epileptic seizure* and *epilepsy* amongst physicians, people with epilepsy, and researchers, the International League Against Epilepsy (ILAE) and the International Bureau for Epilepsy (IBE) defined an epileptic seizure as:

> a transient occurrence of signs and/or symptoms due to abnormal excessive or synchronous neuronal activity in the brain

and epilepsy as:

> a disorder of the brain characterised by an enduring predisposition to generate seizures and by the neurobiological, cognitive, psychological, and social consequences of this condition [42].

However, very recently a new definition of epilepsy was proposed by the same organization, in an attempt to clarify some anomalies created by their earlier definition [43]. In this new framework, a person is considered to have epilepsy if they meet one of the following conditions:

1. they have experienced at least two unprovoked (or reflex) seizures separated by a time-span of more than 24 h,
2. they have experienced at least one unprovoked (or reflex) seizure and have a probability of further seizures similar to the general recurrence risk for people with epilepsy (at least 60%),
3. they have been diagnosed with an epilepsy syndrome.

This revised definition was developed mainly for clinical use, and the authors recognise that other groups such as researchers might prefer to use the older definition, or devise a new definition individually.

Epilepsy has a global prevalence of around 1% (roughly 50 million people worldwide) and, despite a common misconception as a condition that is relatively easily treated with medication, has a high mortality rate [40]. Indeed, people with epilepsy fall into three groups: those who enter remission spontaneously, those whose seizures are effectively controlled through the use of anti-epileptic drugs (AEDs), and those whose seizures cannot be controlled using standard therapeutic interventions. About 5–10% of all people will suffer a single seizure before the age of 80, with a 40–50% probability of experiencing a second seizure if the first encounter remains untreated [9]. Much current research aims to find ways to predict and control seizures, as there currently is no structural way of predicting whether a particular AED will be effective [59, 112]. For those people with epilepsy who are unresponsive to AEDs, surgical treatment might provide an optional solution to affect the frequency and severity of the seizures [146].

A critical advancement in our understanding of brain structure and function has resulted from the development of the electroencephalogram (EEG) by Berger in the 1920s [12]. The EEG is a measure of the aggregated electrical activity, predominately from cortical pyramidal neurons, although these may be modulated by inhibitory interneurons, as well as neurons from deeper brain areas. This results in characteristic patterns of activity, that correlate strongly with the state of arousal of the subject [96]. EEG is used clinically, alongside a study of psychological and behavioural symptoms, to determine the initiation and termination of seizures in people with epilepsy.

Seizures fall into two main categories: generalised and focal. The Commission on Classification and Terminology of the ILAE recently proposed a framework for classifying seizures [11]. It is assumed that focal seizures consistently originate in networks located in one particular hemisphere, whereas generalised seizures engage bilateral networks. Generalised seizures are then further distinguished into tonic-clonic, absence, myoclonic, tonic, and atonic seizures. In turn, these may be further extended on the basis of the underlying cause, which can be either genetic, structural/metabolic, or unknown. This has resulted due to older categorisation (e.g. idiopathic, symptomatic, and cryptogenic) being considered insufficient and problematic from a clinical perspective. It should be noted however, that the current framework of existing classifications is prone to future reinterpretation due to advances in clinical and computational neuroscience, using imaging techniques such as EEG, electrocorticograms (ECoG), magnetoencephalographic (MEG), functional magnetic resonance imaging (fMRI), and diffusion tensor imaging (DTI). An extensive overview on the many applications of EEG and its importance on studying seizures in particular can be found in the book of Niedermeyer and Lopes da Silva [96]. Whilst the majority of cases of epilepsy have no known cause, a set of features including the age of onset, typical seizure types, seizure frequency, and patterns observed in the EEG, generally allow clinicians to describe specific epilepsy syndromes.

Much current research conceptualises epilepsy as a complex and dynamical disease which naturally promotes the use of mathematical and computational models to study the condition. For example, the dynamics of EEG during seizures is presumed to reflect patterns of hypersynchronous electrical activity of neuronal networks, which are thought to be mediated by a disruption to the dynamic balance between excitation and inhibition leading to hyperexcitable networks [84]. It is generally thought that developing integrated approaches to bridge the many spatial and temporal scales of brain activity (see Fig. 7.1) could lead to prediction and control of seizures. The neuronal systems and processes involved in epilepsy implicate dynamic mechanisms from the sub-cellular level (ion channels, neurotransmitters), to the cellular level (individual neurons connected through axons and dendritic trees), up to the level of networks of neurons (specific modules, nuclei), and finally cortical areas and hemispheres. The shortest timescale at the microlevel are the spike-trains of individual neurons, comprises brief spikes and bursts in the order of several milliseconds. At the macroscale an EEG recording of a spike-wave discharge (SWD) will typically show a spike of average activity in the 20–30 Hz range and a slower

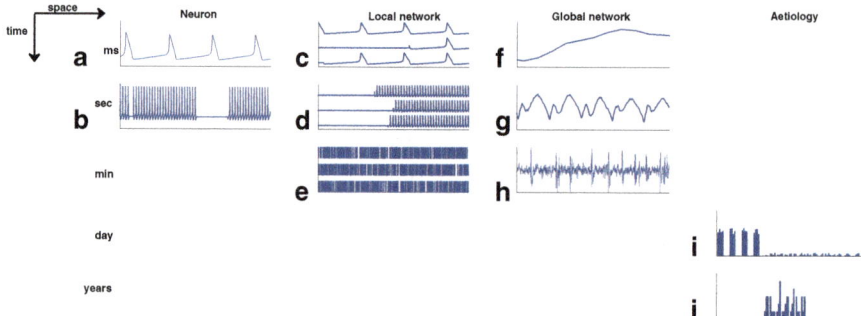

Fig. 7.1 Schematic representation of spatial and temporal scales relevant in epilepsy and seizures. (**a–b**) Consider the activity of a single neuron generating a spike-train. (**c–e**) Consider a network of coupled neurons. The generated spike-trains show corresponding episodes of firing, introducing the notion of synchronicity. (**f–g**) Consider a single channel of EEG from a person with absence epilepsy displaying SWD characteristic of absence seizure EEG. (**h**) The seizure activity arises from healthy background activity, generating seizures lasting for several seconds. (**i**) Schematic histogram counting the number of SWDs in bins over 24 h. (**j**) Schematic representation of absence epilepsy seizures over 20 years where the person grows out of the condition

wave component at 2–4 Hz. Individual seizures will typically initiate and terminate within seconds or minutes. The onset of a seizure is generally believed to be related to specific changes in the macroscopic brain activity prior to the seizure. Circadian factors, such as the level of stress [4, 66], and the lack of sleep can play a crucial role in the frequency and severity of seizures over days. Finally, seizure frequency varies over the time course of months and years, as people with epilepsy might grow out of their condition (in particular in juvenile and childhood types of epilepsy).

Juvenile absence epilepsy (JAE) and childhood absence epilepsy (CAE) are part of a larger group called idiopathic generalised epilepsy (IGE), which is a particularly interesting type of epilepsy as they are generally considered to have an underlying genetic cause. Although absence seizures are very common in IGE, they are not a necessary and sufficient condition. For an overview of the concepts and classifications relevant to IGE, we refer to a review by Mattson [83]. By combining computational, human, and animal experimental studies there is now a clearer understanding of the pathophysiology of absence seizures and the associated SWDs. Absence seizures are usually seen in children and adolescents and are characterised by impairments in consciousness. Particular strains of genetically epileptic rats (WAG-Rij or GAERS) or animals treated with $GABA_A$-antagonists in the feline generalised penicillin model of epilepsy (FGPE) can reproduce patterns of activity similar to SWDs, as well as display behavioural symptoms resembling absence seizures and response to medication [14, 23, 101]. These experimental and animal models are crucially important in identifying the different spatial brain structures involved in absence seizures. Further, they provide a testing environment for investigating the influence of new types of medication and other treatments of epilepsy and seizures.

Although the generation of SWDs and absence seizures are not clearly understood physiologically, the spike component of the SWD is generally associated with firing of cortical neurons, whereas the slower wave component is thought to be related to hyperpolarisation, likely caused by inhibition. A large body of experimental results point to a critical role of the interplay between the thalamus and cortex in the generation of absence seizures and SWDs, and there are many modelling attempts investigating probable causes, such as excessive corticothalamic feedback or inhibitory rebound potentials. For a detailed overview of these type of models and how they relate to activity patterns during sleep, we refer to the work by Destexhe and Sejnowski [34].

The purpose of the present review is to provide an overview of established and influential computational models of generalised and focal epilepsies, alongside a presentation of more recent approaches, that attempt to integrate either computational and experimental, or computational and clinical work respectively. Given seizures arise from the same structures that govern normal brain function, we first provide an introduction to the many scales relevant in modelling neural activity more generally.

Multilevel Computational Models of Neural Activity

When building a mathematical or computational model of neural activity it is important to consider the constituent building blocks required. As described in the introduction, it is presumed (due to the nature of behavioural manifestations associated with seizures) that a malfunction of dynamical interactions between large-scale brain regions plays a critical role. Unpicking this statement suggests that seizures may be thought of as an emergent property, dependent upon the dynamics within a brain region and the connectivity between regions. In this context, the term brain connectivity may refer to many different things, including anatomical links (structural connectivity), statistical dependencies (functional connectivity), or causal interactions (effective connectivity) across different scales (from synapses to whole brain areas). Whilst ultimately large-scale neural activity arises from interactions between neural populations at many levels of description, the challenge of building and analysing fully multiscale computational models means that the focus is instead often constrained to a single scale of description that is typically governed by the availability of data. However, it is worth considering what are the fundamental building blocks of neural activity across multiple spatial scales, and how might function at one level inform and constrain function at another?

At the very highest level we might consider behaviour to be an emergent property of the interactions between macroscopic brain regions, including the cerebral cortex, hippocampus and subcortical nuclei such as the thalamus and basal ganglia. Brain regions are then typically subdivided into smaller functional compartments. For example, the cortex consists of around 10^5–10^6 cortical columns which on their turn consists of around 50–100 smaller minicolumns (see Table 7.1) [93, 118, 130].

Table 7.1 Detailed organisation of the cortex

Structure	Thickness	Neurons/synapses	Scale	Data
Cortex	2–4 mm	1011/1014	Global	EEG/fMRI
Cortical column	200–1000 μm	10^4–10^8/	Meso	LFPs
Minicolumn	20–40 μm	100–200/10^4	Micro	Single-unit recordings

This combinatorial explosion in connectivity is what makes a fully multiscale model of the brain difficult to develop. Further, complexity theory teaches us that even if such a model could be developed it may be of limited value. Across complex networks the emergent dynamics of the network are not typically predictable from a study of the component parts. Further, how far should our reduction process continue? For example, beyond the single neuron level there are several smaller scales of description such as cell biophysics or the molecular biology of gene regulation.

Rather than an exhaustive process of reduction we instead focus on three distinct spatial scales of description for which computational models have been most commonly used. The microscale, where we consider the properties of individual neurons and their dendritic and synaptic connections. The mesoscale where we consider the properties of networks of circuits consisting of larger numbers of neurons, such as cortical columns, and refers to the spatial scale at which we can approximate the statistical properties of the population, without having to study the properties of all the individual neurons. Finally, the macroscale where we consider the brain as described by specific regions (grey matter), connected by fibre pathways (white matter). We first give a brief overview of how different modelling approaches operate across these scales, before describing their specific contributions to the understanding of seizures in terms of neural dysfunction at these different scales of description. For an alternative, more illustrative, summary of these challenges, we refer the reader to the recent review by Tejeda and colleagues [124], which compares the possible advantages and disadvantages of using specific type of models (i.e., deterministic, stochastic, phenomenological, physiological).

Microscale

At the microscale, anatomical and physiological studies have revealed many of the main characteristics and interconnections of cortical microcircuits. Fundamentally, a neuron consists of three main parts: the cell body, the axon, and the dendritic tree. Information comes into the neuron through the dendritic tree, where it is integrated at the cell body and the transformed signal is eventually sent as output through the axon. The dynamical behaviour of neurons is closely related to the evolution of the potential difference between the inside and the outside of the cell. The dynamics of the membrane potential depend on the electrophysiological properties of the neuron, such as the specific ion currents and conductances, and the overall connectivity-structure including the synaptic inputs. More specifically, it is mediated by the inflow and outflow of ionic currents across the membrane caused by the ion-pumps and ion-channels within the membrane. Neurotransmitters are responsible for opening and closing the ion channels by binding to the receptors on the cell membrane and thereby increase (depolarise) or decrease (hyperpolarise) the membrane potential. For example, GABA can both inhibit and excite neurons in the brain, whereas glutamate is the main excitatory neurotransmitter in the brain. When a neuron receives sufficiently strong input through its synapses, the spikingthreshold is exceeded and the membrane potential undergoes a very fast transient change of the membrane voltage called a spike or action potential. Action potentials propagate via the axon to other neurons, which then on their turn can be excited if their synaptic inputs are sufficiently strong. Spikes are the main means of communication between neurons and spike-trains and specific spike-timings are generally considered as the basic building blocks for encoding and transmitting information [30, 70]. It is important to point out that information flow is not constant or instantaneous as conductances cause transmission delays, thereby further complicating the dynamical complexity of the system, even at this small spatial scale. This complexity is confounded by the many different types of neurons, characterised by different ionic properties and neurotransmitters [104].

Within a mathematical framework, we may characterise a neuron as an excitable, non-linear, dynamical system arising as the result of being close to a bifurcation from resting state to spiking activity. One can study neural excitability in a deterministic setting through a separation of time-scales, with typically a slow recovery variable and a fast voltage variable. The resting state of a neuron corresponds to a stable equilibrium, and large enough inputs can push the system into a stable period orbit, corresponding to periodic spiking activity. The intrinsic properties of the neuron, such as the number and type of currents, their conductances, the number of ion-channel and their kinetics, affect the location, the shape, and the period of the stable limit cycle. Transitions between resting and spiking states can occur in different dynamical settings as the equilibrium and limit cycle might coexist (bistable setting) or one of the states might disappear due to external inputs (bifurcation setting).

The Hodgkin-Huxley model of the squid giant axon is one of the most widely utilised and accepted description of neurons with voltage-sensitive currents [58]. Using patch-clamp techniques, three main currents in the squid axon were determined: a voltage-gated K^+-current, a Na^+-current, and an Ohmic leak current. The Hodgkin-Huxley equations model the dynamics of the membrane potential by describing the evolution of the activation variables with Markovian kinetics, assuming that the proportion of open channels is dependent on the number of activation gates and inactivation gates and the probability of an activation gate being open or closed. The evolution of the activation variables of the voltage-gated channels depends in turn on the voltage-sensitive steady-state activation function and a time constant, which are determined using voltage-clamps. Positive and negative feedback loops in the model can lead to the generation of an action potential as the system might be perturbed from its attractor. To illustrate this, assume that an external current causes a slight depolarisation in the cell-membrane. Consequently the sodium-conductance will increase, thereby depolarising the cell even further.

After a short time-period, the slower sodium-channels activate, resulting in an increase in the potassium-conductance, causing the cell to repolarise and eventually hyperpolarise.

Although the structure of the squid giant axon is somewhat simplistic in comparison to cortical neurons, the Hodgkin-Huxley equations form the basis for many deterministic conductance-based models. As experimental techniques have advanced, new details of the mechanisms governing ion channels have emerged, such as spike frequency adaptation or synaptic depression, and these insights can be incorporated into new conductance-based models thereby providing a closing agreement between model output and experimental data (such as single-unit recordings). One famous extension of the Hodgkin-Huxley equations is the Connor-Stevens model which takes into account an additional transient current within the framework (a K^+-conductance) [24]. As an alternative, rather than increasing the complexity of the system, some researchers have sought to reduce the model to its most critical parts. For example, by using a fast-slow decomposition, lower-dimensional models such as the FitzHugh-Nagumo model or the Morris-Lecar equations are derived [44, 92]. These reductions allow mathematical treatment such as phase-plane analysis, bifurcation analysis, and fast slow-analysis while still capturing the essential dynamic properties of the full Hodgkin-Huxley equations. As such, conductancebased neurons provide a successful mechanistic understanding of crucial phenomena such as neuronal excitability and spike-generation [60].

Although morphological studies have revealed much insight in the detailed structure of neurons and inspired detailed compartmental models using cable theory (see [17] for example), most simple models characterise a neuron as a single compartment or point corresponding to the cell-body. In these point-models, current flows strictly inside and outside of the cell, not between different regions within the cell. One of the earliest and simplest model of a neuron is the leaky integrate-and-fire neuron, a simple resistance-capicitator circuit with an Ohmic leakage [1]. By restoring the membrane potential of a neuron to a resetting-value after a specified threshold is reached, the model is able to mimic the generation of action potentials. Although integrate-and-fire models are piece-wise continuous and the spikes are not resultant from any realistic biological dynamics or kinetics, their reduced formulation makes them particularly suitable for mathematical analysis and for proving several theorems. As such, they allow a particular simple setting that includes spikes, excitability, refractory periods, and the difference between excitation and inhibition. This balance between excitation and inhibition has been a crucial dynamical concept in the study of neural activity, especially in relation to synchronisation of networks, and as such often provides a direct relation to the modelling of seizures [128].

An alternative approach for modelling microscale neuronal activity is to focus exclusively on the firing rate properties of neurons [39]. Firing rate models are built on the assumption that, on average, the input from a presynaptic neuron is proportional to its firing rate, and that the total synaptic input is obtained by summing the contributions of all the presynaptic neurons. These approaches specify a relationship between the firing rate and the input a neuron receives from other neurons in the network, or external or sensory inputs (such as an applied current). Note that firing rate models explicitly include the interactions between neurons, whereas the previously discussed conductance-based models were mainly studied in isolation without specified input-relationships. These approaches naturally extend the scale from an individual neuron to the level of networks of interacting neurons [134].

Mesoscale

Computational models of single neurons are often used to describe the underlying dynamics of realistic neuronal networks consisting of interconnected neurons. By coupling neurons together into larger ensembles or (sub)populations, networks of variable size are constructed as sets of coupled differential equations. Simulating these networks then gives the evolution of the state-variables of every individual neuron and reveals the emergent spatiotemporal patterns at the network level. Networks at the microscale are in the order of micrometers, whereas at the mesoscopic scale larger networks of cortical (mini)columns operate at the scale of hundreds of micrometers. The mesoscale is a relevant level of observation in the context of integration of information from the microscale towards whole brain areas, as it extends the level from single neurons to interacting local neural groups. Multi-unit recordings or local field potentials (LFPs), measuring summated dendritic current, can reveal activity of the brain at the mesoscopic level [130]. As shown in the work of Destexhe, networks of single compartment models are used to calculate field potentials, thereby relating both experimental data as well as models from the microscale to the mesoscale [34].

At this level of description, we can consider different levels and types of connectivity or coupling. First there is the level of neuronal connections, where coupling will either have an increasing (excitatory) of decreasing (inhibitory) effect on the membrane potential of the receiving neuron. Further, there might be a hierarchal structure within the overall network, as information might flow from one area to another in a feedback (top-down) or feedforward (bottom-up) structure within the network [130]. Once the type of neuronal coupling (i.e., strength, delay, periodicity) is established and the connectivity-structure described (i.e., all-to-all, sparse connectivity, specific areas), the output of the simulated network can be used to study the collective behaviour of the network as well as the individual neuronal responses.

Current detailed, biophysical modelling attempts typically incorporate more biophysical details into multi-compartmental models using specified software (i.e., NEST, Genesis, Neuron) with the computational power of supercomputers. A particularly striking example of this is the Blue Brain Project [80] and its sequel, the Human Brain Project. It should be highlighted that these large-scale simulations and models limit a global understanding of the dynamical behaviour of the underlying model, in the sense that one typically cannot decide whether the observed simulation displays converged activity or whether there are any other possible attractors within the system. Additionally, given the large numbers of physiological parameters and the fact that it is often notoriously hard to establish reliable estimates for them, a thorough understanding of the dynamical behaviour of these models may require a thorough sensitivity-analysis, as the model might critically depend on the exact settings of the parameters. An interesting future avenue of work here would be to develop uncertainty quantification techniques (see for example [98]), which have proved popular in large-scale climate models, for quantifying uncertainty associated with parameters of these large-scale simulations.

Networks of interacting neurons can exhibit collective behaviour that is not intrinsic to the individual activity patterns, such as synchronisation. Synchronisation is a crucially important concept in the computational modelling of the brain and seizures. For example, synchronisation within large-scale networks of interacting neurons causes the emergence of oscillatory activity in the thalamocortical system, such as the alpha and gamma rhythms [20]. The existence of distinct oscillatory frequencies in LFPs and EEG suggest that the brain, despite its dynamical complexity, produces patterns and trajectories that could be projected onto much lowerdimensional subspaces and studied accordingly. This approach, often termed meanfield approximations, form the basis for lumped models that typically operate at both the meso- and macroscale, by describing the evolution of neural activity with collective variables, such as the proportions of active neurons at a given time in a population, the mean membrane potential, or average firing rate. Simulating networks of thousands of individual conductance-based neurons at such a larger scale is computationally expensive because of the many dimensions, and furthermore, the large number of neurons makes it impracticable to study the influence of each element individually. Lumped approaches offer an alternative for modelling neuronal networks by coupling large collections of individual neurons, by characterising the activity of the neuronal populations in an aggregated manner. An important example of this approach is the Kuramoto-model, which describes the behaviour of large groups of neurons as near identical phase-coupled oscillators [71]. Here we consider the action potential of the neuron to reflect a periodic oscillation and focus on the study of the phase of this oscillation rather that the membrane potential directly. In the case of weak coupling, the amplitudes of the oscillations remain approximately constant, such that interactions can be described by focusing on the phase alone. Synchronisation within this framework typically relates to phase-locking (or coherency), and then often studied in the thermodynamic limit (e.g. the network-size growing to infinity). Mathematical analysis of the order-parameter can then reveal that the dynamics of the network of oscillators can be described by studying its statistical properties [60] and treating the network as a field.

Another motivation for considering mean-field activity of neural populations results from the nature of experimental data recorded at the mesoscale. As LFPs reflect the summated dendritic current at a scale of around 1 mm, they highlight the common action of local networks of neurons rather than the activity of individual neurons [130]. Another popular approach for describing neural population activity is based on a mean-field approximation of the ensemble density function. The ensemble-approach is explained in detail in a review by Deco and others [31], and is based around methods from statistical mechanics to formulate a probability density function that captures the distribution of all neuronal states of a population.

Instead of following the evolution of thousands of individual neurons, the probability density function describes the time-dependent average activity level of the whole ensemble. By using appropriate assumptions, the stationary solutions of the probability density function might be analysed in a generic setting. However, one can simplify the probability density approach by relating the evolution of the density to a single variables, instead of a collection. It is this specific mean-field assumption forms the basis of neural mass models [45].

A neural mass model treats large-scale activity as a point process, and as such can equally reflect that activity of neuronal ensembles or EEG-sources. This comes at the cost of throwing away higher-order moments, such that interacting populations can only influence each other through their expected population rate. Despite this limitation, this simplification of the dynamics of the neuronal populations allows one to study interacting subpopulations with small computation times. A particular popular neural mass formulation is the Jansen-Rit model [61] which forms the basis of many of the epilepsy models we consider later.

Macroscale

We now consider models at the macroscale of millimetres, where very large numbers of neurons and neuronal populations form distinct brain regions which are interconnected by inter-regional pathways. It is the brain activity arising from macroscopic populations that we observe directly from EEG or MEG recordings. The activity patterns recorded by EEG are widely regarded as the summation of interactions of large populations of cortical pyramidal neurons, which due to their dendritic organisation align perpendicularly to the surface of the cortex [97]. Whilst EEG reflects the extracellular output of pyramidal cells, these are generated as a consequence of receiving both excitatory and inhibitory postsynaptic potentials.

One approach to study the emergent rhythms of these large-scale brain regions are neural field models. Neural field models are effectively extensions of neural mass models describing the average or coarse-grained activity of populations of interacting neurons by treating the cortex as a continuous excitable medium. Consequently, the spatially extended sheet is modelled as a function of space and time, and the macroscopic ensemble dynamics are described by a set of partial differential equations or integro-differential equations. Using fundamental methods from statistical mechanics, Wilson and Cowan developed the basis of neural field models by extending the earlier work of Beurle by including both excitation, inhibition, and a refractory period within their approach [143, 144]. A review by Destexhe shows how the Wilson-Cowan equations have been used to include more sophisticated and realistic mechanisms such as bursting, adaptation, and synaptic depression, and how their work predicted neural oscillations and stimulus-evoked responses [35]. Naturally, neural field models have been used to model the generation of EEG rhythms, and a key feature of neural field models in the context of seizures is related to the functional significance of the macroscopic variables in relation to EEG-recordings, making it easier to intuitively relate the computational model to the observed data.

The terminology of mean-field approaches can be somewhat confusing in the overall context of neural fields, as neural field modelling is not necessarily confined to only mean-field activity, but can often result from other assumptions and approximations, such as the probability density approach or from an integro-differential approach [31]. Neural field models have been extensively used to study a wide variety of topics including memory storage, pattern-formation, and travelling waves [2, 18, 25, 28], where they are typically formulated as a PDE, with the assumption of some underlying Green's function describing the connectivity kernel [62]. Alternatively, many mean-field approaches are defined as a set of integro-differential equations describing the coarse-grained activity of a population of neurons, where particular choices of the integral kernel describe the synaptic connectivity structure of the cortex. Common examples of these connectivity structures include short-range (local) excitation and long-range (lateral) inhibition (so-called Mexican hat), long-range excitatory to excitatory connections and short-range inhibitory to inhibitory connectivity (the inverse Mexican hat), and global excitation [45, 62, 97]. The choice of weight function enables specific dynamical patterns to emerge such as oscillatory behaviour, travelling waves, or bumps [18]. For example, Liley et al. [75] and Robinson et al. [107] include all main types of connectivity at the local scale (e-e, e-i, i-e, i-i) but only long-range excitatory connections that interact with both the excitatory and the inhibitory populations. Liley further includes high-order neurotransmitter kinetics and synaptic reversal potential to describe the excitatory and inhibitory cortico-to-cortical interactions more accurately, finding alpha oscillations to be crucially dependent on the local inhibitory-inhibitory interactions [75]. Furthermore, these models show how different biophysical assumptions regarding the dendritic and axonal structure and dynamics affect the nature of the equations, as the voltage-equations might be either first-order (Liley) or second-order (Robinson) depending on whether the population response has (in)finite rise times affecting the propagation of the response through the cortical sheet. In summary, these modelling choices and assumptions on anatomical and physiological properties have a critical influence in determining the overall dynamical repertoire of the models as well as the suitability of analytical tools.

Currently, there is great interest in understanding the relationship between biophysically detailed spiking models at one scale, and neural field models at the other.

In relating different levels of description, the question arises how microscopic properties of individual neurons relate to mesoscopic local networks or the macroscopic behaviour of large-scale networks and brain areas. Given the wide variety of different model approaches (see Table 7.2), how do these models relate to each other, i.e., how they complement or differ, and how we can describe the complex dynamical distributed activity of the brain aided by these modelling approaches? This is a particularly complex task because of the range of spatial scales (from micrometers to centimetres) as well as the variety in temporal scales, with dynamical changes taking place at the range of milliseconds to years within the brain. Under some very specific constraints, the relationship can be inferred [110], but this relationship is neither unique, nor true in generality [31]. Indeed, as highlighted by Bressloff, there is currently no structural multi-scale analysis of conductance based

7 Multilevel Computational Modelling in Epilepsy

Table 7.2 Summary of classical neurocomputational models

Type	Model class	Model	References
Neuronal	Conductance-based	Connor-Stevens	[24]
		FitzHugh-Nagumo	[44]
		Hodgkin-Huxley	[58]
		Morris-Lecar	[92]
	Phase-models	Izhikevich	[45]
		Kuramoto	[71]
	Phenomenological	Firing-rate	[39]
		Integrate-and-fire	[1]
Neural field	Phenomenological	Amari	[2]
		Wilson-Cowan equations	[143, 144]
Neural mass	Physiological	Jansen-Rit	[61]

neural networks that allows a rigorous derivation of neural field equations [18]. An alternative approach is to include higher-order moments beyond the first order mean-field approximation, for example including the variance of activity [41]. Very recently, work by Visser and colleagues has focussed on the addition of delays into the framework of neural fields. These authors argue that conduction delays are likely to critically effect the synchronising effects of brain networks and should therefore be incorporated [127, 133]. For a more detailed overview of the development and application of both neural mass and neural field models, we refer to a review by Coombes [26].

The Application of Multilevel Models in Epilepsy

Given that epilepsy is a pathological condition whose hypothesised causes have been characterised both experimentally and clinically across a wide range of spatial and temporal scales, computational modelling approaches to epilepsy have grown rapidly (see for example [137] for a review of articles predating 2005). Computational models provide an additional tool with which to interrogate experimental and clinical data, and when appropriately utilised, enable a reiterative cycle whereby models can be used to identify underlying candidate mechanisms from data recordings, which may then in turn be tested, and thus validated, through new experiments. It is hoped that this understanding may ultimately lead to new techniques for seizure prediction, treatment and control [91, 114, 115, 138]. Further, the textbook "Computational neuroscience in epilepsy" edited by Soltesz and Staley [117], provides a structural overview of some aligned approaches to computational study seizures and epilepsy and demonstrates the critical advances that the pursuit of a multidisciplinary approach to the problem can enable.

There are further excellent reviews that focus on the overall process of multilevel computer modelling in epilepsy; discussing the basic mathematical concepts (e.g.

attractors, nonlinearity, stability), underlying multiscale models (e.g. deterministic versus stochastic, microscopic versus lumped), as well as their application to various types of epilepsy, including both focal and generalised epilepsies [72, 78, 88, 120]. In a further recent review, Badawy and colleagues [4] present a summary of experimental and modelling evidence for dynamic changes of excitability within epileptic brain networks both during (ictal) as well as away from (interictal) seizures. They discuss how the underlying physiology influences the balance between inhibition and excitation in the interictal state before the onset and evolution of a seizure, highlighting several candidate mechanisms, including changes in blood sugar levels and hormones, that could change the level of cortical excitability of these epileptic brain networks over time (see also [105]).

From these publications, we have highlighted a selection of what we consider to be the key publications (see Table 7.3) which have strongly influenced the current modelling approaches (see Table 7.4) that are the focus of the remainder of this chapter.

One approach to studying epilepsy considers the hypothesis that seizures could be the result of the structural failure of mechanisms within the brain that are reflected by scale-invariant power laws [27]. If the brain follows such power laws, this means that dynamical changes and characteristic behaviour can be found at every spatial and temporal scale within the brain. Furthermore, systems that operate near a critical state often follow a power law distribution, leading to the notion of self-organised criticality as a potential mechanism relevant to epilepsy [7, 99]. Some studies discuss the possibility that the dynamics emerging from an observance to power law distributions might interfere with neural computation, thereby disrupting the healthy network activity and forcing the brain in a state of hyperexcitability [87, 145].

Table 7.3 Classical papers in computational modelling of epilepsy

Seizure type	Model type	Main findings references
General	Field	Enflurane and isoflurane induce epileptiform activity [76]
		Focal/absence seizures caused by physiological [108] parameter-changes
Focal	Neuron	Weak excitatory synapses generates seizure-like activity [37]
		Gap junctions could underlie fast population oscillations [126]
Absence (SWD)	Neuron	Thalamocortical network mechanisms generates SWD [14]
		GABA-receptors in thalamocortical circuits generates [32, 33] SWD
		Cortical LTS cells important in genesis of SWD [36] SWD caused by dynamical bifurcations in bistable [121] framework
	Phen	Noise governs transitions to seizure-state in bistable [122] framework
Tonic-clonic	Field	Bifurcation-analysis reveals difference between absence [16] and tonic-clonic seizures
Astrocytes	Neuron	Intracellular oscillation patterns in epileptic astrocytes [5]

7 Multilevel Computational Modelling in Epilepsy

Table 7.4 Recent papers in computational modelling of epilepsy

Seizure type	Model type	Main findings references
General	Neuron	Inhibitory synapses crucial in generating seizures [56] Failure of adaptive self-organised criticality causes [87] seizures
		Control of seizures through depolarising periodic [100] stimulation
		Computational improvements in multi-scale [102] epilepsy modelling
		Slow depression mechanism enforces seizure ter- [131] mination
		Addition of gap junctions suppresses seizure-like [135] activity
		Changes in balance of conductances explain [148] seizures
	Mass	General framework for generating seizure-like pat- [47, 48, 49] terns
Focal	Neuron	Conditions on prediction of focal seizures [3]
	Mass	Importance of balance between excitatory and in- [89] hibitory feedback
	Phen	Synchronisation of LFP causes afterdischarges [67]
Absence (SWD)	Neuron	Non-ictal phases in bistable model affect likeli- [69] hood of seizures
	Phase	Abnormal white fibre connections facilitate gener- [147] ation of SWD
	Mass	SWD result from intermittency caused by spatial [51, 52, 53, 54] heterogeneities
	Field	SWD caused by bifurcation as well as bistability [81, 82] Genetic algorithm reveals clinical differences be- [95] tween seizures
	Phen	Network structure influences escape times in [8] bistable model
		Bistable model describes seizure-onset [68]
NCSE/SWD	Field	Phase-locking delta-oscillations stochastic precur- [57] sor SWD

Extending these concepts from a dynamical systems perspective, Lopes da Silva and colleagues [77] highlight two fundamental mathematical mechanisms that could lead to the generation of a seizure as reflected in electrophysiological data recordings: the bifurcation and the noise-induced transition in a multistable system. Both scenarios are shown to give rise to the abrupt transition from background activity to seizure dynamics. Perhaps the most well characterised of these is the spike-wave discharge (SWD), where a low-amplitude signal transitions near instantaneously into a high amplitude, approximately 3 Hz signal, each oscillation consisting of a sharp spike and a slow wave. In the bifurcation setting, a smooth change in an underlying system parameter or parameters, leads to a sudden transition in the system dynamics. Alternatively, in a multistable setting there are several stable attractors, and the initial conditions and noisy inputs are determining factors for understanding the dynamical behaviour of the system. For example, one could consider the healthy interictal state as well as a synchronised seizure-state as stable attractors that co-exist. In normal subjects, the basin of attraction of the seizure-state may be considered to be vanishingly small. On the other hand, in people with epilepsy, the basin of attraction of the seizure state has positive measure and thus an appropriate perturbation may lead the dynamics into the seizure regime. These types of ideas have formed the basis of a number of physiological studies of SWDs at both the micro- and macro-scales of spatial description. Many of these studies have explored in detail the role of the neocortex in generating SWDs, especially in conjunction with thalamic networks. Motivated by experimental evidence, a compartmental microscopic model is presented in the work of Destexhe which provides a detailed approach for understanding the mechanisms underlying SWDs [32, 36]. The authors summarised in vivo evidence regarding the behaviour of individual neurons and brain regions obtained from the feline generalised penicillin model. The networks consists of thalamocortical projection cells (TC), local inhibitory cortical interneurons (IN), reticular projection cells (RE), and corticothalamic projection cells (CT). It is shown that $GABA_B$-mediated inhibition plays a crucial role within the model, and SWDs are initiated and terminated by slow time-scale currents in the TC cells while the cells are quiescent between SWDs [33]. During these periods of rest, several TC cells will slowly lose stability thereby causing an initial burst of activity that will then further recruit the rest of the network, eventually causing a SWD at the macroscale. Similar experimental and computational models focussing on the cellular and network structures underlying SWDS are discussed in reviews by Blumenfeld and McCormick [14, 84].

Building on these fundamental ideas, a mean-field model introduced by Robinson and colleagues distinguishes the short timescale dynamics associated with the oscillatory phase of the SWD from the longer timescales implicated in the initiation and termination of SWDs at the macroscale [16, 108]. The model describes the averaged activity of three homogeneously synchronised populations of TC, RE and cortical cells and is thus closely related to the microscale neural work model of Destexhe. Seizure-onset results from a dynamical bifurcation of short timescale dynamics, as the change from the inter-ictal to the ictal state appears due to a change of parameters. Bifurcation analysis of this model demonstrates how an increase in the cortical input into the TC population can lead to spike-and-wave dynamics. This model was extended to explicitly account for the slow $GABA_B$-mediated inhibition of the TC population by RE neurons [81, 109] and this slow inhibitory process was also shown to be capable of leading to a transition to SWDs. These two studies highlight two critical mechanisms necessary for the transition to SWDs. First, a variation in the dynamic balance between excitation and inhibition within the TC nucleus of the thalamus. Second, a separation of time-scales resulting in a slowfast system [111]. In the work of Robinson and colleagues, the time-delay within the cortico-thalamic loop gives rise to a separation of scales, whereas in the work of Marten et al. the separation is caused by the difference mechanisms of action of GABAergic inhibition. A critical difference between the two models is the ability of the Marten et al model to generate poly-spike and wave complexes, which are typically observed in the EEG of people with absence epilepsy. Using this model, the same authors developed a multi-objective evolutionary algorithm to fit parameters of the model from clinical EEG recordings [95], from which they hypothesise that the resulting path through parameter space of the generative model may provide a mechanistic characterisation of different types of epilepsy syndrome.

A possible limitation of these studies is that they only considered space-clamped solutions, justifying this assumption by the observation that the abnormal dynamics associated with SWD occur near simultaneously across the entire cortex. However, there is a growing body of experimental evidence for the involvement of large-scale networks in the generation of SWDs. For example the classical work of Meeren and colleagues demonstrated that SWDs in the WAG/Rij genetic model originate in a specific region of neocortex—the somatosensory cortex—before spreading to other cortical regions [85]. Similarly, Buszaki and colleagues investigated the genesis of SWDs in Fischer-344 rats, and proposed the discharges might be the result of rhythmic cell populations in the thalamus [19]. A historic review of the different paradigms on the genesis of generalised absence seizures can be found in [86].

Motivated by this, Taylor and colleagues consider a spatially extended neural field model, based upon the Wilson-Cowan model with two inhibitory populations and observe that spike and wave oscillations can generate and propagate in the model [123]. Once more, the presence of two inhibitory time-scales being critical to generating the activity. Whilst this study was one of the first to consider spatial-extent in the context of generalised epilepsy, it might be considered somewhat limited, as the spatial-extent is homogenous, whereas the human cortex is fundamentally heterogeneous. To address this, Goodfellow and colleagues extend the Jansen-Rit neural mass model to incorporate spatial heterogeneity [51, 52, 53]. They demonstrate that epileptic activity could arise transiently as a consequence of connectivity structure of the tissue. These intermittent transitions between background and spike-wave like activity provide a third candidate mechanism for generating seizure-like activity in brain networks. By introducing a discrete-time map, the authors further present a mathematical analysis of the role of heterogeneity. They study both spontaneous and stimulation-induced activity patterns, and relate these findings to clinically recorded rhythms. In their most recent work, they demonstrate the fundamental mechanisms by which intermittent transitions can arise through a reduction of their detailed neural mass model to a phenomenological set of equations, that might be considered as a normal form of the more detailed model [54].

The use of these reduced approaches (that we term phenomenological models) provides the opportunity to study (often analytically) the fundamental mechanisms that may contribute to seizures. Typically these studies consider a bistable mechanisms rather than the bifurcation mechanisms considered thus far. For example, recent work by Kalitzin and colleagues, and Benjamin and colleagues [8, 69] describe the bistability using a subcritical Hopf bifurcation, permitting a bistable regime in which noise governs the transitions between the interictal and ictal stages. This phenomenological approach provides an alternative framework for exploring the role of heterogeneity, whereby interactions between internal node dynamics and the macroscale network structure can lead to transitions to seizures, either emerging from a single brain region or as a consequence of the network in general.

A particularly exciting recent study by Jirsa and colleagues [64] presents a structural taxonomy of seizures and derives a 5-dimensional model (the "epileptor") containing the necessary elements for describing invariant seizure activity and temporal features including spike-wave discharges, fast oscillations, and a slower permittivity variable [64]. This model can be considered as an extension of these previous studies, which focus on seizure initiation, to incorporate normal forms for the bifurcation mechanisms postulated to underlie both seizure initiation and seizure termination, through a saddle-node bifurcation and a homoclinic bifuraction respectively. Model based predictions have been effectively validated through several experimental and clinical observations. In particular, a baseline shift at seizure-onset (due to the saddle-node bifuraction) and logarithmic scaling of interspike intervals at offset (due to the homoclinic bifurcation). This baseline shift is particularly significant, as it is typically removed as an artefact by clinicians studying epilepsy, yet may play a critical role in understanding the onset of seizures.

Building on this theme, Richardson [105] presents experimental and clinical evidence suggesting that multi-frequency oscillations may play an important role in the generation of seizures, and that these oscillations may emerge from a particular brain region—the classical concept of a "seizure onset zone"—or from properties of the network structure more generally. This alternative concept for the emergence of rhythms is well described in the work of Buszaki [20], yet its potential application to epilepsy has only very recently been considered. In this setting, Terry and colleagues demonstrated how the interplay between excitability, network structure, and noise can generate emergent dynamics reflecting both generalised and focal seizures [125]. Intriguingly, the study demonstrates settings where the emergent dynamics of both seizure types can result from either a single brain region (the classical concept of "seizure onset") or from alterations to the more general network structure. Further, the influence of removing an edge from the network is studied in detail, revealing an interesting explanation of experimental results (where loss of white matter has been associated with the generation of seizures [22]) in that the system becomes less sTable (and thus more prone to transitions to seizures) as connections are removed. This may at first seem counterintuitive, in that the spread of seizure activity across the whole cortex might be felt to occur most rapidly if all regions of the cortex are strongly connected.

The role of disrupted networks in focal epilepsies has also received much recent attention. For example, a recent study by Wendling and colleagues, describes an approach for combining computational modelling with multiscale experimentally recorded data (specifically in the hippocampus (both in vitro and in vivo)) and clinical MTLE data [142]. The study reveals mechanisms underlying oscillatory activity from inter-ictal spikes to high-frequency oscillations and seizures, including lumped-parameter approach as well as a detailed microscopic model of coupled networks of hippocampal CA1 and CA3 cells. In line with this, the role of the gap junctions, the electrical coupling between neurons, should be considered as a possible key player in the generation of (focal) seizures, as reported by Traub and Volman [126, 135]. Jiruska and colleagues discuss evidence suggesting excessive synchronisation of large neuronal populations could lead to hypersynchronous states during seizures [65]. The authors describe how seizures can result from dynamical changes in the synchronisation patterns during the ictal and interictal stages, both at the microlevel as well as the macrolevel. For example, desynchronisation is often observed before the onset of the seizure whereas high levels of synchronisation are typically observed when the seizure terminates. The onset of seizures is often related to a characteristic area of brain networks, and this concept of seizure onset zones is particularly relevant in relation to focal epilepsy, where the use of electrical stimulation has attracted recent interest [48, 129]. Further, Anderson describes how the prediction of focal seizures might be complicated by the various different mechanisms that could deregulate the healthy background state by using a detailed neocortical model [3]. In a related approach, the work of Blenkinsop is focussed on the time evolution of macroscopic epileptic rhythms in focal-onset epilepsy, by using a neural mass model based on the work by Wendling [13, 139]. These neural mass models are an extension of the standard Jansen-Rit model, incorporating additional inhibitory populations, and describe the relationship between commonly observed wave-forms in clinical intracranial EEG recordings. The bifurcation routes from background to seizure activity of these models have been characterised in the work of Grimbert and Faugeras [55]. More recently, van Drongelen and Visser describe an integrational approach to relate microscopic detailed models to mesoscopic and macroscopic models [37, 132].

To understand the differences between a singular pathological region or a pathological dynamical network disrupting normal neural activity, Richardson argues that combining connectomics with brain dynamics is necessary to provide further insight into the relationship between individual node dynamics and the collective behaviour of the network [106]. The relationship between structure and dynamics and the emergence of function has generated much recent interest across the field of biology more generally, particularly to study the role of complex interactions on the global behaviour of networks [15]. Specific to epilepsy, there have been several studies based on data from people with epilepsy that aim to find differences within these recorded time-series between people with epilepsy and healthy subjects. By using graph theory, signal analysis, and time-series analysis, these approaches aim to derive a graph (a specific set of nodes and edges) which is then typically characterised by a collection of measures such as the small world index (SWI) or the characteristic path length. Determining structural, effective, or functional connectivity between different regions is usually based on regression technique or Bayesian methods. With the development of several useful toolboxes, one can easily use a wide variety of network-derivation techniques such as Granger causality, dynamic causal modelling, transfer entropy, and partial directed coherences to derive an (un)directed, (un)weighted graph, and then analyse the graph properties by using the Brain Connectivity Toolbox [6, 21, 50, 113, 116]. Within this context, one can study the resting state connectivity structure as well as how networks change dynamically over time, analysing how the graph properties evolve during seizures or just before the seizure-onset [73, 74]. One study explores the significant differences in functional connectivity structures in IGE in resting state activity [79], while an additional study reveals how networks become more orderly during absence seizures [103]. These and other studies are reviewed in more detail by Engel and colleagues [38].

Clinical Applications

Much of the focus on the use of computational models of seizures (and epilepsy more generally) is in their ultimate application in the clinic. A method that could reliably identify whether a person with epilepsy was in remission, or had responded to medication would revolutionise the clinical management of the condition. Consequently, much research has focussed on the identification of features in neural activity that can differentiate people with epilepsy and normal controls. For example, Wendling and colleagues provide a spatial analysis of iEEG in case of partial epilepsy by using non-linear regression analysis [141], whereas Freeman conducts a similar spatial analysis of electrocorticograms (ECoG) to find optimal conditions for analysing the frequency content of the recorded signals [46]. Additionally, computational models are developed to describe and reproduce the recorded signals, thereby providing the possibility of discerning the mechanisms underlying the observed dynamical phenomena. Wang and colleagues describe how epileptic EEG recordings can typically be categorised as one of four dominant wave forms (fast oscillations, large slow waves, fast spiking, spike waves) [136]. By focussing on the interplay between excitation and inhibition, they formulate a minimal model that reproduces the essential wave forms, thereby inferring a set of necessary interacting processes underlying epileptic neural activity. Jirsa presents a neural field framework for describing EEG and MEG patterns in the context of electrical stimulation [63], whereas David and Friston use a neural mass formulation to focus on the coupling structure and propagation delays underlying these EEG and MEG signals [29]. Dynamic causal modelling differs from the bivariate or multivariate regression techniques in the sense that DCM is a Bayesian method, constructing a generative model describing the propagation of neural activity through brain networks, estimating the causal interactions using Bayesian inference techniques. As such, DCM may be thought of as an alternative formulation to many of the dynamical systems approaches to studying seizures and epilepsy considered in earlier sections. DCM has been used both in the context of neural mass modelling more generally [90] and has been applied to EEG/fMRI data from people with epilepsy as a result of brain tumour near the hypothalamus [94].

Computational models may also have a critical role to play in informing surgical resection that is used as a treatment strategy for medically intractable epilepsies. Understanding the different candidate mechanisms of seizure-onset may be critically important in informing suitable treatment. A particular treatment option (for example removal of a specific brain region) may not be effective if the underlying mechanism is a disruption within the overall network structure. Furthermore, seizures onset in people with intractable epilepsy may well have different, or even multiple, underlying dynamic mechanisms and hence require different treatment.

Future Directions

Whilst the main focus of our chapter has been to review relevant computational approaches for studying seizures and epilepsy, we conclude with a few remarks on the future direction of the field. As experimental techniques (particularly methods for recording neural activity at very small spatial scales and optogenetic methods for manipulating specific neural populations) evolve, we can expect that physiological parameters and functions within models of microlevel activity will significantly advance. For example, we envisage that computational models will have a critical role to play in advancing our understanding of so-called "microseizures" [119]. Further, by incorporating more details of the actual dynamical processes at the genetic, molecular and cellular level, these models may ultimately enable the genetic contribution to a wide variety of epilepsy syndromes to be identified. Additionally, as we move towards exascale computing, very detailed models (such as promoted within the Human Brain Project) may enable the multiscale interrelationships within the brain to be described, and consequently the identification of mechanisms by which their disruptions lead to neurological disease.

At the macroscale, further work to delineate the contribution of network structure and dynamics in the emergence of seizure activity may result in new methods for distinguishing people with epilepsy from controls, or to classify people with epilepsy according to their likely response to available anti-epilepsy drugs (AEDs). A pragmatic approach to achieving this could be based on a combination of computational models and anatomical or functional networks derived from people with epilepsy and trying to identify how the internal brain dynamics as well as the brain connectivity structure are modulated by AEDs over time. Databases of longitudinal EEG recordings could be utilised to inform the estimation of parameters of computational models before and after treatment, and changes in these parameters may determine whether patients have appropriately responded to treatment. In this regard, we might think of seizures as emerging from either normal dynamics with abnormal connectivity, or abnormal dynamics with normal connectivity, or maybe even a combination of the two.

To summarise, we have presented an overview of the wide variety of computational approaches for understanding the dynamic mechanisms of seizures and epilepsy. We have attempted to balance classical studies, with a focus on more recent advances. In particular, we promote the concept of computational models of epilepsy and in particular their careful integration with experimental or clinical studies across multiple levels of description as critical to advance our understanding of this significant neurological disorder.

Acknowledgments Wessel Woldman was supported by a PhD studentship awarded by the College of Engineering Mathematics and Computer Science. Financial support of the Medical Research Council of the United Kingdom via grant MR/K013996/1 and Epilepsy Research UK via grant P1203 is acknowledged. We thank Simon Todd for useful comments on an earlier version of the manuscript.

References

1. Abbott LF. Lapique's introduction of the integrate-and-fire model neuron (1907). Brain Res Bull. 1999;50:303–4.
2. Amari S. Homogeneous nets of neuron-like elements. Biol Cybern. 1975;17:211–20.
3. Anderson WS, Azhar F, Kudela P, Bergey GK, Franaszczuk PJ. Epileptic seizures from abnormal networks: why some seizures defy predictability. Epilepsy Res. 2012;99:202–13.
4. Badawy RAB, Freestone DR, Lai A, Cook MJ. Epilepsy: ever-changing states of cortical excitability. Neuroscience. 2012;222:89–99.
5. Balazsi G, Cornell-Bell A, Neiman AB, Moss F. Synchronization of hyperexcitable´ systems with phase-repulsive coupling. Phys Rev E. 2001;64:041912.
6. Barnett L, Seth AK. The MVGC multivariate Granger causality toolbox: a new approach to Granger-causal inference. J Neurosci Methods. 2014;223:50–68.
7. Benayoun M, Cowan J.D, van Drongelen W, Wallace E. Avalanches in a stochastic model of spiking neurons. PLoS Comput Biol. 2010;6:e1000846.
8. Benjamin O, Fitzgerald THB, Ashwin P, Tsaneva-Atanasova K, Chowdhury F, Richardson MP, Terry JR. A phenomenological model of seizure initiation suggests network structure may explain seizure frequency in idiopathic generalised epilepsy. J Math Neurosci. 2012;2:1.
9. Berg AT. Risk of recurrence after a first unprovoked seizure. Epilepsia. 2008;49:13–8.
10. Berg AT, Berkovic SF, Brodie MJ, Buchhalter J, Cross JH, van Emde Boas W, Engel J, French J, Glauser TA, Mathern GW, Moshe SL, Nordli D, Plouin P, Scheffer´ IE. Revised terminology and concepts for organization of seizures and epilepsies. Epilepsia. 2010;51:676–85.
11. Berg AT, Scheffer IE. New concepts in classification of the epilepsies: entering the 21st century. Epilepsia. 2011;52:1058–62.
12. Berger H. Uber das Elektrenkephalogramm des Menschen. Archiv f¨ur Psychiatrie und Ner-¨venkrankheiten. 1929;87:527–70.
13. Blenkinsop A, Valentin A, Richardson MP, Terry JR. The dynamic evolution of focalonset epilepsies: combining theoretical and clinical observations. Eur J Neurosci. 2012;36:2188–200.
14. Blumenfeld H. Cellular and network mechanisms of spike-wave seizures. Epilepsia. 2005;46:21–33.
15. Boccaletti S, Latora V, Moreno Y, Chavez M, Hwang D-U. Complex networks: structure and dynamics. Phys Rep. 2006;424:175–308.
16. Breakspear M, Roberts JA, Terry JR, Rodrigues S, Mahant N, Robinson PA. A unifying explanation of primary generalized seizures through nonlinear brain modeling and bifurcation analysis. Cereb Cortex. 2006;16:1296–313.
17. Bressloff PC, Coombes S. Physics of the extended neuron. Int J Modern Phys B. 1997;11:2343–92.
18. Bressloff PC. Spatiotemporal dynamics of continuum neural fields. J Phys A Math Theor. 2012;45:033001.
19. Buszaki G. The thalamic clock: emergent network properties. Neuroscience. 1991;41:351–64.
20. Buszaki G. Rhythms of the brain. Oxford: Oxford University Press; 2006.
21. Chavez M, Martinerie J, Le Van Quyen M. Statistical assessment of nonlinear causality:´ application to epileptic EEG signals. J Neurosci Methods. 2003;124:113–28.
22. Chahboune H, Mishra AM, DeSalvo MN, Staib LH, Pucaro M, Scheinost D, Papademetris X, Fyson SJ, Lorincz ML, Crunelli V, Hyder F, Blumenfeld H. DTI abnormalities in anterior corpus callosum of rats with spike-wave epilepsy. NeuroImage. 2009;47:459–66.
23. Coenen AM, van Luijtelaar EL. Genetic animal models for absence epilepsy: a review of the WAG/Rij strain of rats. Behav Genet. 2003;33:635–55.
24. Connor JA, Stevens CF. Prediction of repetitive firing behaviour from voltage clamp data on an isolated neurone soma. J Physiol. 1971;213:31–53.
25. Coombes S. Waves, bumps, and patterns in neural field theories. Biol Cybernet. 2005;93:91–108.

26. Coombes S. Large-scale neural dynamics: simple and complex. NeuroImage. 2010;52:731–9.
27. Coombes S, Terry JR. The dynamics of neurological disease: integrating computational, experimental and clinical neuroscience. Eur J Neurosci. 2012;36:2118–20.
28. Coombes S, Venkov NA, Shiau L, Bojak I, Liley DTJ, Laing CR. Modeling electrocortical activity through improved local approximations of integral neural field equation. Phys Rev E. 2007;76:051901–8.
29. David O, Friston KJ. A neural mass model for MEG/EEG: coupling and neuronal dynamics. NeuroImage. 2003;20:1743–1755.
30. Dayan P, Abbott LF. Theoretical neuroscience: computational and mathematical modeling of neural systems. Cambridge: MIT Press; 2001.
31. Deco G, Jirsa VK, Robinson PA, Breakspear M, Friston K. The dynamic brain: from spiking neurons to neural masses and cortical fields. PLoS Comput Biol. 2008;4:e1000092.
32. Destexhe A. Spike-and-wave oscillations based on the properties of GABAB receptors. J Neurosci. 1998;18:9099–111.
33. Destexhe A. Can GABAA conductances explain the fast oscillation frequency of absence seizures in rodents? Eur J Neurosci. 1999;11:2175–81.
34. Destexhe A, Sejnowski TJ. Thalamocortical assemblies. Oxford: Oxford University Press; 2001.
35. Destexhe A, Sejnowski TJ. The Wilson-Cowan model, 36 years later. Biol Cybernet. 2009;101:1–2.
36. Destexhe A, Contreras D, Steriade M. LTS cells in cerebral cortex and their role in generating spike-and-wave oscillations. Neurocomputing. 2001;38:555–63.
37. Drongelen W. van, Lee HC, Hereld M, Chen Z, Elsen FP, Stevens RL. Emergent epileptiform activity in neural networks with weak excitatory synapses. IEEE Trans Neural Syst Rehabil. 2005;13:236–24.
38. Engel J, Thompson PM, Stern JM, Staba RJ, Bragin A, Mody I. Connectomics and epilepsy. Curr Opin Neurol. 2013;26:186–94.
39. Ermentrout GB, Terman DH. Mathematical foundations of neuroscience. Berlin: Springer; 2010.
40. Fact Sheet 999 Epilepsy. World Health Organization. 2012. Available via World Health Organization. http://www.who.int/mediacentre/factsheets/fs999/en/ Cited 13 Jan 2014
41. Faugeras O, Touboul J, Cessac B. A constructive mean field analysis of multi population neural networks with random synaptic weights and stochastic inputs. Front Comput Neurosci. 2009;3:0808.1113.
42. Fisher RS, van Emde Boas W, Blume W, Elger C, Genton P, Lee P, Engel J. Jr. Comment on epileptic seizures and epilepsy: definitions proposed by the International League Against Epilepsy (ILAE) and the International Bureau for Epilepsy (IBE). Epilepsia. 2005;46:1698–9.
43. Fisher RS, et al. A practical clinical definition of epilepsy. Epilepsia. 2014;55:475–82.
44. FitzHugh R. Mathematical models of threshold phenomena in the nerve membrane. Bull Math Biophys. 1955;17:257–78.
45. Freeman WJ. Mass action in the nervous system. New York: Academic; 1975.
46. Freeman WJ, Rogers LJ, Holmes MD, Silbergeld DL. Spatial spectral analysis of human electrocorticograms including the alpha and gamma bands. Epilepsia. 2000;95:111–21.
47. Freestone DR, Aram P, Dewar M, Scerri K, Grayden DB, Kadirkamanathan V. A data-driven framework for neural field modeling. NeuroImage. 2011;56:1043–58.
48. Freestone DR, Kuhlmann L, Grayden DB, Burkitt AN, Lai A, Nelson TS, Vogrin S, Murphy M, DSouza W, Badawy R, Nesic D, Cook MJ. Electrical probing of cortical excitability in patients with epilepsy. Epilepsy Behav. 2011;22:110–8.
49. Freestone DR, Kuhlmann L, Chong MS, Grayden DB, Aram P, Postoyan R, Cook MJ. Patient-specific neural mass modeling: stochastic and deterministic methods. In: Tetzlaff R, Elger CE, Lehnertz K, editors. Recent advances in predicting and preventing epileptic seizures. Singapore: World Scientific Publishing Co. http://www.worldscientific.com/worldscibooks/10.1142/8886; 2013.

50. Friston K, Moran R, Seth AK. Analyzing connectivity with Granger causality and dynamic causal modelling. Curr Opin Neurobiol. 2013;23:1–7.
51. Goodfellow M, Schindler K, Baier G. Intermittent spike-wave dynamics in a heterogeneous, spatially extended neural mass model. NeuroImage. 2011;55:920–32.
52. Goodfellow M, Taylor P, Wang Y, Garry D, Baier G. Modelling the role of tissue heterogeneity in epileptic rhythms. Eur J Neurosci. 2012;36:2178–87.
53. Goodfellow M, Schindler K, Baier G. Self-organized transients in a neural mass model of epileptogenic tissue dynamics. NeuroImage. 2012;59:2644–60.
54. Goodfellow M, Glendinning P. Mechanisms of intermittent state transitions in a coupled heterogeneous oscillator model of epilepsy. J Math Neurosci. 2013;3:17.
55. Grimbert F, Faugeras O. Bifurcation analysis of Jansen's neural mass model. Neural Comput. 2006;18:3052–68.
56. Hall D, Kuhlmann L. Mechanisms of seizure propagation in 2-dimensional centre-surround recurrent networks. PLoS ONE. 2013;8:e71369.
57. Hindriks R, Meijer HGE, van Gils SA, van Putten MJAM. Phase-locking of epileptic spikes to ongoing delta oscillations in non-convulsive status epilepticus. Front Syst Neurosci. 2013;7:111.
58. Hodgkin AL, Huxley AF. A quantitative description of membrane current and its application to conduction and excitation in nerve. J Physiol. 1952;117:500–44.
59. Howard P, Twycross R, Shuster J, Mihalyo M, Remi J, Wilcock A. Anti-epileptic´ Drugs. J Pain Symptom Manage. 2011;42:788–804.
60. Izhikevich EM. Dynamical systems in neuroscience: the geometry of excitability and bursting. Cambridge: MIT Press; 2007.
61. Jansen BH, Rit VG. Electroencephalogram and visual evoked potential generation in a mathematical model of coupled cortical columns. Biol Cybernet. 1995;73:357–66.
62. Jirsa VK, Haken H. A derivation of a macroscopic field theory of the brain from the quasimicroscopic neural dynamics. Physica D. 1997;99:503–26.
63. Jirsa VK, Jantzen KJ, Fuchs A, Kelso JAS. Spatiotemporal forward solution of the EEG and MEG using network modeling. IEEE Trans Med Imaging. 2002;21:493–504.
64. Jirsa VK, Stacey WC, Quilichini PP, Ivanov AI, Bernard C. On the nature of seizure dynamics. Brain 2014;137(Pt 8):2210–30. doi:10.1093/brain/awu133.
65. Jiruska P, de Curtis M, Jefferys JGR, Schevon CA, Schiff SJ, Schindler K. Synchronization and desynchronization in epilepsy: controversies and hypotheses. J Physiol. 2013;591:787–97.
66. Joels M. Stress, the hippocampus, and epilepsy. Epilepsia. 2009;50:586–97.
67. Kalamangalam GP, Tandon N, Slater JD. Dynamic mechanisms underlying afterdischarge: a human subdural recording study. Clin Neurophysiol. 2014;125:1324–38.
68. Kalitzin SN, Velis DN, Lopes da Silva FH. Stimulation-based anticipation and control of state transitions in the epileptic brain. Epilepsy Behav. 2010;17:310–23.
69. Kalitzin SN, Koppert M, Petkov G, Velis DN, Lopes da Silva FH. Computational model prospective on the observation of proictal states in epileptic neuronal systems. Epilepsy Behav. 2011;22:102–9.
70. Kandel E, Schwartz J, Jessel TM. Principles of neural science. New York: Elsevier; 1991.
71. Kuramoto Y. Chemical oscillations, waves and turbulence. Berlin: Springer; 1984.
72. Lehnertz K. Epilepsy and nonlinear dynamics. J Biol Phys. 2008;34:253–66.
73. Lehnertz K, Ansmann G, Bialonski S, Dickten H, Geier C, Porz S. Evolving networks in the human epileptic brain. Physica D. 2014;267:7–15.
74. Liao W, Zhang Z, Mantini D, Xu Q, Ji G, Zhang H, Wang J, Wang Z, Chen G, Tian L, Jiao Q, Zang Y, Lu G. Dynamical intrinsic functional architecture of the brain during absence seizures. Brain Struct Funct. 2014. doi: 10.1007/s00429-013-0619-2

75. Liley DTJ, Cadusch PJ, Dafilis MP. A spatially continuous mean field theory of electrocortical activity. Netw Comput Neural Syst. 2002;13:67–113.
76. Liley DTJ, Bojak I. Understanding the transition to seizure by modeling the epileptiform activity of general anesthetic agents. J Clin Neurophysiol. 2005;22:300–13.
77. Lopes da Silva FH, Blanes W, Kalitzin SN, Parra J, Suffczynski P, Velis DN. Dynamical diseases of brain systems: different routes to epileptic seizures. IEEE Trans Biomed Eng. 2003;50:540–8.
78. Lytton WW. Computer modelling of epilepsy. Nat Rev Neurosci. 2008;9:626–37.
79. Maneshi M, Moeller F, Fahoum F, Gotman J, Grova C. Resting-state connectivity of the sustained attention network correlates with disease duration in idiopathic generalized epilepsy. PLoS ONE. 2012;7:e50359.
80. Markram H. The blue brain project. Nat Rev Neurosci. 2006;7:153–60.
81. Marten F, Rodrigues S, Benjamin O, Richardson MP, Terry JR. Onset of polyspike complexes in a mean-field model of human electroencephalography and its application to absence epilepsy. Philos Trans A Math Phys Eng Sci. 2009;367:1145–61.
82. Marten F, Rodrigues S, Suffczynski P, Richardson MP, Terry JR. Derivation and analysis of an ordinary differential equation mean-field model for studying clinically recorded epilepsy dynamics. Physical Review E. 2009;79:021911.
83. Mattson RH. Overview: idiopathic generalized epilepsies. Epilepsia. 2003;44:2–6.
84. McCormick DA, Contreras D. On the cellular and network bases of epileptic seizures. Ann Rev Physiol. 2001;63:815–46.
85. Meeren HKM, Pijn JPM, Van Luijtelaar ELJM, Coenen AML, Lopes da Silva FH. Cortical focus drives widespread corticothalamic networks during spontaneous absence seizures in rats. J Neurosci. 2002;22:1480–95.
86. Meeren H, van Luijtelaar G, Lopes da Silva FH, Coenen A. Evolving concepts on the pathophysiology of absence seizures. Arch Neurol. 2005;62:371–6.
87. Meisel C, Storch A, Hallmeyer-Elgner S, Bullmore E, Gross T. Failure of adaptive self-organized criticality during epileptic seizure attacks. PLoS Comput Biol. 2012;8:e1002312.
88. Milton JG. Epilepsy as a dynamic disease: a tutorial of the past with an eye to the future. Epilepsy Behav. 2010;18:33–44.
89. Molaee-Ardekani B, Benquet P, Bartolomei F, Wendling F. Computational modeling of high-frequency oscillations at the onset of neocortical partial seizures: from 'altered structure' to 'dysfunction'. NeuroImage. 2010;52:1109–22.
90. Moran R, Pinotsis DA, Friston K. Neural masses and fields in dynamic causal modeling. Front Comput Neurosci. 2013;7:57.
91. Mormann F, Andrzejak RG, Elger CE, Lehnertz K. Seizure prediction: the long and winding road. Brain. 2006;130:314–33.
92. Morris C, Lecar H. Voltage oscillations in the barnacle giant muscle fiber. Biophys J. 1981;35:192–213.
93. Mountcastle VB. The columnar organization of the neocortex. Brain. 1997;120:701–22.
94. Murta T, Leal A, Garrido MI, Figueiredo P. Dynamic causal modelling of epileptic seizure propagation pathways: a combined EEG-fMRI study. NeuroImage. 2012;62:1634–42.
95. Nevado-Holgado AJ, Marten F, Richardson MP, Terry JR. Characterising the dynamics of EEG waveforms as the path through parameter space of a neural mass model: application to epilepsy seizure evolution. NeuroImage. 2012;59:2374–92.
96. Niedermeyer E, Lopes da Silva FH. Electroencephalography: basic principals, clinical applications, and related fields. London: Williams and Wilkins; 2005.
97. Nunez PL, Srinivasan R. Electric fields of the brain: the neurophysics of EEG. 2nd ed. New York: Oxford University Press; 2006.
98. O'Hagan A. Bayesian analysis of computer code outputs: a tutorial. Reliab Eng Syst Saf. 2006;91:1290–300.
99. Osorio I, Frei MG, Sornette D, Milton J, Lai Y.-C. Epileptic seizures: quakes of the brain? Phys Rev E. 2010;82:021919.

100. Owen JA, Barreto E, Cressman JR. Controlling seizure-like events by perturbing ion concentration dynamics with periodic stimulation. PLoS ONE. 2013;8:e73820.
101. Pellegrini A, Musgrave J, Gloor P. Role of afferent input of subcortical origin in the genesis of bilaterally synchronous epileptic discharges of feline generalized epilepsy. Exp Neurol. 1979;64:155–73.
102. Pesce LL, Lee HC, Hereld M, Visser S, Stevens RL, Wildeman A, van Drongelen W. Large-scale modeling of epileptic seizures: scaling properties of two parallel neuronal network simulation algorithms. Comput Math Methods Med. 2013;2013:182145.
103. Ponten SC, Douw L, Bartolomei F, Reijneveld JC, Stam CJ. Indications for network regularization during absence seizures: weighted and unweighted graph theoretical analyses. Exp Neurol. 2009;217:197–204.
104. Purves D, Augustine G, Fitzpatrick D, Hall W, LaMantia, A-S, White L. Neuroscience, 5th ed. Sunderland: Sinauer Associates; 2012.
105. Richardson MP. New observations may inform seizure models: very fast and very slow oscillations. Prog Biophys Mol Biol. 2011;105:5–13.
106. Richardson MP. Large scale brain models of epilepsy: dynamics meets connectomics. J Neurol Neurosurg Psychiatry. 2012;83:1238–48.
107. Robinson PA, Rennie CJ, Wright JJ. Propagation and stability of waves of electrical activity in the cerebral cortex. Phys Rev E. 1997;56:826–40.
108. Robinson PA, Rennie CJ, Rowe D. Dynamics of large-scale brain activity in normal arousal states and epileptic seizures. Phys Rev E. 2002;65:041924.
109. Rodrigues S, Barton D, Szalai R, Benjamin O, Richardson MP, Terry JR. Transitions to spike-wave oscillations and epileptic dynamics in a human cortico-thalamic mean-field model. J Comput Neurosci. 2009;27:507–26.
110. Rodrigues S, Chizhov AV, Marten F, Terry JR. Mappings between a macroscopic neural mass model and a reduced conductance-based model. Biol Cybernet. 2010;102:361–71.
111. Rodrigues S, Barton D, Marten F, Kibuuka M, Alarcon G, Richardson MP, Terry JR. A method for detecting false bifurcations in dynamical systems: application to neuralfield models. Biol Cybernet. 2010;102:145–54.
112. Rogawski MA, Loscher W. The neurobiology of antiepileptic drugs. Nat Rev Neurosci. 2004;5:553–64.
113. Rubinov M, Sporns O. Complex network measures of brain connectivity: uses and interpretations. NeuroImage. 2010;52:1059–69.
114. Schelter B, Timmer J, Schulze-Bonhagel A. Seizure prediction in epilepsy. Weinheim: Wiley; 2008.
115. Schiff SJ. Neural control engineering: the emerging intersection between control theory and neuroscience. Cambridge: MIT Press; 2011
116. Seth AK. A MATLAB toolbox for Granger causal connectivity analysis. J Neurosci Methods. 2010;186:262–73.
117. Soltesz I, Staley K. (eds). Computational neuroscience in epilepsy. San Diego: Academic; 2008.
118. Sporns O, Tononi G, Kotter R. The human connectome: a structural description of the¨ human brain. PLoS Comput Biol. 2005;1:e42.
119. Stead M, Bower M, Brinkman BH, Lee K, Marsh WR, Meyer FB, Litt B, Van Gompel J, Worrell GA. Microseizures and the spatiotemporal scales of human partial epilepsy. Brain. 2010;133:2789–97.
120. Stefanescu R, Shivakeshavan RG, Talathi SS. Computational models of epilepsy. Seizure. 2012;21:748–59.
121. Suffczynski P, Kalitzin SN, Lopes Da Silva FH. Dynamics of non-convulsive epileptic phenomena modeled by a bistable neuronal network. Neuroscience. 2004;126:467–84.
122. Suffczynski P, Lopes da Silva FH, Parra J, Velis DN, Bouwman BM, van Rijn CM, van Hese P, Boon P, Khosravani H, Derchansky M, Carlen P, Kalitzin SN. Dynamics of epileptic phenomena determined from statistics of ictal transitions. EEE Trans Biomed Eng. 2006;53:524–32.

123. Taylor PN, Baier G. A spatially extended model for macroscopic spike-wave discharges. J Comput Neurosci. 2011;31:679–84.
124. Tejada J, Costa KM, Bertti P, Garcia-Cairasco N. The epilepsies: complex challenges needing complex solutions. Epilepsy Behav. 2013;26:212–28.
125. Terry JR, Benjamin O, Richardson MP. Seizure generation: the role of nodes and networks. Epilepsia. 2012;53:166–9.
126. Traub RD, Whittington MA, Buhl EH, LeBeau FE, Bibbig A, Boyd S, Cross H, Baldeweg T. A possible role for gap junctions in generation of very fast EEG oscillations preceding the onset of, and perhaps initiating, seizures. Epilepsia. 2001;42:153–70.
127. van Gils SA, van Janssens SG, Kuznetsov YA, Visser S. On local bifurcations in neural field models with transmission delays. J Math Biol. 2013;66:837–87.
128. van Vreeswijk C, Abbott LF, Ermentrout GB. When inhibition not excitation synchronizes neural firing. J Comput Neurosci. 1994;1:313–21.
129. Valent´ın A, Alarcon G, Garc´ıa-Seoane JJ, Lacruz ME, Nayak SD, Honavar M, Selway RP, Binnie CD, Polkey CE. Single-pulse electrical stimulation identifies epileptogenic frontal cortex in the human brain. Neurology. 2005;65:426–35.
130. Varela F, Lachaux JP, Rodriguez E, Martinerie J. The brainweb: phase synchronization and large-scale integration. Nat Rev Neurosci. 2001;2:229–39.
131. Vincent RD, Courville A, Pineau J. A bistable computational model of recurring epileptiform activity as observed in rodent slice preparations. Neural Netw. 2011;24:526–37.
132. Visser S, Meijer HGE, Lee HC, van Drongelen W, van Putten, MJAM, van Gils SA. Comparing epileptiform behavior of mesoscale detailed models and population models of neocortex. J Clin Neurophysiol. 2010;27:471–8.
133. Visser S, Meijer HGE, van Putten MJAM, van Gils SA. Analysis of stability and bifurcations of fixed points and periodic solutions of a lumped model of neocortex with two delays. J Math Neurosci. 2012;2:8.
134. Vogels TP, Rajan K, Abbott LF. Neural network dynamics. Annu Rev Neurosci. 2005;28:357376.
135. Volman V, Perc M, Bazhenov M. Gap junctions and epileptic seizures—two sides of the same coin? PLoS ONE. 2011;6:e20572.
136. Wang Y, Goodfellow M, Taylor PN, Baier G. Phase space approach for modeling of epileptic dynamics. Phys Rev E. 2012;85:1–11.
137. Wendling F. Neurocomputational models in the study of epileptic phenomena. J Clin Neurophysiol. 2005;22:285–7.
138. Wendling F. Computational models of epileptic activity: a bridge between observation and pathophysiological interpretation. Expert Rev Neurotherapeut. 2008;8:889–96.
139. Wendling F, Bartolomei F, Bellanger JJ, Chauvel P. Interpretation of interdependencies in epileptic signals using a macroscopic physiological model of the EEG. Clin Neurophysiol. 2001;112:1201–18.
140. Wendling F, Bartolomei F, Bellanger JJ, Chauvel P. Epileptic fast activity can be explained by a model of impaired GABAergic dendritic inhibition. Eur J Neurosci. 2002;15:1499–508.
141. Wendling F, Bartolomei F, Senhadji L. Spatial analysis of intracerebral EEG in the time and frequency domain: identification of epileptogenic networks in partial epilepsy. Philos T Roy Soc A. 2009;367:297–316.
142. Wendling F, Bartolomei F, Mina F, Huneau C, Benquet P. Interictal spikes, fast ripples and seizures in partial epilepsies: combining multi-level computational models with experimental data. Eur J Neurosci. 2012;36:2164–77.
143. Wilson HR, Cowan JD. Excitatory and inhibitory interactions in localized populations of model neurons. Biophys J. 1972;12:1–24.
144. Wilson HR, Cowan JD. A mathematical theory of the functional dynamics of cortical and thalamic nervous tissue. Kybernetik. 1973;13:55–80.

145. Worrell GA, Stephen CA, Cranstoun SD, Litt B, Echauz J. Evidence for selforganized criticality in human epileptic hippocampus. Expert Rev Neuroreport. 2002;13:2017–21.
146. Wyllie E, Comair YG, Kotagal P, Bulacio J, Bingaman W, Ruggieri P. Seizure outcome after epilepsy surgery in children and adolescents. Ann Neurol. 1998;44:740–8.
147. Yan B, Li P. The emergence of abnormal hypersynchronization in the anatomical structural network of human brain. NeuroImage. 2013;65:34–51.
148. Ziburkus J, Cressman JR, Schiff SJ. Seizures as imbalanced up states: excitatory and˜ inhibitory conductances during seizure-like events. J Neurophysiol. 2013;109:1296–306.

Chapter 8
Computational Modeling of Neuronal Dysfunction at Molecular Level Validates the Role of Single Neurons in Circuit Functions in Cerebellum Granular Layer

Shyam Diwakar

Abstract Using mathematical modelling, we attempted to reconstruct the information transmission at the granular layer of the cerebellum, a circuit whose functions and dysfunctions remain yet to be explored in detail. Information transmission at the Mossy Fiber (MF)—Granule cell (GrC) synaptic relay is crucial to understand mechanisms of signal coding in the cerebellum and related impacts of connectivity mechanisms. Using biophysically detailed multi-compartmental models, simple spiking neurons we reconstructed granular layer micro-circuitry and estimated both single neuron behaviour and network activity in terms of center-surround patterns, as observed during sensory and tactile stimulation. The chapter also includes local field potential reconstructions to show plasticity mechanisms at the molecular level is reflected at the network activity level, indicating network LFP in the granular layer is a regulated activity signal arising from the underlying granule cells and the feed-forward inhibition from the Golgi cells. The role of selective inhibition by Golgi cells for coincidence detection is presented. Exploring the EPSP-spike complex in granular neurons revealed potential mechanisms for sparse recoding in cerebellum and quantification of information encoding in individual neurons of the cerebellar granular layer. We also look into two specific forms of neuronal dysfunction with ataxia-like behaviour in knockout mice models and in NMDAR-related autism. While network activity was severely affected, the amplitude of damage is critical of the mechanisms at the cellular or molecular level. The study further enhances our understanding of specific coding geometries in the cerebellum and spatio-temporal processing in a primary circuit of the cerebellum.

Keywords Cerebellum · Granular layer · Computational modelling · Local field potential · EPSP-spike complex · Feedforward inhibition · Knockout mutations

---

S. Diwakar (✉)
Amrita School of Biotechnology, Amrita Vishwa Vidyapeetham (Amrita University), Amritapuri, Clappana P.O., Kerala 690525, India
e-mail: shyam@amrita.edu

Introduction

Marr, Albus, Ito and others have shown cerebellum as a region of motor learning [1–3]. A large microcircuit, the cerebellar granular layer has been known to perform functions related to sensory and tactile signal processing [4]. Such circuits are reliable for studies as they are fairly conserved among species [5].

Analysing neuronal processing helps determine the possible role and function of a neuron in a particular neural microcircuit. Information processing in neural circuit can be analysed in terms of spikes or EPSPs, i.e. by quantifying how much information the neural responses convey about the input stimuli [6–9]. At the cellular scale, single-neurons process information mainly through spikes or action potentials in addition to excitatory post-synaptic potentials or EPSPs.

Circuits show various functional roles compared to the underlying neurons [10]. The cerebellar input layer circuitry has been suggested to perform spatio-temporal processing of signals coming through the mossy fibers. The combination of excitatory and inhibitory inputs have been not only shown to serve as inputs but also as mechanisms of function such as motor control, which is only one of the known functions of such circuits [11, 12]. Therefore the study of such circuits and their dynamical aspects along with their behaviour and relationship to internal model control theories have become important.

Since an early time and especially during the years of Camillo Golgi and Santiago Ramon y Cajal, the cerebellum was known as a structure related to motor control. However much later several cognitive functions have been attributed to the cerebellar structures and circuits [13]. Cerebellum has been observed to show homogeneity in circuit organization and hence the "modules" or various circuits in the cerebellum are assumed to contribute to the diversity of the functions attributed to the cerebellum [14]. Previous studies assume signal decorrelation as one of the main functions of the cerebellum input layer, namely the granule cells due to the apparent combinatorial organization [15]. However it is suggested now that the granular layer can lend to many more functions than just signal decorrelation [16].

In this chapter, we look into how single neuron models play a role in network circuit functions and how predictions of circuit functions can be elucidated from known experimental validations of neural activity via computational models of a cerebellar circuit.

Cerebellum Input Layer Circuit

The cerebellum [15] is divided into different functionally- distinct regions (see Fig. 8.1. There is a need for analysis on the functionalities and behaviour of the two independent processing regions: granular layer and Purkinje cell layer [17]. Understanding how the granular layer of the cerebellum processes incoming information is crucial because its output forms a major part of the output of cerebellar cortex [18]. Specific connection geometry exists between Mossy Fibers (MF), Granule Cells (GrC) and Golgi cells (GoC) in the granular layer [19].

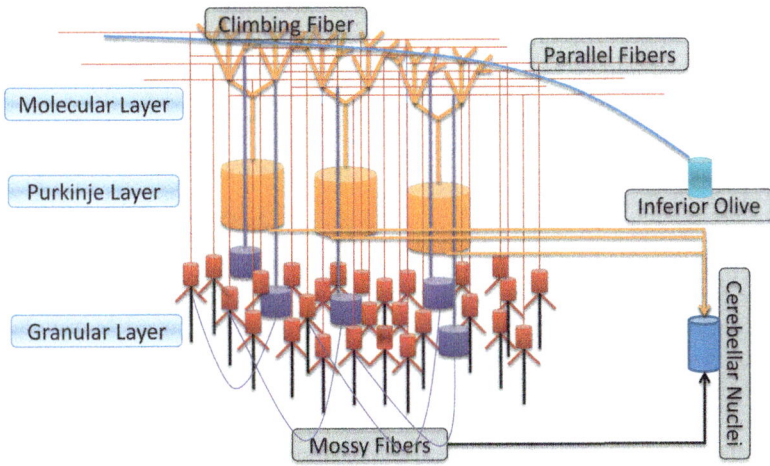

Fig. 8.1 Schematic representation of cerebellar circuits. At the sub-cortical levels, afferent fibers carrying sensory information activate the deep cerebellar nuclei and inferior olive cells. In the cerebellar cortex, there are granule cells, Golgi cells, Purkinje cells, stellate cells, basket cells, Lugaro cells and unipolar brush cells. The granular layer (in this work) is a feed-forward excitatory network with several inhibitory loops; the main inputs to granule cells come from mossy fibers that diverge from deep cerebellar nuclei. Golgi cells provide inhibitory inputs to granule cells and granule cell ascending axons form parallel fibers that activate the Purkinje cells. The cerebellum and its circuits are arranged as modules, microzones and multizonal complexes [89]

Granular layer is hypothesized to perform recording of the MF inputs into a sparse representation that permits noise reduction by lateral inhibition [20], generated by Golgi cell. Due to the limited set of combinations of MF excitatory and GoC inhibitory inputs, there are a discrete number of synaptic connections for studying the relationship between input and output in the MF-GrC relay.

Long-term changes in synaptic strength named as synaptic plasticity is one of the important neurochemical foundations of learning and memory [21, 22]. Introduced by Donald Hebb in 1949, spike time dependent plasticity is supposed to provide the cellular basis for learning and memory and takes the form of either Long Term Potentiation (LTP) or Long Term Depression (LTD) [1, 23]. A high frequency signal (typically 100 Hz) induces a strong glutamate release into the synapse and depolarizes the postsynaptic neuron [24]. The depolarization induced by successive excitatory postsynaptic potentials (EPSPs) overlap and cumulatively produces action potentials or spikes. Amount of neurotransmitters released into the synaptic cleft activates postsynaptic receptors to elicit a postsynaptic response. In the cerebellum granular layer, induction of LTP at MF-GrC synapses enhances the spike train response of GrCs [25]. During MF- GrC LTP, there is an enhancement in neurotransmitter release and intrinsic excitability of GrC [25, 26]. LTD expression was associated with a decrease in release probability of the mossy fiber, showing changes opposite to those characterizing LTP [27]. Therefore induction of plasticity by changing the intrinsic excitability of the GrC and excitatory release probability of MF [23, 26] at MF-GrC synapses changes the spike/EPSP train response of GrC [17].

At the cellular scale, single-neurons process information mainly through spikes or action potentials [28]. The mechanisms involving the responses produced by the granular layer implicates changes in release probability and intrinsic excitability over mossy fiber-granule cell relay [29–31].

Granule Neuron

Granule cell is one of the most numerous of neurons in the cerebellar cortex [32, 33]. The soma in the rat cerebellum is spherical and is usually of ~ 5 µm in diameter [34]. The cells have 3–5 dendrites on an average and are about ~ 13 µm in length. The axon is usually very thin ~ 0.1 µm has two known components, ascending axon which then extends as the parallel fiber [35, 36]. The axon starts from the granular layer and extends into the molecular layer. The granule neuron is one of the best known neurons where the biophysical properties have been characterized extensively and modelled using both detailed and simple compartmental models [13, 32, 34, 36, 37].

Cerebellar granule cells have been known to recode information coming from the mossy fibers via expansion recoding due to large convergence-divergence ratio [13, 38, 39]. Because of the location, such cells are considered critical since all operations of cerebellum depend on the elaboration of incoming sensory and tactile signals into the cerebellum. The connectivity with the granular layer, makes granular layer signal recoding unique. Each granule cell dendrite receives an input from a different mossy fiber and an inhibitory synapse from a Golgi cell. The granule cell dendrites also form a specialized structure called glomerulus, where the postsynaptic densities are observed [40, 41]. The glomeruli act as microenvironments for neurotransmitter related mechanisms.

Golgi Neuron

Golgi cells are inhibitory interneurons of the cerebellar granular layer, which receive inputs from Mossy fibers and respond to afferent stimulation *in vivo* with a burst-pause sequence interrupting their irregular background low-frequency firing [42–44]. Golgi neurons show fastest transient in soma, intermediate in dendrites and slowest in axon [43]. Golgi cells are known to inhibit granule neurons via two loops [45, 46].

Methods

We have used NEURON [47] as the modelling tool. The study carried out in this paper involved the use of computational models of neurons based on experimental data from p19-23 Wistar rat cerebellum [33, 43]. Mathematical neuron models of granule cell [33, 48], and Golgi cell model [43, 44] were used in this network study. Modeling reliability for spiking models was based on the extensive characterization of membrane currents and the compact electrotonic structure of cerebellar granule cells [33, 34]. The models used AMPA (2-amino-3-(5-methyl-3-oxo-1,2- oxazol-4-yl) propanoic acid) and NMDA(N-Methyl-D-aspartate) receptor components as excitatory MF-GrC synapses and GABA (γ-Aminobutyric acid)-ergic synapses for the Golgi cell- GrC relay [33]. On an average, each granule cell receives excitatory connections from 4 to 5 mossy fibers [25, 33, 34].

Multi-Compartmental Granule Neuron Model

Detailed multi-compartmental granule cell (GrC) model [33] was used and simulations included responses by varying the excitatory (E) and inhibitory (I) synaptic inputs. The model of the granule cell was based on multi-compartmental cable theory and included soma, axon, hillock and dendritic compartments. The model consisted of 52 active compartments connected to each other via the 3/2-power law [18]. For each of the compartments, membrane voltage V_m had to be estimated separately.

$$\frac{dV_m}{dt} = \frac{1}{\tau_m}\left(V - \frac{\sum_i g_i(V-V_i) + \sum_{syn} g_{syn}(V-V_{syn}) + \sum_{br} g_{br}(V-V_{br})}{g_{tot}}\right)$$

Where g is the conductance corresponding to i (ion channel), syn (synaptic dynamics), br (neighbouring attached branch) and tot (total). Here $\tau\_m = R\_mC\_m$ which is the time constant of oscillation of the membrane based on its membrane resistance, R_m and membrane capacitance, C_m. The calcium current in the model was included as

$$\frac{d[Ca]}{dt} = -\frac{I_{Ca}}{(2F.A.d)} - \left(\beta_{Ca}\left([Ca]-[Ca]_O\right)\right)$$

where d is the depth of a shell adjacent to the cell surface of area A, β_Ca determines the loss of calcium ions from the shell approximating the effect of fluxes, ionic pumps, diffusion, and buffers, $[Ca]_O$ is resting [Ca] and F is the Faraday's constant. [Ca] is the calcium channel dynamics as reported in [33].

Granule Cell Synapses

The model GrC has 1–4 excitatory (one for each dendrite) and 0(no inhibition)–4 inhibitory connections (one for each dendrite) [33]. The detailed explanations of ionic channel dynamics, compartmental localization of ion channels and electrotonic structure of this granule neuron model are described elsewhere [25, 33, 34]. Since granule cell is one of the rarest neurons where the ionic channel densities can be accurately determined using whole-cell patch clamp, the ion channel dynamics that were modelled previously [33–35] and is not repeated here. Also, excitatory and inhibitory synaptic inputs to the dendrites were located in dendritic tips, although in neighbouring dendritic compartments. Presynaptic dynamics for the MF-GrC was modelled separately as in [25, 33] due to components such as facilitation and depression.

Excitatory post-synaptic mechanisms were shown as AMPA and NMDA post-synaptic receptor components as seen in granule neurons. AMPA receptor dynamics was modelled using a three-state scheme and a 2D diffusion model whereas the NMDA receptors used Boltzmann equation as seen in [49]. Both the excitatory pre-synaptic and excitatory post-synaptic mechanisms are described in detail elsewhere [25, 37]. The GoC-GrC inhibitory synapse model was based on the following pre-synaptic dynamics: release probability = 0.35, $\tau REC = 36$ ms, $\tau facil = 58.5$ ms and $\tau I = 0.1$ ms, respectively and as described in [50], Effects of blocking inhibition by adding gabazine were also simulated by setting GABAergic conductance in inhibitory fibers to zero.

Golgi Cell Model

The Golgi cell model (GoC) was adapted from [43]. The model produced responses that was used to drive the inhibitory connections onto the granule cell via the Golgi cell-Granule neuron feed-forward circuit.

Granule Layer Circuitry

Mossy fibers activate granule neuron via AMPA and NMDA receptors at the post-synaptic site [25, 33]. Mossy fiber also activated Golgi cells with a divergence ration of 1:3.6 [13]. In most of our circuit models, we ignored the feedback loop from granule cells to Golgi cells since that circuit takes part at a later stage of the post-synaptic response and was not characteristic of the models in this study.

Mutant Neuron Models

To model mutant models, we adapted the sodium channel related intrinsic excitability [35, 36] and synaptic dysfunctions [51, 52] as seen during certain disorders. The sodium channel used in control and mutant models was a 13-state, allosteric sodium channel model with values as described elsewhere [35]. For the synaptic sparseness, we selectively disabled NR2A-related NMDA receptors and reconstructed the synaptic response as indicated.

Center-Surround Excitation

Stimulating mossy fibers with an electrode at a particular point activates granule cells in the network in a center-surround activation pattern [53]. Within a 'spot', cells which are in close proximity to the electrode will receive high excitation and the periphery layer cells receive less excitation. Our model of the granular network shows the pattern activated as a 'spot'. The center-surround pattern showing the decreasing strengths of excitation can be noticed from the center to the periphery.

Simulating LTP-LTD

By modifying intrinsic excitability and release probability [25, 54], we simulated plasticity in the granule cells. We modified intrinsic excitability by changing ionic current density or gating. We modified the on-off gating characteristics of sodium channel to modify sodium activation and inactivation parameters [36] for higher and lower intrinsic excitability. Combining intrinsic excitability and varying release probability of synapses we were able to model long-term potentiation (LTP) and long-term depression (LTD) for simulating spike-time dependent plasticity (STDP) in the MF-GrC synapses.

Local Field Potential Reconstruction

Network local field potentials (LFP) was reconstructed using techniques based on convolution of jittered single extracellular potentials [36] and by linear summation of compartmental components [12]. In both methods, NEURON's extracellular mechanism [47] was used for calculating membrane ionic currents from the compartments. The network model consists of 1382 Multi-compartmental granule cells, 5 Golgi cell, 88 Mossy fibers rosettes (MF) and 16,500 synapses arranged in 35 µm cubic slice of cerebellar cortex model. Convergence and divergence ratios used to build the network model were based on earlier studies [39, 55].

Artificial Spiking Neuron Models

We also employed another group of models for their simplicity in reproducing known neuronal behaviour patterns, namely, the artificial spiking models. These models are computationally inexpensive and can be used to simulate large scale network models with relatively less time. There are many existing spiking models starting from leak integrate-and-fire models to modified versions of spike response model. Simplified spiking models depend on two equations, one regulating the membrane potential and the other regulating the adaptation or subthreshold dynam-

Table 8.1 AdEx parameter values used for modeling neuronal firing

	C (pF)	g_L (nS)	E_L (mV)	ΔT (mV)	V_T (mV)	I (pA)	V_R (mV)	a (nS)	b (pA)	τ_w (ms)
DCN	200	10	−70	2	−50	0	−58	−20	675	8
IO	350	10	−65	2	−50	0	−58	−13	1200	1
Golgi cell	511	13.1	−58	7	−60	0	−50	−20	1033	14.65
Purkinje cell	100	10	−65	2	−50	0	−58	−13	260	1
Granule cell	1	10	−70	2	−50	20	−58	−10	265	0.71

ics. The membrane potential will rise till threshold and then exponential mechanism triggers rapid rise and resets back to the resting potential. We used the adaptive exponential integrate and fire model [56] which is well known to replicate realistic neuronal firing patterns. This model replaces strict threshold with more realistic smooth threshold [57] as observed in exponential integrate and fire neuron [58].

$$C\frac{dV}{dt} = -g_L(V - E_L) + g_L * \Delta T * \exp\left(\frac{V - V_T}{\Delta T}\right) - w + I$$

$$\tau_w \frac{dw}{dt} = a(V - E_L) - w$$

Where 'C' is the membrane capacitance, 'g_L' is leak conductance, E_l is the resting potential, 'ΔT' is the slope factor and 'V_T' is the threshold potential. Variable 'w' determines the level of adaptation of the neuron and 'a' modifies sub-threshold adaptation [57]. Exponential term determines the early activation of voltage-gated sodium channels in the model (Table 8.1).

Using spiking neuron models, we reproduced different neuronal firing patterns as observed in rat cerebellum specifically those of deep cerebellar nuclei (DCN), inferior olivary nucleus (IO), Golgi cell, Purkinje cell and granule cell (see Table 8.1).

Validating In Vitro and In Vivo Behavior

In vitro like behaviors were studied in the granule cells by giving single spike as input via MF terminals. *In vivo* like behaviors were characterized by short bursts (5 spikes per burst). First EPSP/spike latency was measured from the time of input stimulus to time of the peak of the output EPSP/spike. In plots latency is indicated synonymously as occurrence times of spike/EPSP. The excitatory stimulus was applied at t=20 ms and inhibitory stimulus at t=24 ms. The EPSP/spike amplitude was measured from the resting voltage of −70 mV to the voltage of the peak of the EPSP/spike generated. *In vivo* GrCs tend to discharge bursts in vivo [60] whereas in vitro GrCs produced spikes or doublets with high excitation and EPSPs with lower excitation [63].

Results

Granule Cell Post-Synaptic Responses

Using a mathematical model, we explored the role of EPSPs as a major information carrying component in addition to spikes. Most studies consider spikes as the major information carrying component in neurons. Studying both spikes and EPSPs and how EPSP-spike complex modulate information encoding in single neurons can possibly reveal the importance of the large number of granule neurons in the cerebellum and how sparse recoding happens in the cerebellar input layer.

Variation in number of synaptic inputs affected both types of post-synaptic responses in granule neurons. Typically excitation with minimum number of inputs generates excitatory post-synaptic potentials (EPSPs) while larger number of inputs causes spikes. Granule neuron model [33], under control condition in vitro with one or two active mossy fiber (MF) inputs generated EPSPs. Three excitatory synaptic inputs generated spikes and four inputs generated spike doublets (Fig. 8.2a). *In vivo* synaptic inputs generated more spikes (Fig. 8.2b) due to burst-like inputs in MFs.

The EPSPs and spikes generated were found to be sensitive to the inhibitory-excitatory balance of the MF-GrC relay. Increase in MF excitatory inputs tend to reduce the number of EPSPs generated while it increased the number of spikes generated. When inhibition was included, EPSPs tend to increase but the spiking activity of the GrC decreased (see Fig. 8.3, EPSPs are not seen due to *in vivo* like inputs along MF). Single mossy fiber input combined without or with inhibitory connections always generated zero spikes but produced EPSPs.

Over all possible combinations of excitation (E) and inhibition (I) i.e. summing all responses for combinations I0E1, I0E2 etc. until I4E4 generated an estimate of the maximum possible EPSPs and Spikes over all possible release probabilities (See Fig. 8.4).

Varying MF Release Probabilities Affect Spike-EPSP Ratio

To understand whether increase in release probability could determine changes in EPSP, the release probability (U) of excitatory synapse was modified from control ($U = 0.416$) [25] while the release probability of inhibitory fiber was set at its control value ($U_{inh} = 0.34$) [59]. Various synaptic activation patterns were applied as inputs via MF and for each of the activation patterns, number of spikes; first spike latency (from the peak of the spike) and amplitude of the initial spike were measured. Here, EPSPs were measured from the initial membrane potential of -70 mV. With inputs

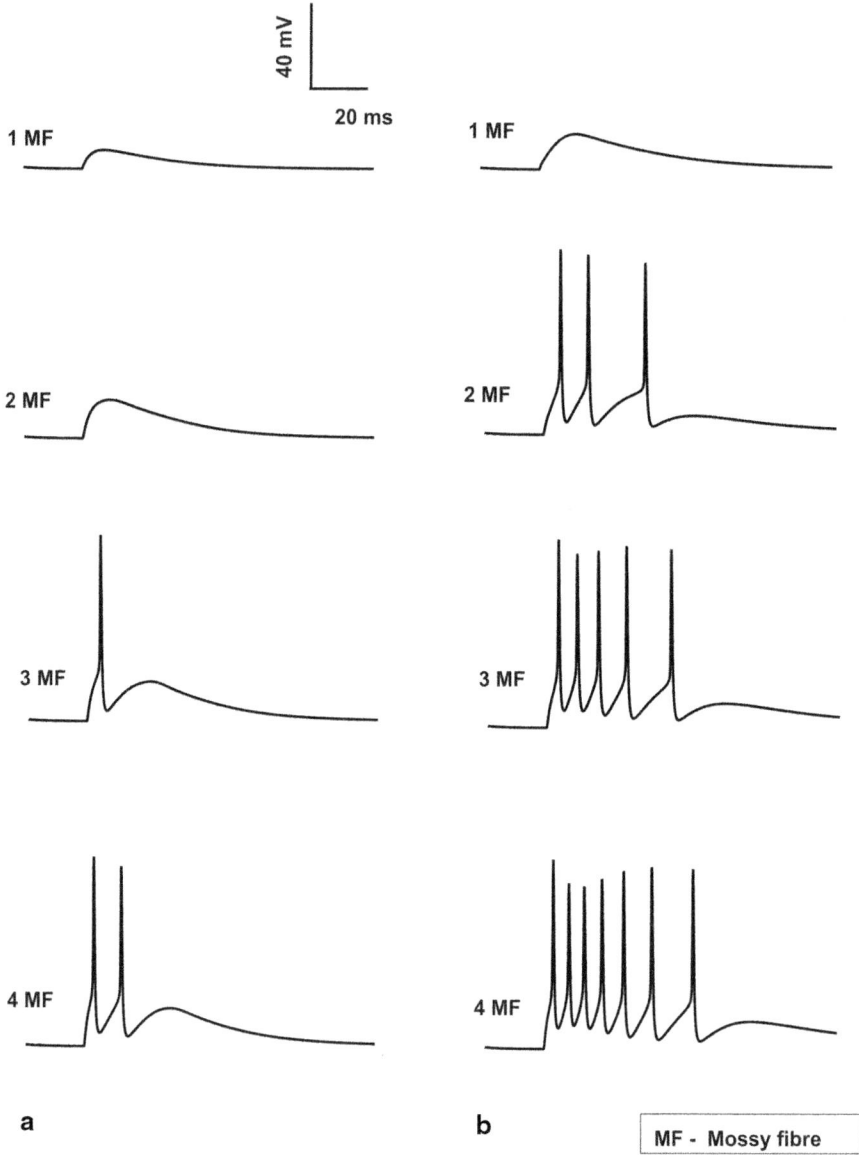

Fig. 8.2 Spiking properties of granule cell. **a** *In vitro* Granule cell shows spikes only if *3 MF* inputs are activated which elicits a single spike and when all *4 MF* inputs are activated a doublet is seen. No inhibitory inputs are provided to the cell. **b** *In vivo* Granule cell without inhibition gives 3, 5 and 7 spikes for *2, 3* and *4 MF* activation

from 4 MF excitatory synapses and 2 inhibitory synapses (see Fig. 8.5a) single spike was observed and the spike measured 15.52 mV at 23.225 ms under control condition (U=0.416).

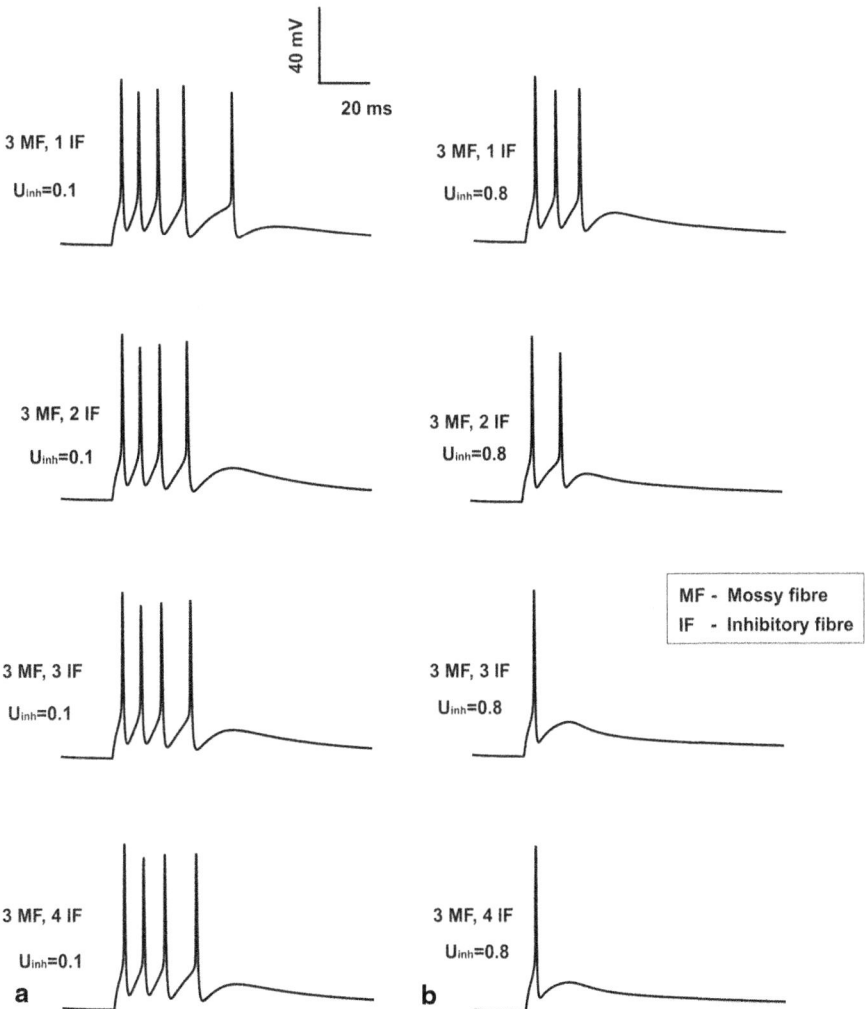

Fig. 8.3 Feed forward inhibition and granule cell. **a** Granule cell having with low inhibition (low release probability on inhibitory synapses). **b** Granule cell firing with high inhibition (high release probability on inhibitory synapses). For comparison sake, the number of MF inputs is set to three and the release probability of excitatory synapses was set at control (0.416) value. One can note from *top* (1 inhibitory fiber input) to *bottom* (4 inhibitory inputs) varying inhibitory synaptic inputs are applied

When release probabilities of the excitatory synapses were increased above 0.416, number of spikes increased, spike amplitude increased and the first spike latency decreased. A decrease in number of spikes was observed when the release probability of excitatory synapse was decreased below 0.416. First spike latency decreased and amplitude increases in all cases (see Fig. 8.6a). The same protocol was repeated with other activation patterns (data not shown) comprising combination of

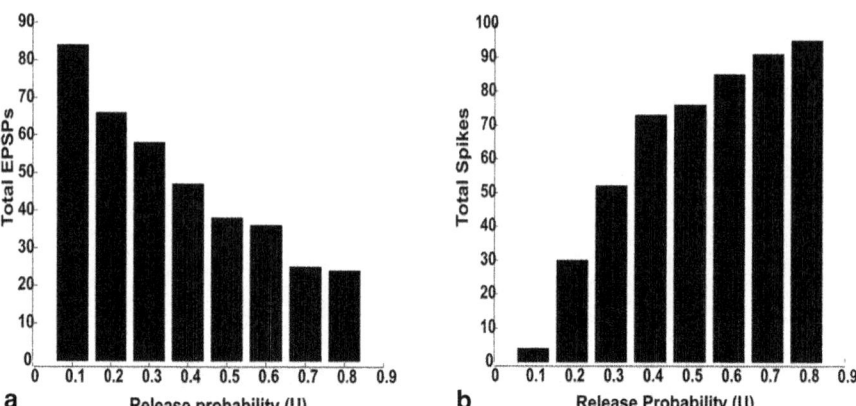

Fig. 8.4 Total EPSPs **a** and spikes, **b** generated while varying the MF release probability. All spikes and EPSPs for all possible combinations of excitatory (E) and inhibitory (I) synaptic inputs such I0E1 to I4E4 were summed to estimate total spikes. *In vivo* like inputs—bursts along MF were simulated. Input bursts favored spikes due to temporal summation of inputs facilitated by NMDA channels

excitatory and inhibitory synapses. *In vitro* model behavior with increased release probability of excitatory synapse was associated with decreased first spike latency, increase in amplitude and increase in number of spikes.

GrCs tend to discharge bursts *in vivo* [60]. With input combination of 4MF synapses and 2 inhibitory synapses, *in vivo* model showed 5 spikes, the timing and amplitude of initial spike was 15.55 mV and 23.15 ms respectively under control condition (U = 0.416). The release probability of MF synapse was first reduced and then increased from the control condition to study how the changes in release probability of MF synapse affects granule cell firing *in vivo* (Fig. 8.5b). At U = 0.1 release probability the GrC model showed 2 spikes with initial spike amplitude and timing as 9.28 mV and 31.775 ms respectively. Release probability of the excitatory synapse was increased to 0.3, 5 spikes were observed, amplitude and first spike latency measured 14.82 mV and 24 ms respectively. With increase in release probability above 0.416 (control), number of spikes remained unchanged (at 5 spikes), but small changes in spike amplitude and spike timing was observed (see Table II in [61]). *In vivo* model with the reduction in release probability of MF synapses below 0.416 (control) was associated with decrease in number of spikes, decrease in spike amplitude and increase in first spike latency. When the release probability was increased above the control (U = 0.416), number of spikes remained unchanged and small changes in spike amplitude and spike latency were observed.

Varying MF release probabilities and measuring the EPSP-spike response for the different excitatory- inhibitory input combinations *in vivo* involving an increased MF

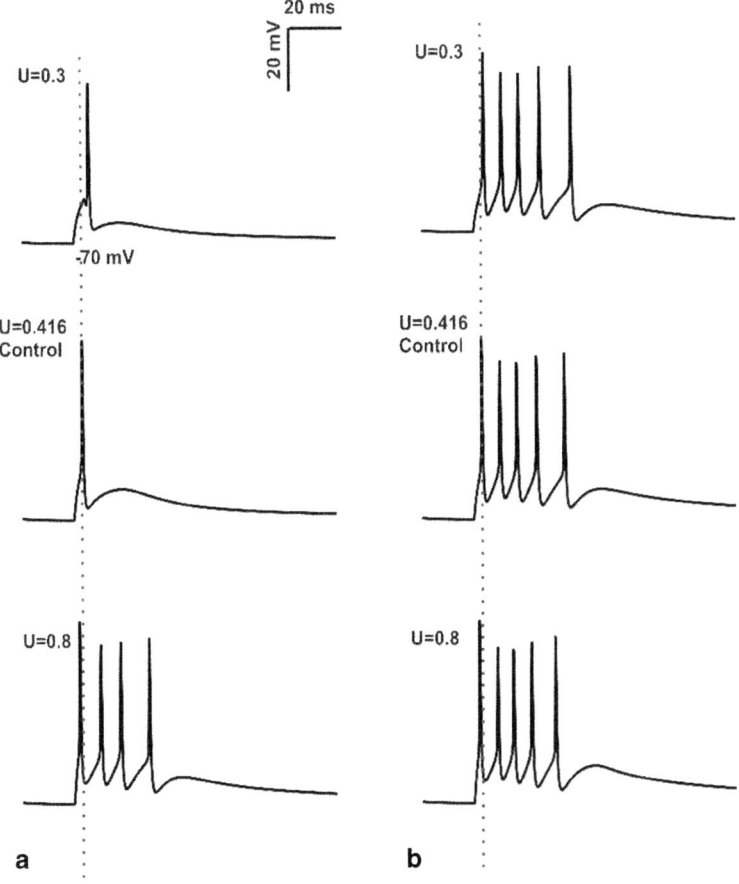

Fig. 8.5 Changes in Mossy Fiber release probability. **a** Spike response obtained using *in vitro* like input through 4 MF synapses, 2 inhibitory synapses of granule cell with varying release probability. First spike latency was reduced (see the *grey dotted line*) and number of spikes was increased by the increase in release probability. **b** 4 MF synapses, 2 inhibitory synapses activation of granule cell *in vivo* with changes in release probability (U). At U=0.3, change in initial spike delay and decrease in number of spikes was clearly observed. During control condition (U=0.416), increase in number of spikes and shortening of first spike latency was seen (the *grey dotted line* passing through the center of first spike). When the release probability of excitatory synapses was increased to 0.8, number of spikes remains constant, small change in initial spike latency was seen

release probability decreased the number of EPSPs and increased spiking (Fig. 8.4a, b), as expected. The percentage of EPSPs generated during LTD was 67.5% while during LTP was 32.5%. The percentage of spikes produced during LTP was 31.4% while during LTD was 68.6%. The increase—decrease pattern of EPSP–spike coupling was found to follow a seemingly symmetric behavior. Release probability change affected GrC firing only if a favorable excitatory—inhibitory balance was maintained in the cell (data not shown). The simulations suggest that in vivo, spikes

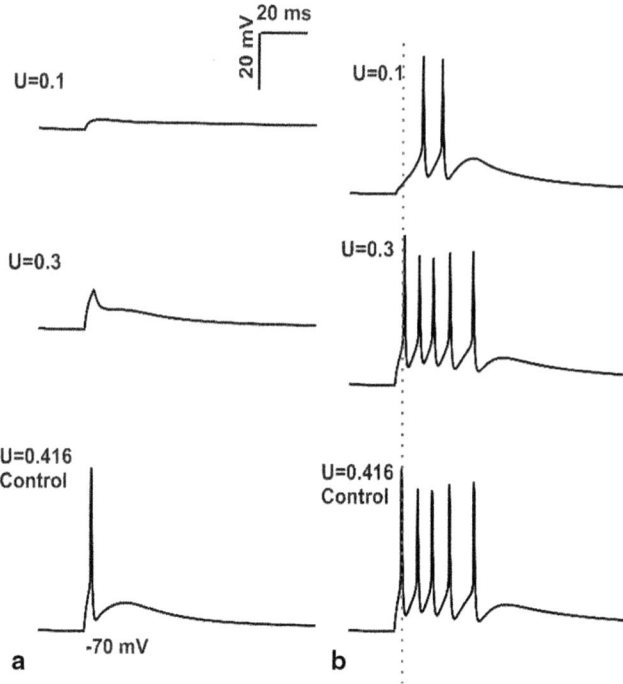

Fig. 8.6 Simulations of changes in mossy fiber release probability and intrinsic excitability. **a** *In vitro* behavior of granule cell model with low release probabilities of MF synapses **b** *In vivo* behavior of granule cell model during low release probability of mossy fiber. First spike latency was increased when there was a reduction in MF synapse release probability coupled with low intrinsic excitability of granule cell (see *grey dotted line* passing through control). It can be noted that the decrease in release probability of MF synapses was associated with decrease in number of spikes

were more common than EPSPs. A full estimate of the EPSP-Spike counts during control, LTP and LTD is shown in Fig. 8.7. The contributions with MF release probabilities between EPSPs and spikes in the EPSP-spike complex were not linear (as indicated by total numbers in Fig. 8.4).

A study on the effects of varying of inhibitory release probabilities on granule neuron firing was performed (see Fig. 8.3). Spike amplitude was changed significantly when the release probability of inhibitory synapse was increased from 0.1 to 0.8, but the number of spikes and timing was preserved. Spike amplitude was decreased for both increases in number of inhibitory inputs as well as the release probability of inhibitory synapses during *in vitro* simulations. *In vivo* behavior of granule cell model indicated the increase or decrease in release probability of inhibitory synapse did not affect the timing of granule cell firing due to delayed inhibition. However, decreased release probability of inhibitory synapses resulted in increased number of spikes, but there was no effect on amplitude or latency.

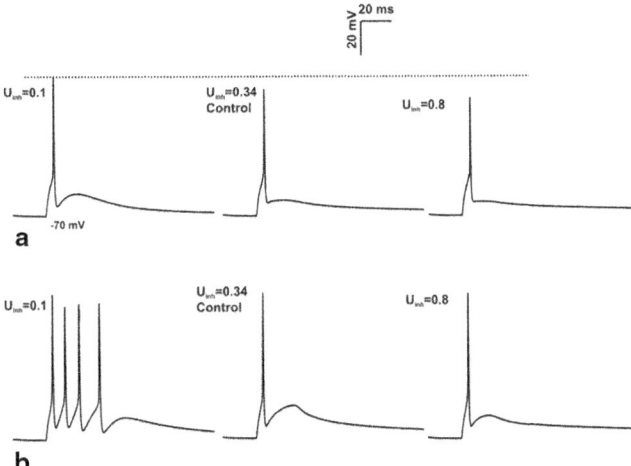

Fig. 8.7 Effect of inhibition on spike amplitude. **a** 4 inhibitory synapses, 3 excitatory synapses activation of granule cell with varying the release probability of inhibitory synapse for single spike input. Increase in the release probability of inhibitory synapse cause the reduction in spike amplitude (see the *grey dotted line*). **b** Response produced by burst-like input through 4 inhibitory synapses, 3 excitatory synapse in granule cell. *In vivo* behavior showed significant changes not in amplitude but in number of spikes

Predicting the Effects of LTD

Plasticity in MF-GrC has been known to be bidirectional [23]. The expression of LTD was associated with decrease in release probability of the excitatory synapses, thereby showing directionally opposite characteristics of LTP [27]. LTD arises due to weak, asynchronous, sporadic activity in pre and postsynaptic neurons [54]. To test effects of LTD induction, intrinsic excitability of the model was reduced and the release probability of MF synapses was reduced below 0.416(control) and its effect in granule cell firing was observed for various synaptic activation patterns. The release probability of inhibitory synapse was kept constant ($U_{inh} = 0.34$).

With low release probability of MF synapses ($U = 0.1$) *in vitro* granule cell model with low intrinsic excitability did not produce spikes (see Fig. 8.6a). In the presence of low levels of inhibition (0 or 1 inhibitory synapses), and during co-activation of 4 MF synapses, the model produced single spike at $U = 0.3$ release probability. The effect of low release probability of MF synapses and low intrinsic excitability of *in vivo* granule cell was simulated and studied.

When the release probability of MF synapse was increased to 0.3, ($U = 0.3$) number of spikes was increased to 5, first spike latency was decreased while amplitude was increased (Fig. 6b, also see Table V in [61]). *In vivo* like inputs to granule cell model under control condition produced 5 spikes and the initial spike measured 15.55 mV at 23.125 ms.

Low intrinsic excitability of granule cell affected spike amplitude and first spike latency, but it did not show significant effect on number of spikes (see [61]). Decrease in release probability of MF synapses affects spike latency, number of spikes and spike amplitude. MF-GrC LTD was associated with decrease in spike amplitude and increase in first spike latency.

Selective Inhibition and Spiking During Plasticity

Inhibition had modulatory effect on granule cell spiking. Spike amplitude was changed significantly when the release probability of inhibitory synapse was increased from 0.1 to 0.8, even though number of spikes and timing was preserved (see Table VI in [61]). With increasing release probability in the inhibitory synapses, spike amplitude decreased. This was observed for various combinations of inhibitory and excitatory synapses.

For *in vivo* like inputs, amplitude and latency of the first spike remained unchanged (see Table VII in [61]). Golgi cells regulate the induction of long-term synaptic plasticity at the mossy fiber–granule cell synapse. The main output from Golgi cells to granule neurons is GABAergic and inhibits the granule cells [50]. The spike amplitude and first spike latency remained constant for all release probabilities of inhibitory synapses for any particular combination of excitatory-inhibitory synapses. Inhibition did not affect first spike latency or spike amplitude during burst like *in vivo* inputs along MF (see Fig. 7b). For single spike (as seen *in vitro*) inputs along excitatory and inhibitory synapses, varying inhibition (both number of active inhibitory synapses and varying inhibitory release probabilities) regulated spike amplitude (see Fig. 7a).

The simple spiking network models were used to abstract rate coding information and produced similar results (data not shown).

Local Field Potential Reveals Role of Single Neuronal Functions in Populations

Population signals of neural circuits reveal emergent behavior as patterns of information flow in the underlying neural circuits [12, 36, 62]. The role of precise timing between intracellular events in a single neuron and its encoding within population signals such as local field potentials has been explored with attention to understand the rate code correlation between neurons in such signals [63]. This attention to spike encoding is also because of information being encoded in noise [36, 64]. Local field potentials (LFPs) are recorded as waveforms of extracellular activity that arise from complex interactions of spatial distribution of current sources, temporal dynamics, spatial distribution of dipoles and underlying conductive properties of the extracellular medium [65, 66].

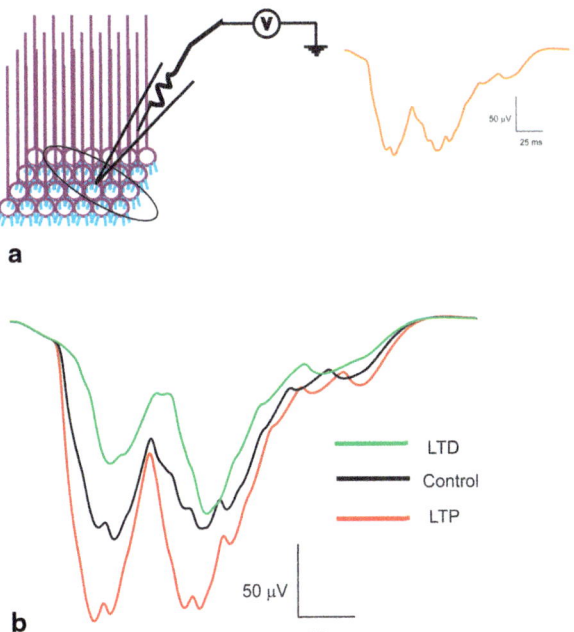

Fig. 8.8 Local Field Potential reconstructions. **a** Schematic of a LFP being recorded from a population of granule neurons. *Orange* trace on the *right* shows T-C wave generated as post-synaptic LFP for *in vivo*-like burst inputs. Two bursts through the granule neurons generate the T (short burst input) and C wave (long burst). **b** Induced plasticity reconstructions at the LFP level [36]. With induced plasticity, amplitude and lag varied compared to the control. Amplitude increased and lag decreased for LTP while lag increased and amplitude decreased for LTD

Local field potentials from experimental data are known to reflect the neural activity of the cells in the vicinity [12, 36, 67]. Granular layer LFP generated by tactile stimulation [23] was reproduced in the network model (see Fig. 8.8a) by giving short and long pulses through mossy fiber bundles [60, 68, 69]. The in vivo LFP was composed of 'T' and 'C' waves. The Trigeminal (T) wave corresponds to afferent inputs and the following cortical (C) wave is generated from cerebral cortex and pontine nuclei. In our network, the 'T' wave was generated by applying 5 spikes at 500 Hz along the mossy fiber synapse at 20 ms, and the 'C' wave was produced by 9 spikes at 500 Hz via mossy fiber synapse at 60 ms. The model was able to reconstruct population responses from the network model (see Fig. 8.8).

Spike-time dependent plasticity inductions were simulated by altering the intrinsic excitability and the synaptic release probability as mentioned in methods of [12, 36]. During induced LTP, simulations showed the LFP wave [23, 36] with an increased amplitude and width for both T and C waves compared to the control conditions (see Fig. 8.8b). With induced LTD–like conditions, the T and C wave of LFP showed a decreased amplitude and width compared to control (see Fig. 8.8b). Although modification of single neuron properties affected the spike propagation to continued inputs, varying such properties affected local field signals. The amplitude and lag variations of reconstructed post-synaptic LFP seems to be affected by the excitation pattern, synaptic variations and nature of stimulus [36].

Spatio-Temporal Dynamics are Characteristic of Network Geometry

Studies resolving single cell activity in multiple neuron responses have been indicating granule cells produced various spikes reflecting the variability of neurotransmission process [70, 71]. The pattern of excitation and role of inhibition elicited by delivering a single stimulus to white matter was reported using maps generated by voltage sensitive dye-based imaging [10]. Granular layer responses was regulated by inhibition [72]. To test the frequency-dependent transmission in granular layer, we reconstructed the map using 730 granule cells, 2 Golgi cells, 40 mossy fiber rosettes to pack 35 µm^3 of cerebellar cortex (see Fig. 8.9).

Golgi cells could control both temporal dynamics and the spatial distribution of information transmitted through the cerebellar granular layer network [73]. The strength of the inhibition depends on the number of inhibitory connections and

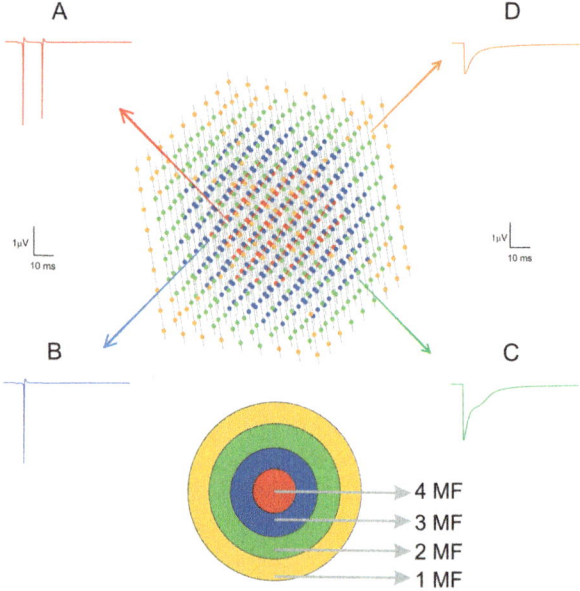

Fig. 8.9 Spatial geometry-based fluctuations of LFP. The circuit model is shown in 3D in the center. The colors show the excitation inputs to the circuit model. All cells were excited with *in vitro* like inputs (single spikes through MFs). Network model was simulated with center-surround "spot" activation. Dots (*red, blue, green, orange*) represent the soma of cells in the network. Type-1 cells (5% of total cells) located in the center of the network received *4 MF* excitation (shown in *red*), Type-2 cells (45% of total cells) in the network received *3 MF* excitation (shown in *blue*), Type-3 cells (35% of total cells) in the network received *2 MF* excitation (shown in *green*) and Type-4 cells (15% of total cells) in the network received *1 MF* excitation (shown in *orange*). Cells which received *4 MF* inputs (core 1) reproduced a doublet **a**. Cells that received *3* and *2 MFs* showed single spike. **b, c**. Cells with *1 MF* showed an EPSP as a response **d**. **a, b, c, d** are extracellular responses

synaptic release probability. The dynamics of the granule cells–Golgi cell circuit were explained by the simultaneous activation of both neurons through the mossy fibers, followed by activation of the feed-forward and feed-back inhibitory loops [72, 74]. Local field reconstructions also suggested the modulatory role of inhibitory inputs on the activities of granule cells (See Fig. 8.9) via the change in potential shapes of N_{2a} and N_{2b} waves in the network. Reconstructing extracellular properties indicated plasticity has similar mechanisms of regulation of granule cell burst initiation and thereby could implement an adaptable delay affecting downstream activation into circuitry.

To study the role of single compartments of the neuron in the contributing network, it was essential to know the spatio-temporal properties of the single cell and its relation to the network LFP. The electrical potential generated at different parts of the neuron mainly depended on the morphology of the compartment, ion channel density and electrotonic compactness of the neuron. Reconstructing by calculating the extracellular potential generated from compartments of cerebellar granule cell model [75]. The axonal initial segment showed the most contribution of all compartments to the single neuron LFP. Although, spiking and post-synaptic potentials were influenced by the somato-dendritic compartment, the amplitude of extracellular potentials was most emphasized in the axonal initial segment. Somato-dendritic compartments governed the time-course of the extracellular potential. The spatial geometry was reproduced using simple spiking models (described in Methods) and results suggest such models are reliable to study spike-time and temporal activity of fast microcircuits [48].

Spatial Reach of Granule Layer LFP

Since the nature of LFP is known to arise from the nature of circuits, it was primarily known to reflect the synaptic activity of the neural populations near the recording electrode [36, 67, 76]. Here the term, spatial reach is the region of few micrometers in a volume of conducting tissue where the signal can be measured. In order to know whether electrotonically compact neurons show variability of extracellular current with reference to spatial changes of point of recording, we simulated extracellular field around a single granule neuron. Simulations validated the decrease of amplitude and width of the extracellular wave form when the recording electrode point was moved away from the somato-dendritic compartments in the neuron (see Fig. 8.10a). Although large neurons show significant differences in shape and amplitude of the extracellular potential estimated [67, 77], cerebellum granular layer showed similarity due to compactness of the somato-dendritic compartments (see Fig. 8.10a).

We reconstructed the spatial reach of LFP by increasing the distance of recording from cells. When the recording electrode was moved away from the soma, the simulated extracellular wave amplitude decreased exponentially at an average rate of -0.011 mV (Fig. 8.10b). This is significant for the granular layer since the

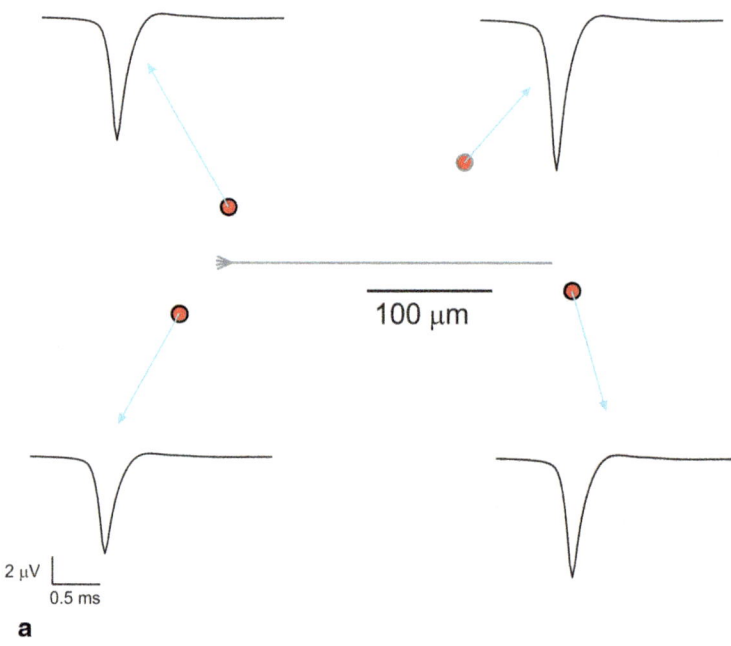

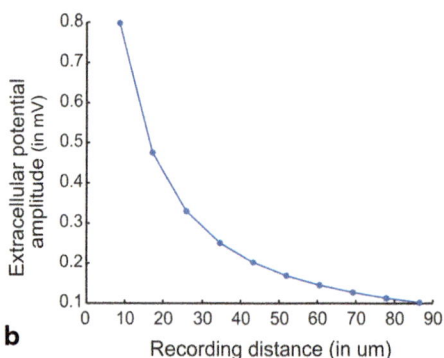

Fig. 8.10 Attenuation of LFP signal. **a** Single neuron extracellular potential reconstruction shows the cerebellar granule cell model [6] and reconstructed extracellular potential at different points around the neuron. The simulated recording calculated away from soma showed decreases in amplitude and width of the wave. **b** Spatial spread of LFP. Plot shows attenuation of post-synaptic LFP at several recording points with increasing distance from the soma of the granule neuron. Recording distance from the soma was plotted on the x-axis whereas amplitude of the simulated extracellular potential at each point on y-axis

neurons are small, electrotonically compact [78] and do not have extracellular field components as larger neurons that have qualitative distributions due to synaptic distribution along the neuronal morphology [67]. The granular layer post-synaptic LFP seemed to be dependent on the position of recording electrode due to the sink-source dipole effect observed in the granular layer [36]. Additionally, these attenuation studies seem relevant to indicate that the underlying mechanisms of induced plasticity acting within clusters are similar for the mossy fiber-granule neuron pathway. From an electrode-centric point of view, both trigeminal (T wave) and thalamo-cortico-pontine pathways (C wave) reconstructed use the same molecular mechanisms for generating the extracellular potential, although one may suggest additional plasticity may affect the C wave in the in vivo-like behavior due to circuit properties.

Mutual Information and Entropy

For validating the geometry of our excitation patterns, we used Shannon's mutual information to extract average amount of information from granule neuron responses and then use the values to map the quantified information in the pattern (see Fig. 8.11). The capability of afferent system to process signals and transmit meaningful accounts of their inputs to downstream neural stages was calibrated. The goal in our study was to estimate the role of spike-EPSP complex that was contributing to the LFP population signal. The spatial geometry of the cells were mapped with a center-surround organization via a color-map based on mutual info values (see Fig. 8.11). In this, the maximum value quantifying 16,000 responses were used to estimate the number of bits. Our values for single granule neurons were similar to [31].

VSD-based recordings had shown similar spot activations [71]. The geometry of colored intensive spots matches with observations seen using laser two-photon techniques [70]. The reconstruction helped quantify the sparse activation of granule neurons allowing the averaging nature of reconstructed local field potentials in such spots to the characteristics such as shape, amplitude and lag [23, 36].

Modeling Neuronal Dysfunction

Modeling neuronal dysfunctions allows study of cellular and molecular alterations and thereby help predict functional properties and validate hypotheses. In this section, we introduce two cases of molecular-level alterations and their cellular altercations based on mathematical modeling.

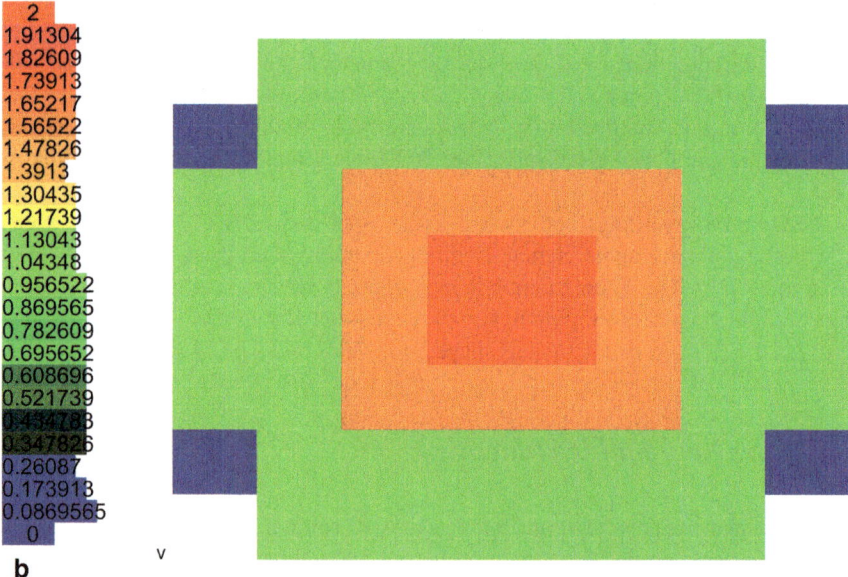

Fig. 8.11 Averaged mutual information for center-surround pattern. **a** Simulated mutual information map representation for *in vitro*- like inputs. **b** Simulated mutual information map representation for *in vivo*- like inputs. In both cases, the number of bits was averaged from calculations based on 16,000 responses of a single granule neuron. The map was made by assigning the estimated value to the neuron with the particular number of inputs. Bursting responses via *in vivo*-like inputs generated higher mutual info values and hence the seemingly additional regions in the color map (**b**). The color map of excited neurons show a geometry like 'spot' observations seen using laser two-photon techniques [70]

Fhf$^{1-/-}$ Fhf4$^{-/-}$ Knock-Out Mutation and Ataxia-Like Behavior

Severe ataxia-like conditions have been shown by Fhf$^{1-/-}$ Fhf4$^{-/-}$ knock-out mice among other neuronal deficits [35]. While wild-type mice slice recordings showed repetitive firing, the mutant mice fired only once and that was at an elevated voltage spike threshold. Fibroblast growth factor homologous factors (FHFs) are a family of proteins, which constitute/support voltage-gated sodium channel to recover from the inactivated state in repeated firing. Any mutation to this protein causes the neuron to fail, to fire an action potential completely or partially by slowdown the recovery period of their sodium channels which causes the neuron to fire only at higher threshold [35]. Behavior-wise the translation of these effects was observed in mice and the mutant animals showed inability to retain their gait and balance on narrow edges while wild-type mice could do it easily. Our previous study [35] involved the modeling of sodium channel modifications as seen in intrinsic excitability of wild type and mutant animals. Indeed, the changes in model at the sodium channel activation and inactivation parameters produced depressed firing dynamics as seen in Fhf$^{1-/-}$ Fhf4$^{-/-}$ knock-out mice [35]. Mathematical modeling suggested change of rate constants during inactivation and activation reproduced the firing behavior seen in mutant models.

To test the reliability of LFP on spike information, we reconstructed the nature of Fhf$^{1-/-}$ Fhf4$^{-/-}$ mutations in LFP signals (see Fig. 8.12). Circuit properties may be significantly limited since spike shapes were altered. With spikes disabled at normal spike threshold levels, excitation only generated a stunted action potential (Fig. 8.12a). At the single cell extracellular level, single neuron extracellular potential was considerably reduced (Fig. 8.12b). Modeling the LFP generated from 700 cells *in vitro* was consequential of the diminished extracellular potential. The local field potential (Fig. 8.12d) lacked intactness of the wave components unlike control (see Fig. 8.8b) also on the in vivo-like behavior. Although the single neuron field potential components were reconstructed from ion channel behavior and electrotonic morphology of individual neurons, the varied channel distribution in compartments attributed to nonlinearity in generated local field potential components. This also suggest that local field potential may be used as an indicator in cases where cognitive deficits may be seen without actual cell death.

Modeling NMDAR Related Vestibulo-Cerebellar Learning Problems

Blocking NMDA receptors [52] in granule neurons showed reduced excitation. Selective disabling of NMDA receptors is reported to be seen during NR2A/NR2B mutations [52]. In order to predict on the nature of such mutations affecting network computation (in addition to affecting the number of spikes), we randomly disabled (26, 36% of total cells) NMDA receptors in the network and reconstructed the local field response. Simulations with in vitro-like inputs showed decreased number of spikes is seen via a change in N_{2a} amplitude compared to control [48] even with a smaller percentage of cells with such mutations.

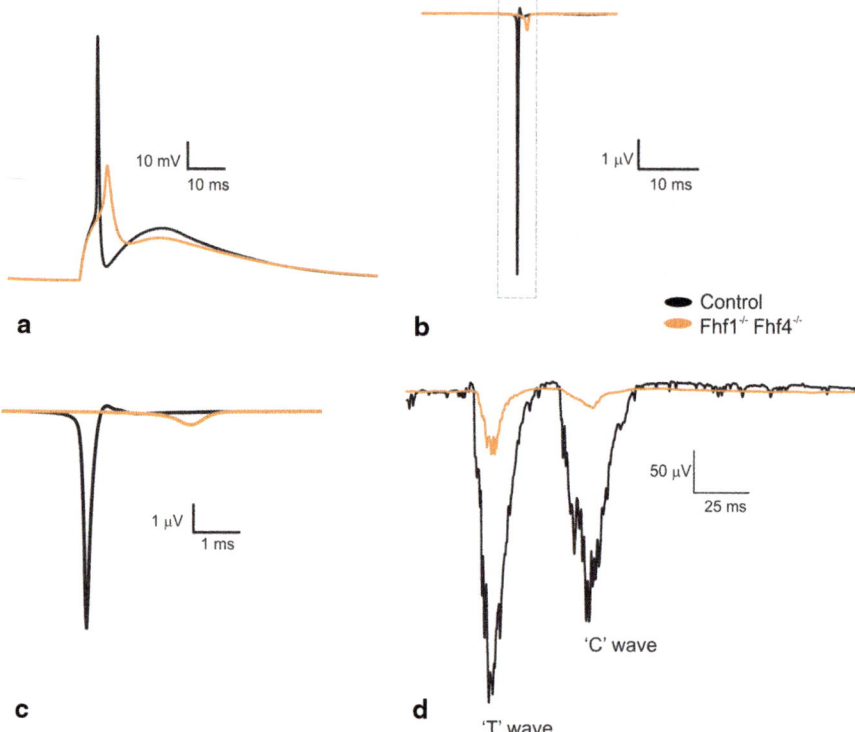

Fig. 8.12 Modeling ataxia like FHF mutations in LFP. **a** Action potential in control or wildtype *(black)* and FHF mice *(orange)*. FHF mice were modeled with changes to sodium channel-related intrinsic electroresponsiveness [35]. **b** Extracellular response to 3 MF inputs for control *(black)* and mutant models *(orange)*. **c** LFP for *in vitro* like inputs. Note the almost flat trace in FHF mice compared to the N_{2a} and N_{2b} waves for the control [36, 72]. **d** LFP for *in vivo* like inputs. Since bursts are usually stronger excitation than spikes, there is a small deflection in the FHF LFP trace *(orange)* compared to control which shows T and C waves [12, 23, 36]

Disabling NMDA receptors in 26 (Fig. 8.13c)–36% (Fig. 8.13d) of cells showed a significant decrease in N_{2b} wave with *in vitro* like inputs. LFP reconstruction from the model clearly showed a 'seemingly linear' outlook in propagating the nonlinearities of individual neurons in population code (evoked LFP). NMDA knockout mice show errors in cerebellar motor learning [52]. Induced plasticity changes were also reflected in the evoked LFP waves.

Comparing this with ataxia like condition, while ataxia like conditions involving spiking neurons show a single peak (Fig. 8.13a) with a reduction in amplitude (compared to control), populations with selectively-disabled NMDA receptors generated both peaks (Fig. 8.13c, d) although N_{2B} peak was reduced [48]. For both simulations, we used the same network with similar excitation patterns. LFP reconstructions indicated the role of molecular mechanisms of induced plasticity were similar

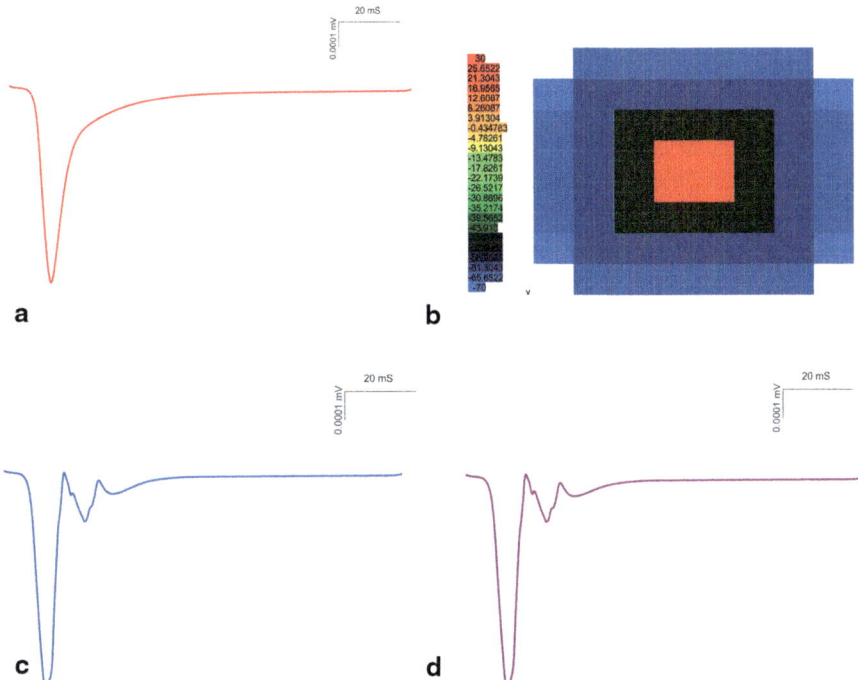

Fig. 8.13 Modeling NMDAR- related dysfunctions in LFP. **a** LFP of a neuronal population modeled with Ataxia-like symptoms by changing sodium channel activation and inactivation properties [35]. Notice the shape and amplitude. **b** Activation pattern used in network model with a percentage of cells with disabled NMDA channels to simulate NR2A—related dysfunctions. Conductance of NMDA receptor was set to zero in the model for a percentage of cells shown in (**b**). **c** LFP response with *in vitro* like inputs with 26% cells *(blue trace)* having NMDA conductance set to zero. **d** LFP response with *in vitro* like inputs with 36% cells *(purple trace)* having NMDA conductance set to zero. Note that in **c** and **d**, LFP trace shows no significant changes in both waves. However in comparison to control N_{2b} wave is reduced in amplitude

and regulated of granule cell burst initiation and could implement an adaptable delay affecting downstream signaling [79].

Discussion

This chapter uses an eclectic set of mathematical tools to extensively analyze information transmission in granule neuron using detailed biophysical models. We used detailed multi-compartmental models, simple spiking models to implement the neuronal behavior of cerebellar input layer neurons and to reconstruct the roles and mechanisms in network operations. The input-output parameter space analysis beyond current experimental techniques has been explored. Part of neural information is transmitted as rate, part as time precision and that high correlation among the MF inputs was characteristic of most informative stimuli. Importantly, the transmitted information was regulated by inhibition especially during induction of LTP and LTD.

The molecular and cellular parameters analyzed eventually help us to quantify the effects of LTP and LTD in the output firing response of GrC [54, 80]. LTP or LTD can alter several functional aspects of GrC firing. In this attempt, we showed that specific temporal dynamics and geometry of connections eventually determine the circuit output [81]. We propose that the topology of connections play a critical role in the expressions of long-term plasticity in the granule cells.

The simulations indicated that the first spike latency is modulated with the change in intrinsic excitability and release probability of the excitatory synapses. Mossy fibers stimuli train determined EPSP temporal summation and spike activation in the granule cell. Simulations also supported the hypothesis [82] that EPSP temporal summation is critical for reaching spike threshold but, once firing begins, it is efficiently regulated by postsynaptic ionic conductance.

The results based on single neuron simulations indicate bidirectional plasticity is regulated differently by spatial inhibition. While EPSPs are less favored during LTP in vitro, LTD favors generation of EPSPs. The sensitivity of initial EPSP timing towards GoC inhibition was clearly demarked during LTD. The modulation role of Golgi cell (GoC) inhibition was supported by both first spike and EPSP latency which could provide a clue to information encoding in the cerebellar granule layer.

Granule neuron spiking increases with release probability changes along excitatory synapses. EPSP-Spike ratio changes for both *in vitro* and *in vivo* like inputs. *In vitro* inputs favor EPSPs while *in vivo* inputs show more changes in spike count. This behavioral difference may be due to temporal summation of EPSPs leading to generation of spikes. Simulations therefore estimates the contributions of EPSP-spike complexes during information coding. By favoring selected combinations of mossy fiber inputs, MF-GrC LTP would then improve pattern recognition, the primary function attributed to the cerebellar MF-GrC relay.

The granular layer network model is significant as spatio-temporal patterns in the cerebellar granular layer has been not well studied. Center-surround activation of granule cells have been recently shown experimentally [53, 83] and reconstructions have critical significance on downstream encoding of cerebellar inputs [55, 84]. Mutual information based maps seems to suggest VSD-like readouts may have quanta correlations to synaptic activity-elicited behavior. Reconstruction of such clusters allows extending the time-window hypothesis [16] and theories related to the sparse coding in the cerebellum [79]. Granular layer with a gain regulation mechanism modulated by tonic inhibition, implements an enhanced transmission of spike bursts as suggested by the model-based LFP reconstructions.

Modeling of neuronal dysfunction as in the case of FHF mice and in the case NR2A changes elicits highlight on two important mechanistic explanations into disorders. Ataxia-like conditions, induced by intrinsic excitability changes, disable spiking or bursts and thereby limit the quanta of downstream information. The shape of LFP reconstructions showed defects in shape, amplitude and lag. However,

NR2A-based NMDAR dysfunctions show temporal summation issues and thereby lose network information in the second wave (N_{2B}). Such knock-out behaviors are observed during some types of autism although the exact role of the effects on spatio-temporal processing remains yet to be ascertained [85]. Cell death in some cases of ataxia-telangiectasia may evoke such selective loss of information-transmission mechanisms [86].

The granular layer simulations has its due importance as it explores how cerebellar granule neurons store and process information in terms of EPSP-spike combinations [60] and the role of plasticity indicating why a large number of granule cells are present. Such a study is potentially an important estimation of cerebellar function in encoding sensory and tactile inputs. Theoretical network models predict that information in the MF-GC relay is a relevant parameter that could optimize cerebellar performance for certain tasks and under appropriate learning rules [20, 87]. Hence, plasticity at this relay may be an important element of tremendous storage capacity in the learning of coordination of actions, sensorimotor or cognitive, in which the cerebellum participates.

Given the pervasiveness of temporal information in external stimuli and the generality of the time-dependent mechanisms, all temporal responses needed for forming coarse temporal code can be stipulated. This, in turn, will help estimate overall behavior in firing in the underlying granular layer network. Also the role of selective inhibition in granular layer suggests mechanisms of coincidence detection and spatial pattern separation as described in Motor learning theory [3].

Conclusion

Several aspects of cerebellar information processing remains yet to be explored. While experimental evidence is being sought, mathematical reconstructions suggest significant roles of operation and mechanisms affecting cellular and network functions. Geometry based reconstructions suggest center-surround and specific activation properties rather than 'sparse' estimations as suggested in Marr-Albus theories [3, 88]. This study illustrates some of the complex dynamics shown by the granular layer cells in reconstructing network activity. From a broad perspective, it may very well be possible, while granule cells form the input layer, Golgi cells a hidden layer [45] and Purkinje neurons the role of perceptron in a gain-adaptation circuit for fine tuning movement precision.

Acknowledgments This work derives direction and ideas from the Chancellor of Amrita University, Sri Mata Amritanandamayi Devi. SD would like to acknowledge Egidio D'Angelo and Sergio Solinas of University of Pavia, Thierry Nieus of IIT Genova and Bipin Nair, Krishnashree Achuthan, Harilal Parasuram, Chaitanya Medini, Nidheesh Melethadathil, Manjusha Nair, Asha Vijayan, Afila Yoosef, Chaitanya Kumar, Sandeep Bodda of Amrita University for their work and support in making this manuscript. This work is supported by Grants SR/CSI/49/2010, SR/CSI/60/2011 and Indo-Italy POC 2012–2014 from DST and BT/PR5142/MED/30/764/2012 from DBT, Government of India.

References

1. Ito M. Mechanisms of motor learning in the cerebellum. Brain Res. 2000;886(1–2):237–45.
2. Tyrrell T, Willshaw D. Cerebellar cortex: its simulation and the relevance of Marr's theory. Philos Trans R Soc Lond B Biol Sci [Internet]. 1992;336(1277):239–57. http://www.ncbi.nlm.nih.gov/pubmed/1353267.
3. Albus J. A theory of cerebellar function. Math Biosci. 1971;10:25–61.
4. D'Angelo E, Koekkoek SKE, Lombardo P, Solinas S, Ros E, Garrido J, Schonewille M, De Zeeuw CI. Timing in the cerebellum: oscillations and resonance in the granular layer. Neuroscience [Internet]. 2009;162(3):805–15. http://www.ncbi.nlm.nih.gov/pubmed/19409229.
5. Raymond JL, Lisberger SG, Mauk MD. The cerebellum: a neuronal learning machine? Science [Internet]. 1996;272(5265):1126–31. http://www.ncbi.nlm.nih.gov/pubmed/8638157.
6. Nemenman I, Bialek W, de Ruyter van Steveninck R. Entropy and information in neural spike trains: progress on the sampling problem. Phys Rev E [Internet]. 2004;69(5):1–6. http://link.aps.org/doi/10.1103/PhysRevE.69.056111.
7. Bialek W, Rieke F, de Ruyter van Steveninck R, Warland D. Reading a neural code. Science (80-) [Internet]. 1991;252(5014):1854–7. http://www.sciencemag.org/content/252/5014/1854.short.
8. Strong S, Koberle R, de Ruyter van Steveninck R, Bialek W. Entropy and information in neural spike trains. Phys Rev Lett [Internet]. 1998;80(1):197–200. http://link.aps.org/doi/10.1103/PhysRevLett.80.197.
9. Shew WL, Yang H, Yu S, Roy R, Plenz D. Information capacity and transmission are maximized in balanced cortical networks with neuronal avalanches, J. Neurosci. 31;2011. pp. 55–63.
10. Mapelli J, Gandolfi D, D'Angelo E. High-pass filtering and dynamic gain regulation enhance vertical bursts transmission along the mossy fiber pathway of cerebellum. Front Cell Neurosci [Internet]. 2010;4:14. http://www.pubmedcentral.nih.gov/articlerender.fcgi?artid=2889686&tool=pmcentrez&rendertype=abstract.
11. D'Angelo E. Neural circuits of the cerebellum: hypothesis for function. J Integr Neurosci [Internet]. 2011;10(3):317–52. http://www.ncbi.nlm.nih.gov/pubmed/21960306.
12. Parasuram H, Nair B, Naldi G, Angelo ED, Diwakar S, D'Angelo E. A modeling based study on the origin and nature of evoked post-synaptic local field potentials in granular layer. J Physiol Paris [Internet]. Elsevier Ltd. 2011;105(1–3):71–82. http://www.ncbi.nlm.nih.gov/pubmed/21843640.
13. D'Angelo E, Mazzarello P, Prestori F, Mapelli J, Solinas S, Lombardo P, Cesana E, Gandolfi D, Congi L. The cerebellar network: from structure to function and dynamics. Brain Res Rev [Internet]. Elsevier B.V. 2011;66(1–2):5–15. http://www.ncbi.nlm.nih.gov/pubmed/20950649.
14. Cerminara NL, Apps R. Behavioural significance of cerebellar modules. Cerebellum. 2011;10:484–94.
15. Marr D. A theory of cerebellar cortex. J Physiol. 1969;202(2):437–70.
16. D'Angelo E, De Zeeuw CI. Timing and plasticity in the cerebellum: focus on the granular layer. Trends Neurosci. 2009;32(1):30–40.
17. Kandel ER, Schwartz JH, Jessell TM. Principles of Neural Science. New York: McGraw-Hill; 2000.
18. Ito M. The modifiable neuronal network of the cerebellum. Jpn J Physiol. 1984;34(5):781–92.
19. Purves D, Augustine GJ, Fitzpatrick D, Hall WC, LaMantia AS, McNamara JO Williams MS, Editors. Neuroscience, Third Edition. Sunderland: Sinauer Associates Inc; 2004.
20. Philipona D, O'Regan JK, Nadal J-P. Is there something out there? Inferring space from sensorimotor dependencies. Neural Comput [Internet]. 2003;15(9):2029–49. http://dx.doi.org/10.1162/089976603322297278.
21. Evans GJO. Synaptic signalling in cerebellar plasticity. Biol Cell. 2007;99:363–78.

22. D'Angelo E. The critical role of Golgi cells in regulating spatio-temporal integration and plasticity at the cerebellum input stage. Front Neurosci. 2008;2:35–46.
23. Roggeri L, Rivieccio B, Rossi P, D'Angelo E. Tactile stimulation evokes long-term synaptic plasticity in the granular layer of cerebellum. J Neurosci [Internet]. 2008;28(25):6354–9. http://www.ncbi.nlm.nih.gov/pubmed/18562605.
24. Nicoll RA. Expression mechanisms underlying long-term potentiation: a postsynaptic view. Philos Trans R Soc Lond B Biol Sci. 2003;358:721–6.
25. Nieus T, Sola E, Mapelli J, Saftenku E, Rossi P, D'Angelo E. LTP regulates burst initiation and frequency at mossy fiber-granule cell synapses of rat cerebellum: experimental observations and theoretical predictions. J Neurophysiol [Internet]. 2006;95(2):686–99. http://dx.doi.org/10.1152/jn.00696.2005.
26. Sola E, Prestori F, Rossi P, Taglietti V, D'Angelo E. Increased neurotransmitter release during long-term potentiation at mossy fibre-granule cell synapses in rat cerebellum. J Physiol [Internet]. 2004;557(Pt 3):843–61. http://dx.doi.org/10.1113/jphysiol.2003.060285.
27. D'Errico A, Prestori F, D'Angelo E. Differential induction of bidirectional long-term changes in neurotransmitter release by frequency-coded patterns at the cerebellar input. J Physiol [Internet]. 2009;587(Pt 24):5843–57. http://www.pubmedcentral.nih.gov/articlerender.fcgi?artid=2808544&tool=pmcentrez&rendertype=abstract.
28. Foffani G, Moxon KA. Studying the role of spike timing in ensembles of neurons. Conference Proceedings 2nd Int IEEE EMBS Conference Neural Engineering 2005. Arlington, VA: IEEE; 2005. pp. 206–8.
29. Zucker RS, Regehr WG. Short-term synaptic plasticity. Annu Rev Physiol [Internet]. 2002;64:355–405. http://www.ncbi.nlm.nih.gov/pubmed/11826273.
30. Zhang W, Linden DJ. The other side of the engram: experience-driven changes in neuronal intrinsic excitability. Nat Rev Neurosci. 2003;4:885–900.
31. Arleo A, Nieus T, Bezzi M, D'Errico A, D'Angelo E, Coenen OJ-MD. How synaptic release probability shapes neuronal transmission: information-theoretic analysis in a cerebellar granule cell. Neural Comput [Internet]. 2010;22(8):2031–58. http://www.mitpressjournals.org/doi/abs/10.1162/NECO_a_00006-Arleo.
32. D'Angelo E, Filippi G De, Rossi P, Taglietti V. Ionic mechanism of electroresponsiveness in cerebellar granule cells implicates the action of a persistent sodium current. J Neurophysiol [Internet]. 1998;80(2):493–503. http://jn.physiology.org/content/80/2/493.long.
33. Diwakar S, Magistretti J, Goldfarb M, Naldi G, D'Angelo E. Axonal na + channels ensure fast spike activation and back-propagation in cerebellar granule cells. J Neurophysiol [Internet]. 2009;101(2):519–32. http://dx.doi.org/10.1152/jn.90382.2008.
34. D'Angelo E, Nieus T, Maffei A, Armano S, Rossi P, Taglietti V, Fontana A, Naldi G. Theta-frequency bursting and resonance in cerebellar granule cells: experimental evidence and modeling of a slow k+ -dependent mechanism. J Neurosci. 2001;21(3):759–70.
35. Goldfarb M, Schoorlemmer J, Williams A, Diwakar S, Wang Q, Huang X, Giza J, Tchetchik D, Kelley K, Vega A, Matthews G, Rossi P, Ornitz DM, D'Angelo E. Fibroblast growth factor homologous factors control neuronal excitability through modulation of voltage-gated sodium channels. Neuron [Internet]. 2007;55(3):449–63. http://dx.doi.org/10.1016/j.neuron.2007.07.006.
36. Diwakar S, Lombardo P, Solinas S, Naldi G, D'Angelo E. Local field potential modeling predicts dense activation in cerebellar granule cells clusters under LTP and LTD control. PLoS One [Internet]. 2011;6(7):e21928. http://www.pubmedcentral.nih.gov/articlerender.fcgi?artid=3139583&tool=pmcentrez&rendertype=abstract.
37. D'Angelo E, Solinas S, Garrido J, Casellato C, Pedrocchi A, Mapelli J, Gandolfi D, Prestori F. Realistic modeling of neurons and networks: towards brain simulation. Funct Neurol [Internet]. 2013;28(3):153–66. http://www.pubmedcentral.nih.gov/articlerender.fcgi?artid=3812748&tool=pmcentrez&rendertype=abstract.
38. Ito M. Questions in modeling the cerebellum. J Theor Biol. 1982;99(1):81–6.
39. Eccles JC, Ito M, Szentagothai J. The cerebellum as a neuronal machine. Berlin: Springer-Verlag; 1967.

40. Hamori J, Somogyi J. Formation of new synaptic contacts by Purkinje axon collaterals in the granular layer of deafferented cerebellar cortex of adult rat. Acta Biol Hung. 1983;34(2–3):163–76.
41. Hamori J, Somogyi J. Differentiation of cerebellar mossy fiber synapses in the rat: a quantitative electron microscope study. J Comp Neurol [Internet]. 1983;220(4):365–77. http://dx.doi.org/10.1002/cne.902200402.
42. Vos BP, Maex R, Volny-Luraghi A, Schutter E De. Parallel fibers synchronize spontaneous activity in cerebellar Golgi cells. J Neurosci. 1999;19(11):RC6.
43. Solinas S, Forti L, Cesana E, Mapelli J, Schutter E De, D'Angelo E. Fast-reset of pacemaking and theta-frequency resonance patterns in cerebellar Golgi cells: simulations of their impact in vivo. Front Cell Neurosci. 2007;1:4.
44. Solinas S, Forti L, Cesana E, Mapelli J, Schutter E De, D'Angelo E. Computational reconstruction of pacemaking and intrinsic electroresponsiveness in cerebellar golgi cells. Front Cell Neurosci. 2007;1:2.
45. D'Angelo E, Solinas S, Mapelli J, Gandolfi D, Mapelli L, Prestori F. The cerebellar Golgi cell and spatiotemporal organization of granular layer activity. Front Neural Circuits [Internet]. 2013;7:93. http://www.pubmedcentral.nih.gov/articlerender.fcgi?artid=3656346&tool=pmcentrez&rendertype=abstract.
46. Forti L, Cesana E, Mapelli J, D'Angelo E. Ionic mechanisms of autorhythmic firing in rat cerebellar Golgi cells. J Physiol [Internet]. 2006;574(Pt 3):711–29. http://dx.doi.org/10.1113/jphysiol.2006.110858.
47. Hines ML, Carnevale NT. NEURON: a tool for neuroscientists. Neuroscientist. 2001;7(2):123–35.
48. Medini C, Nair B, D'Angelo E, Naldi G, Diwakar S. Modeling spike-train processing in the cerebellum granular layer and changes in plasticity reveal single neuron effects in neural ensembles. Comput Intell Neurosci [Internet]. 2012;2012:1–17. http://www.hindawi.com/journals/cin/2012/359529/.
49. Rossi P, Sola E, Taglietti V, Borchardt T, Steigerwald F, Utvik JK, Ottersen OP, Köhr G, D'Angelo E. NMDA receptor 2 (NR2) C-terminal control of NR open probability regulates synaptic transmission and plasticity at a cerebellar synapse. J Neurosci. 2002;22(22):9687–97.
50. Mapelli L, Rossi P, Nieus T, D'Angelo E. Tonic activation of GABAB receptors reduces release probability at inhibitory connections in the cerebellar glomerulus. J Neurophysiol [Internet]. 2009;101(6):3089–99. http://www.ncbi.nlm.nih.gov/pubmed/19339456.
51. Lorenz A, Deutschmann M, Ahlfeld J, Prix C, Koch A, Smits R, Fodde R, Kretzschmar HA, Schüller U. Severe alterations of cerebellar cortical development after constitutive activation of Wnt signaling in granule neuron precursors. Mol Cell Biol [Internet]. 2011;31(16):3326–38. http://mcb.asm.org/content/31/16/3326.long.
52. Andreescu CE, Prestori F, Brandalise F, D'Errico A, De Jeu MTG, Rossi P, Botta L, Kohr G, Perin P, D'Angelo E, De Zeeuw CI. NR2A subunit of the N-methyl D-aspartate receptors are required for potentiation at the mossy fiber to granule cell synapse and vestibulo-cerebellar motor learning. Neuroscience [Internet]. 2011;176:274–83. http://www.ncbi.nlm.nih.gov/pubmed/21185357.
53. Gandolfi D, Lombardo P, Mapelli J, Solinas S, D'Angelo E. Theta-frequency resonance at the cerebellum input stage improves spike timing on the millisecond time-scale. Front Neural Circuits [Internet]. Frontiers. 2013;7:64. http://www.pubmedcentral.nih.gov/articlerender.fcgi?artid=3622075&tool=pmcentrez&rendertype=abstract.
54. D'Angelo E, Rossi P, Gall D, Prestori F, Nieus T, Maffei A, Sola E. Long-term potentiation of synaptic transmission at the mossy fiber-granule cell relay of cerebellum. Prog Brain Res [Internet]. 2005;148:69–80. http://dx.doi.org/10.1016/S0079-6123(04)48007-8.
55. Solinas S, Nieus T, D'Angelo E. A realistic large-scale model of the cerebellum granular layer predicts circuit spatio-temporal filtering properties. Front Cell Neurosci [Internet]. 2010;4:12. http://www.pubmedcentral.nih.gov/articlerender.fcgi?artid=2876868&tool=pmcentrez&rendertype=abstract.

56. Naud R, Marcille N, Clopath C, Gerstner W. Firing patterns in the adaptive exponential integrate-and-fire model. Biol Cybern [Internet]. 2008;99(4–5):335–47. http://www.pubmedcentral.nih.gov/articlerender.fcgi?artid=2798047&tool=pmcentrez&rendertype=abstract.
57. Clopath C, Jolivet R, Rauch A, Luscher H-R, Gerstner W. Predicting neuronal activity with simple models of the threshold type: adaptive exponential integrate-and-fire model with two compartments. Neurocomputing. 2007;70(10–12):1668–73.
58. Fourcaud-Trocmé N, Hansel D, van Vreeswijk C, Brunel N. How spike generation mechanisms determine the neuronal response to fluctuating inputs. J Neurosc [Internet]. 2003;23(37):11628–40. http://www.ncbi.nlm.nih.gov/pubmed/14684865.
59. Mapelli L, Solinas S, D'Angelo E. Integration and regulation of glomerular inhibition in the cerebellar granular layer circuit. Front Cell Neurosci [Internet]. Frontiers. 2014;8:55. http://www.frontiersin.org/Journal/10.3389/fncel.2014.00055/abstract.
60. Chadderton P, Margrie TW, Häusser M. Integration of quanta in cerebellar granule cells during sensory processing. Nature [Internet]. 2004;428(6985):856–60. http://dx.doi.org/10.1038/nature02442.
61. Abdulmanaph N, James P, Nair B, Diwakar S. Characterizing information transmission in cerebellar granule neuron. 2010 IEEE Fifth International Conference on Bio-Inspired Computing: Theories and Applications (BIC-TA) [Internet]. IEEE; 2010. 1487–94. http://ieeexplore.ieee.org/articleDetails.jsp?arnumber=5645277.
62. Pettersen KH, Hagen E, Einevoll GT. Estimation of population firing rates and current source densities from laminar electrode recordings. J Comput Neurosci [Internet]. 2008;24(3):291–313. http://www.ncbi.nlm.nih.gov/pubmed/17926125.
63. Mapelli J, D'Angelo E. The spatial organization of long-term synaptic plasticity at the input stage of cerebellum. J Neurosci [Internet]. 2007;27(6):1285–96. http://www.ncbi.nlm.nih.gov/pubmed/17287503.
64. Toyoizumi T, Aihara K, Amari S. Fisher information for spike-based population decoding. Phys Rev Lett. 2006;97(9):98102.
65. Gold C, Henze Da, Koch C, Buzsáki G, Buzsaki G. On the origin of the extracellular action potential waveform: a modeling study. J Neurophysiol [Internet]. 2006;95(5):3113–28. http://www.ncbi.nlm.nih.gov/pubmed/16467426.
66. Holt GR, Koch C. Electrical interactions via the extracellular potential near cell bodies. J Comput Neurosci [Internet]. 1999;6(2):169–84. http://www.ncbi.nlm.nih.gov/pubmed/10333161.
67. Linden H, Tetzlaff T, Potjans TC, Pettersen KH, Grün S, Diesmann M, Einevoll GT. Modeling the spatial reach of the LFP. Neuron. 2011;72:859–72.
68. Rancz EA, Ishikawa T, Duguid I, Chadderton P, Mahon S, Häusser M. High-fidelity transmission of sensory information by single cerebellar mossy fibre boutons. Nature [Internet]. 2007;450(7173):1245–8. http://www.ncbi.nlm.nih.gov/pubmed/18097412.
69. Jörntell H, Ekerot C-F. Properties of somatosensory synaptic integration in cerebellar granule cells in vivo. J Neurosci [Internet]. 2006;26(45):11786–97. http://www.ncbi.nlm.nih.gov/pubmed/17093099.
70. Gandolfi D, Pozzi P, Tognolina M, Chirico G, Mapelli J, D'Angelo E. The spatiotemporal organization of cerebellar network activity resolved by two-photon imaging of multiple single neurons. Front Cell Neurosci [Internet]. 2014;8:92. http://www.pubmedcentral.nih.gov/articlerender.fcgi?artid=3995049&tool=pmcentrez&rendertype=abstract.
71. Mapelli J, Gandolfi D, D'Angelo E. Combinatorial responses controlled by synaptic inhibition in the cerebellum granular layer. J Neurophysiol [Internet]. 2010;103(1):250–61. http://www.pubmedcentral.nih.gov/articlerender.fcgi?artid=536482&tool=pmcentrez&rendertype=abstract.
72. Mapelli J, D'Angelo E. The spatial organization of long-term synaptic plasticity at the input stage of cerebellum. J Neurosci [Internet]. 2007;27(6):1285–96. http://www.ncbi.nlm.nih.gov/pubmed/17287503.
73. D'Errico A, Prestori F, D'Angelo E. Differential induction of bidirectional long-term changes in neurotransmitter release by frequency-coded patterns at the cerebellar input. J Physiol [Internet]. 2009;587(Pt 24):5843–57. http://www.pubmedcentral.nih.gov/articlerender.fcgi?artid=2808544&tool=pmcentrez&rendertype=abstract.

74. Maffei A, Prestori F, Rossi P, Taglietti V, D'Angelo E. Presynaptic current changes at the mossy fiber-granule cell synapse of cerebellum during LTP. J Neurophysiol. 2002;88(2):627–38.
75. Parasuram H, Nair B, Naldi G, Angelo ED, Diwakar S, D'Angelo E. A modeling based study on the origin and nature of evoked post-synaptic local field potentials in granular layer. J Physiol Paris [Internet]. Elsevier Ltd. 2011;105(1–3):71–82. http://www.ncbi.nlm.nih.gov/pubmed/21843640.
76. Mitzdorf U. Current source-density method and application in cat cerebral cortex: investigation of evoked potentials and EEG phenomena. Physiol Rev [Internet]. 1985;65(1):37–100. http://www.ncbi.nlm.nih.gov/pubmed/3880898.
77. Einevoll GT, Kayser C, Logothetis NK, Panzeri S. Modelling and analysis of local field potentials for studying the function of cortical circuits. Nat Rev Neurosci [Internet]. 2013;14(11):770–85. http://www.ncbi.nlm.nih.gov/pubmed/24135696.
78. D'Angelo E, Nieus T, Maffei A, Armano S, Rossi P, Taglietti V, Fontana A, Naldi G. Theta-frequency bursting and resonance in cerebellar granule cells: experimental evidence and modeling of a slow k+-dependent mechanism. J Neurosci. 2001;21(3):759–70.
79. Ito M. Cerebellar circuitry as a neuronal machine. Prog Neurobiol [Internet]. 2006;78(3–5):272–303. http://dx.doi.org/10.1016/j.pneurobio.2006.02.006.
80. Gerstner W, Kistler WM. Spiking neuron models: single neurons, populations, plasticity [Internet]. Cambridge University Press. 2002. http://icwww.epfl.ch/~gerstner/SPNM/SPNM.html.
81. Bengtsson F, Jörntell H, Jo H. Sensory transmission in cerebellar granule cells relies on similarly coded mossy fiber inputs. Proc Natl Acad Sci U S A. 2009;106(7):2389–94. doi:10.1073/pnas.0808428106.
82. D'Angelo E, Filippi G De, Rossi P, Taglietti V. Synaptic excitation of individual rat cerebellar granule cells in situ: evidence for the role of NMDA receptors. J Physiol. 1995;484(Pt 2):397–413.
83. D'Angelo E, Mazzarello P, Prestori F, Mapelli J, Solinas S, Lombardo P, Cesana E, Gandolfi D, Congi L. The cerebellar network: from structure to function and dynamics. Brain Res Rev [Internet]. Elsevier B.V. 2011;66(1–2):5–15. http://www.ncbi.nlm.nih.gov/pubmed/20950649.
84. D'Angelo E, Solinas S, Garrido J, Casellato C, Pedrocchi A, Mapelli J, Gandolfi D, Prestori F. Realistic modeling of neurons and networks: towards brain simulation. Funct Neurol [Internet]. 2013;28(3):153–66. http://www.pubmedcentral.nih.gov/articlerender.fcgi?artid=3812748&tool=pmcentrez&rendertype=abstract.
85. Bidoret C, Ayon A, Barbour B, Casado M. Presynaptic NR2A-containing NMDA receptors implement a high-pass filter synaptic plasticity rule. Proc Natl Acad Sci U S A [Internet]. 2009;106(33):14126–31. http://www.pubmedcentral.nih.gov/articlerender.fcgi?artid=2729031&tool=pmcentrez&rendertype=abstract.
86. Yang Y, Herrup K. Loss of neuronal cell cycle control in ataxia-telangiectasia: a unified disease mechanism. J Neurosci [Internet]. 2005;25(10):2522–9. http://www.ncbi.nlm.nih.gov/pubmed/15758161.
87. Schweighofer N, Doya K, Lay F. Unsupervised learning of granule cell sparse codes enhances cerebellar adaptive control. Neuroscience. 2001;103(1):35–50.
88. Marr D A theory of cerebellar cortex. This information is current as of August 14, 2007. This is the final published version of this article; it is available at: This version of the article may not be posted on a public website for 12 months after publication u. 2007.
89. D'Angelo E, Casali S. Seeking a unified framework for cerebellar function and dysfunction: from circuit operations to cognition. Front Neural Circuits [Internet]. 2012;6:116. http://www.pubmedcentral.nih.gov/articlerender.fcgi?artid=3541516&tool=pmcentrez&rendertype=abstract.

Chapter 9
Modelling Cortical and Thalamocortical Synaptic Loss and Compensation Mechanisms in Alzheimer's Disease

Damien Coyle, Kamal Abuhassan and Liam Maguire

Abstract Confirming that synaptic loss is directly related to cognitive deficit in Alzheimer's disease (AD) has been the focus of many studies. Compensation mechanisms counteract synaptic loss and prevent the catastrophic amnesia induced by synaptic loss via maintaining the activity levels of neural circuits. In this chapter we investigate the interplay between various synaptic degeneration and compensation mechanisms, and abnormal cortical and thalamocortical oscillations based on two studies involving different implementations of a large-scale neural network model. Study 1 involves a large scale cortical model (C-model) and includes 100,000 neurons exhibiting several cortical firing patterns, 8.5 million synapses, short-term plasticity, axonal delays and receptor kinetics. The structure of the model is inspired by the anatomy of the cerebral cortex. Study 2 involves an extended model, a thalamocortical network model which oscillates within the alpha frequency band (8–13 Hz) as recorded in the wakeful relaxed state with closed eyes. The thalamocortical network model (TC-model) includes different types of cortical excitatory and inhibitory neurons recurrently connected to thalamic and reticular thalamic regions with the ratios and distances derived from the mammalian thalamocortical system.

The results of Study 1 suggest that cortical oscillations respond differently to compensation mechanisms. Local compensation preserves the baseline activity of theta (5–7 Hz) and alpha (8–12 Hz) oscillations whereas delta (1–4 Hz) and beta (13–30 Hz) oscillations are maintained via global compensation. Applying compensation mechanisms independently shows greater effects than combining both compensation mechanisms in one model and applying them in parallel. Consequently, it can be speculated that enhancing local compensation might recover the neural processes and cognitive functions that are associated with theta and alpha oscillations whereas inducing global compensation might contribute to the repair of neural (cognitive) processes which are associated with delta and beta band activity.

D. Coyle (✉) · K. Abuhassan · L. Maguire
School of Computing and Intelligent Systems, Faculty of Computing and Engineering,
Magee Campus, Ulster University, Derry, Northland Road, BT48 7JL, UK
e-mail: dh.coyle@ulster.ac.uk

© Springer International Publishing Switzerland 2015
B. S. Bhattacharya, F. N. Chowdhury (eds.), *Validating Neuro-Computational Models of Neurological and Psychiatric Disorders,* Springer Series in Computational Neuroscience 14, DOI 10.1007/978-3-319-20037-8_9

Study 2 focuses on investigating the impacts of four types of connectivity loss on the model's spectral dynamics, namely degeneration of corticocortical, thalamocortical, corticothalamic and corticoreticular couplings, with an emphasis on the influence of each modelled case on the spectral output of the model. Synaptic compensation has been included in each model to examine the interplay between synaptic deletion and compensation mechanisms, and the oscillatory activity of the network. The results of power spectra and event related desynchronisation/synchronisation (ERD/S) analyses show that the dynamics of the thalamic and cortical oscillations are significantly influenced by corticocortical synaptic loss. Interestingly, the patterns of changes in thalamic spectral activity are correlated with those in the cortical model. Similarly, the thalamic oscillatory activity is diminished after partial corticothalamic denervation. Given the results, it can be speculated that thalamic atrophy is a secondary pathology to cortical shrinkage in Alzheimer's disease. In addition, this study finds that the inhibition from neurons in the thalamic reticular nucleus (RTN) to thalamic relay (TCR) neurons plays a key role in regulating thalamic oscillations; disinhibition disrupts thalamic oscillatory activity even though TCR neurons are more depolarized after being released from RTN inhibition. Both study 1 and study 2 indicate that compensation mechanisms may vary across cortical regions and the activation of inappropriate compensation mechanism in a particular region may fail to recover network dynamics and/or induce secondary pathological changes in the network. Both studies provide a better understanding on the neural causes of abnormal oscillatory activity in neurodegeneration.

Keywords Alzheimer's disease · Synaptic loss · Synaptic compensation mechanisms · Electroencephalography (EEG) · Large-scale network model · Izhikevich neuron model · Message passing interface · Validation

Introduction

There is a strong correlation between cognitive deficit and the degree of the Electroencephalographic (EEG) abnormality in Alzheimer's disease (AD). On the lower scale, it is confirmed that the structure and function of cortical circuits is disrupted by neuropathological factors during AD progression [1–4]. These circuits are the sources of the recorded EEG signals from the scalp surface. Further investigations are required to gain a greater understanding of the underlying neurological basis of abnormal EEG dynamics, linking changes at synaptic level to larger networks and their oscillatory activity.

EEG has been used for the clinical diagnosis of AD over the last few decades [5–8]. Due to its low cost and non-invasive nature, it has been utilized in many studies investigating abnormal brain oscillations which are correlates of cognitive dysfunction. EEG slowing is one of the main characteristics of EEG in AD [9, 10]. EEG slowing is associated with an overall power content shift towards the lower bands of the frequency domain [1]. This has been quantified using power spectral analysis [10–14, 156]. Power spectral analysis of AD patients in relaxed wakefulness with eyes closed has shown a decrease in the mean frequency, alpha (8–12 Hz) and beta (13–30 Hz) band powers with a parallel increase in delta (1–3 Hz) and theta (4–7 Hz) band powers compared with those in healthy elderly groups; increases in delta band power occurs at later stages of the disease [2, 3, 14–16]. Alpha and theta frequency bands have local dynamics [17, 18] while beta and gamma frequency bands have global dynamics [18]. The shift from predominantly higher frequency activity to slower activity during disease progression is referred to as EEG slowing.

de Waal et al. [19] have reported severe EEG abnormalities in early-onset AD patients compared to subjects with late-onset AD. The variations in EEG dynamics (obtained in the resting state) have been quantified in a recent EEG study that involves a large sample population of old and young AD participants (320 participants) and elderly controls (247 participants) [10]. The relative power spectra of EEG in old AD patients (> 65 years) have lower delta and higher alpha band powers than the young ($\leq$65 years) AD group. The young AD group also shows enhanced theta band power and reduced beta band power compared to old AD participants [10].

The correlation between EEG abnormality and cognitive decline has been investigated in [12]. A significant correlation has been observed between enhanced relative theta band power (during relaxed wakefulness with closed eyes) and reduced performance on neuropsychological tests, measuring global cognition, memory, language and executive functioning. Alpha reactivity was decreased in AD patients compared to the healthy control group and it was correlated with reduced performance on neuropsychological tests of global cognition, memory and executive functioning. Alpha reactivity has been obtained during the memory activation task and it refers to the percentile decrease in absolute alpha band power during the memorizing period as compared to the resting state. Decreased alpha reactivity has also been reported in a recent study [20]. Alpha coherence during either the resting or memory activation states was not correlated with cognitive decline [12].

EEG slowing in AD can be related to changes in excitatory circuit activity and connectivity loss [21]. Koeing et al. [21] have compared a large cohort of AD and age matched healthy controls using Global Field Synchronization (GFS), a measure to quantify global EEG synchronization. A high GFS index for a certain frequency band reflects increased functional connectivity between brain processes. The AD group showed increased GFS values in the delta band, and decreased GFS values in alpha and beta frequency bands, supporting the disconnection syndrome hypothesis [21–24].

EEG and computational modelling studies have observed a decrease in the mean frequency, alpha (8–13 Hz) and beta (14–30 Hz) band powers with a parallel increase in delta (1–3 Hz) and theta (4–7 Hz) band powers in AD and Mild Cognitive Impairment (MCI) patients compared with those in healthy elderly groups [1, 2, 15, 21, 25] Jelic 2000. The underlying neural causes are still not clearly understood but there is speculation that the interplay between cortical and thalamic structures influences brain oscillation significantly. The thalamocortical network is a substantial structure and is central to brain function [26], consisting of the thalamus and the cortex, recurrently connected to each other with neural pathways. The collective firing activity of the reciprocally connected neuronal populations in the thalamocortical system, referred to as thalamocortical oscillations and influencing heavily the dynamics of EEG patterns, plays a significant role in controlling our functional and cognitive behaviours. Based on a magnetic resonance imaging (MRI) study of 139 memory complainers (MC) and probable AD subjects, [27] have observed reduced volumes of thalamus in AD. The overall brain size and the volume of grey matter in neocortical area were shown to be significantly reduced in probable AD subjects and the volumes of the left side of hippocampus, putamen and thalamus in probable AD subjects are tightly associated with some cognitive test scores. The anterior and lateral parts of the thalamus are surrounded by a thin sheet of inhibitory neurons known as reticular thalamic neurons (RTN) that are essential for the oscillatory activity of thalamic neurons [1, 28]. Computational models of the thalamocortical system have shown that the thalamic reticular fibres contributes to the thalamocortical oscillations [1, 29]. This circuitry is speculated to be impaired in AD [1]. The neuropathology in the RTN can result in abnormal oscillatory activity in the thalamic nuclei.

Various studies have produced several hypotheses to describe the neuropathological mechanisms of AD [30]. The cholinergic hypothesis states that AD is caused by a deficit in producing the neurotransmitter acetylcholine which has an important role in synaptic plasticity. Whilst therapeutic techniques derived from this hypothesis have treated symptoms of AD, it could neither reverse cognitive deficit (i.e., it is not a cure) nor prevent the accumulation of the pathological components, neurofibrillary tangles (NFT) and neurotic plaques (NP) [31]. Neurofibrillary tangles (NFT) are pathological tangles of hyperphosphorylated tau protein. The protein tau links the microtubules in the cytoskeleton in the axons of neurons. The tau hypothesis refers to the neurodegenerative effects after tau disrupts the cytoskeleton causing axonal death and synaptic loss when detached from microtubules which results in a disruption in information flow between neurons in the network [32]. Biological studies have reported a correlation between the number of NFT and the degree of dementia [33, 34].

Amyloid beta (Aβ) is the main component of NP. The amyloid hypothesis proposes that AD pathogeneses are related to the Aβ accumulation in the brains of AD patients either in the form of extracellular amyloid plaques or in an intracellular soluble form, known as soluble Aβ oligomers [35]. These pathological forms of Aβ (together with NFT) cause neural and synaptic loss. A relevant quantitative morphometric research on biopsies from temporal and frontal cortices within an average of 2–4 years of the clinical onset of AD show synaptic decreases of 36% in layer 5 of the temporal cortex, 25% in layers 2–3 and a synaptic decrease of 27% in layer V of the frontal cortex [36]. The results also reveal greater damage to synapses than neurons [36].

Recent cerebrovascular studies have found that microvascular alterations and blood-brain barrier (BBB) dysfunction occur before Aβ accumulation, atrophy and cognitive deficit [37–40]. The BBB is formed normally by a continuous layer of tightly-connected endothelial cells that controls the passage of molecules and cells into the brain, limits the entry of red blood cells and plasma components into the brain and clears the brain from neurotoxic molecules such as AB [38, 41]. Thus, the BBB is crucial for the normal functioning of the brain. The BBB's permeability is regulated by the neurovascular unit (NVU) which is composed of pericytes, vascular smooth muscle cells (VSMC), endothelial cells, glia and neurons [38, 41]. Pericytes are vascular cells that surround the endothelial cells and have contacts with astrocytes. Interactions between pericytes and endothelial cells are essential for the regulation of the BBB structure [42]. Together with the blood–spinal cord barrier (BSCB) and the BBB, pericytes maintains the removal of neurotoxic molecules from the central nervous system (CNS) and block toxic macromolecules in the blood from accessing the CNS [38, 41]

Zlokovic [38] has proposed a neurovascular hypothesis of AD that combines vascular deficits and neuropathological factors. The neurovascular hypothesis of AD states that "primary damage to brain microcirculation (hit one) initiates a non-amyloidogenic pathway of vascular-mediated neuronal dysfunction and injury, which is mediated by BBB dysfunction and is associated with leakage and secretion of multiple neurotoxic molecules and/or diminished brain capillary flow that causes multiple focal ischaemic or hypoxic microinjuries. BBB dysfunction also leads to impairment of amyloid-β clearance, and oligaemia leads to increased amyloid-β generation. Both processes contribute to accumulation of amyloid-β species in the brain (hit two), where these peptides exert vasculotoxic and neurotoxic effects. According to the two-hit vascular hypothesis of Alzheimer's disease, tau pathology develops secondary to vascular and/or amyloid-β injury" [38].

Several experimental studies have found that the neurodegenerative process is accompanied by synaptic compensation mechanisms described as "a homeostatic mechanism which maintains the excitatory response of individual neurons and prevents the catastrophic amnesia associated with synapse loss" [43–45]. From a neurobiological perspective, compensation might result from neuritic outgrowth [46], an increase in neurogenesis processes [47] or increased expressions of the postsynaptic protein PSD-95 and Apolipoprotein D [48]. PSD-95 protein determines the size and strength of the synapse [49]. On an activity level, synaptic compensation senses and regulates the firing rate of the network at the neuron or network level [50]. Stellwagen and Malenka [51] indicated that glia are well placed to sense the overall activity of the network [52, 53] through glutamate spillover [54], direct synaptic contact [55] or from other activity-regulated signals, and then provide a feedback signal to regulate network behaviour. For example, it has been reported that NG2-glia cells contact all of the synapses within their domain (many millions) [56] and a single astrocyte covers up to 2 million synaptic contacts operating within its domain [57]. Such couplings between neurons and astrocytes contribute to global synaptic homeostasis. Individual neurons can also sense changes in their spiking activity through calcium-dependent sensors which in turn adjust the abundance of glutamate receptors at the synapse [44]. Synaptic scaling protocols adjust both the AMPA and the NMDA currents [44].

Studies on deafferentation-induced synaptic scaling can shed some light on the impact of such processes on the network dynamics in AD [58]. Of particular interest to the current study are EEG dynamics and local field potentials (LFPs). An *in vivo* experimental study has recorded EEG signals from postcruciate and anterior areas of deafferented cat suprasylvian gyri immediately following deafferentation as well as after 3 h [59]. Immediately after deafferentation, a decrease in the amplitude of EEG waves was observed. The EEG signals recorded from the middle and posterior regions of the suprasylvian gyrus, where the degree of deafferentation was maximal, were most affected. After 3 h of partial deafferentation, the EEG activity in the posterior and middle regions of the suprasylvian gyrus was recovered to the normal physiological level in 60% of the deafferented cats. Seizure activity in the vicinity of partially deafferented areas was observed in the other 40% of cats' cortices [59]. Paroxysmal activity was detected only in cats that exhibit an oscillation in the sleep slow waves in normal conditions. From a computational modelling perspective, Fröhlich et al. [50] have reported EEG slowing for increasing degrees of denervation in a computational model of global synaptic homeostasis. In the case of severe deafferentation (80–100%), slow periodic EEG discharges have been obtained in response to global synaptic homeostasis.

Homeostatic mechanisms can be implemented through synaptic scaling, increasing the number of receptors or sensitivity to deafferentation [30]. Horn et al. [60, 61] and Rowan [62] have modelled synaptic compensation in artificial neural networks to investigate memory decline in AD. Horn et al. [60] simulated a global compensatory mechanism after performing random synaptic deletion. This global compensatory mechanism is defined as a similar increase (upscale) in all the remaining synapses by multiplying all remaining weights by one common factor. In a recent study, Fröhlich et al. [50] used an electrophysiological network model to implement global synaptic scaling through increasing the AMPA conductance between pyramidal neurons in response to deafferentation. The compensation factor was calculated based on changes in the overall firing rate of excitatory cells in the network [50]. Local synaptic compensation involves each neuron regulating its own compensation factor based on changes to its post-synaptic activity [61]. Computational studies of compensation mechanisms were inspired by the observation of increasing synaptic strength in aged and AD brains [30, 61].

The studies presented here are targeted at investigating the underlying neurological basis of abnormal EEG dynamics in AD using computational models of neuronal networks. In a previous study [63], we investigated the effect of excitatory circuit disruption on the output oscillatory activity of two heterogeneous neuronal network models including a conductance-based local neuronal network oscillating in beta band [64] and a simple larger network model of different types of cortical neurons that oscillates in different frequency bands [65]. The results showed that beta and alpha band powers are significantly affected by excitatory neural and synaptic loss. Secondly, the results of modelling functional impairment in the excitatory circuit showed that beta band power exhibits the most decrease compared with other bands.

In this chapter, we incorporate representations of more complicated neural processes such as compensation mechanisms in a much larger and more plausible network models. As mentioned above, compensation mechanisms have been experimentally confirmed during dementia. The objective is to elucidate the gradual cognitive decline in AD sufferers and what forms of compensation may prevent the pathological EEG correlates (and thus neural network activity) of cognitive decline. An enlargement in the remaining synapse size has been observed in response to synaptic loss in an experimental study in AD [66, 67].

The chapter examines the interplay between various synaptic deletion and compensation mechanisms, and abnormal oscillatory activity in AD using two different model implementations: Study 1 involving a large scale cortical model (C-model) and study 2 involving a large scale thalamocortical model (TC-model). The models are developed in C with Message Passing Interface (MPI) and simulated on a High Performance Computing (HPC) facility.

Study 1 involves a cortical neuronal network model of 100,000 neurons and 8.5 million synapses consisting of distribution of axonal conduction delays, receptor kinetics (AMPA, NMDA, $GABA_A$ and $GABA_B$), and short-term plasticity [68]. Three compensation mechanisms are applied to the cortical model: a global compensation mechanism, a local (neuronal-based) compensation mechanism and a compensation mechanism that combines both global and local compensation mechanisms in parallel which is referred to as combined compensation mechanisms (CCM) as described in Sect. "Synaptic Degeneration and Compensation". Synaptic compensations were implemented by increasing the synaptic weights as in existing studies [61, 62]. This approach influences NMDA and AMPA glutamate receptors based on experimental findings which suggest that synaptic scaling mechanisms regulate both receptors [44]. A mathematical formulation of the receptor kinetics is included in Sect. "Synaptic Kinetics".

Study 2 involves a thalamocortical neuronal network model of 100,000 cortical neurons, 3340 thalamocortical neurons, 3340 reticular thalamic neurons and more than ten million synapses with Amino-3-Hydroxyl-5-Methyl-4-Isoxazole-Propionate (AMPA), γ-Aminobutyric Acid (GABA), N-Methyl-D-aspartate (NMDA) and gap junction (GJ) kinetics, short-term plasticity and a distribution of axonal conduction delays. The model includes different types of cortical excitatory and inhibitory neurons recurrently connected to thalamic and reticular thalamic regions with the ratios and distances found in the mammalian thalamocortical system. The network model oscillates in alpha frequency band as recorded in the wakeful relaxed state with closed eyes. The model has been utilised to study the impacts of four types of connectivity loss on abnormal oscillatory activity in AD, namely degeneration of corticocortical, thalamocortical, corticothalamic and corticoreticular couplings, with an emphasis on the influence of each modelled case on the spectral output of the model. It is believed that the cognitive decline in AD is caused by the impaired connectivity between cortical regions [24].

In order to evaluate the model dynamics, the study links the output oscillatory activity with findings of EEG studies in AD [3, 4, 13, 69]. As mentioned earlier, EEG spectral analysis in AD is characterized by a decrease in the mean frequency, alpha

(8–12 Hz) and beta (13–30 Hz) band powers with an increase in delta (1–3 Hz) and theta (4–7 Hz) band powers compared with those in healthy elderly groups (Jelic 2000; [2, 15].

In study 1 we found that (1) compensation mechanisms can maintain the average firing rate of the lesioned network, (2) the abnormal changes in the output power spectra are minimized after activating compensation mechanisms, (3) the baseline GFS outputs are more stable with local compensation and global compensation models as compared to the CCM and to models 'without compensation', (4) the global compensation mechanism reorganizes the fine temporal structure of spike trains in individual excitatory neurons and the distribution of firing rates across excitatory neurons whereas the local compensation mechanism maintains the activity-level of the network without changing the organization of the network dynamics and (5) the baseline activity at theta and alpha bands is optimally recovered by local compensation whereas the baseline level of delta and beta bands is maintained by global compensation. The results also suggest that cortical oscillations respond differently to compensation mechanisms. Consequently, it is speculated that enhancing local compensation may contribute to the recovery of the neural processes and cognitive functions that are associated with theta and alpha bands whereas global compensation contributes to the repair of neural (cognitive) processes associated with delta and beta bands. The type of compensation mechanism employed in neural processing and maintaining neural function can vary across cortical regions. The activation of inappropriate compensation mechanism in a particular area may fail to recover the network dynamics and/or may induce secondary pathological changes in the network.

In study 2 the following compensation mechanisms are incorporated: (1) corticocortical synaptic compensation in response to thalamocortical and corticocortical synaptic loss, (2) decreased RTN inhibition to thalamocortical relay (TCR) neurons in response to corticothalamic denervation and (3) decreased RTN self-inhibition after corticoreticular connectivity loss. The biological basis for such choices is discussed in Sect. "Synaptic Degeneration and Compensation" below. The thalamocortical model study (study 2) presents the following novel findings: (1) synaptic compensation plays a significant role in preserving the dynamics of the network (after degeneration), (2) the activity of the network is significantly affected by corticocortical synaptic loss, (3) thalamic atrophy can be a secondary pathology to cortical shrinkage and (4) a deficit in the activity of the RTN population (disinhibition) causes a disruption in the oscillatory activity of the TCR population.

The remainder of the chapter is structured as follows: Materials and methods are described in Sect. "Materials and methods". Then, results and in-depth analysis are provided in Sect. "Results". Finally, discussions and conclusions are presented in Sects. "Thalamocortical Connections" and "Corticothalamic Fibers", respectively.

Materials and Methods

Cortical Model Structure

A neuronal network model inspired by the anatomy of the cerebral cortex is employed (see Fig. 9.1). The anatomy and dynamics of the model follow Izhikevich's model presented in [68]. The model is "an anatomically realistic model" [70]. The model preserves important ratios and relative distances found in the mammalian cortex [70].

The cortical model and the cortical part of the thalamocortical model is inspired by the anatomy of the cerebral cortex with an anatomy and dynamics following the study in [68]. The model preserves important ratios and relative distances found in the mammalian cortex [71]. The ratio of excitatory to inhibitory neurons is 4–1. The radius of local non-myelinated axonal arborisations of an excitatory neuron is 1.5 mm, and the length of the myelinated axon is 12 mm as shown in Fig. 9.1. The long axon is sent to a distant region on the sphere and its collaterals span an area of radius 0.5. Each excitatory neuron innervates 75 randomly chosen local targets and 25 distant targets. The span of local non-myelinated axonal collaterals of an inhibitory neuron is 0.5 mm. Each inhibitory neuron innervates 25 randomly chosen neurons within a circle of radius 0.5 mm. As presented in [68], the axonal conduction velocity is 1 m/s for myelinated axons [72] and 0.15 m/s for non-myelinated collaterals [73].

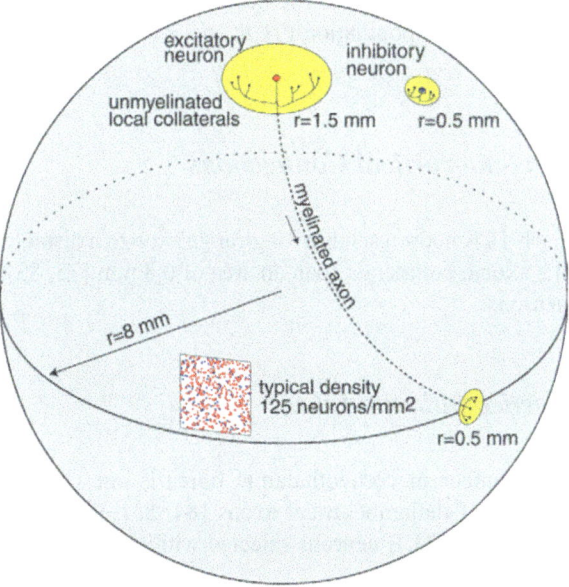

Fig. 9.1 Connectivity of the network model, reproduced with permission from [68]

Thalamic and Reticular Thalamic Network

The relative distribution of thalamic neurons is unknown [68]. It has been observed that there is 350,000 thalamocortical connections per 11,000,000 neurons in layer 4 of the primary visual cortex [74] leading to a ratio of 1/30 [75]. If we assume that each TCR neuron is associated with one fibre, then the distribution of TCR neurons is concluded as a proportion of cortical neurons resulting in 3340 TCR neurons in the thalamic network. The number of reticular thalamic neurons (RTN) is considered equivalent to the number of TCR neurons in the network as modelled in other studies [75–77]. TCR and RTN neurons are randomly allocated on a spherical surface of radius 2 mm.

Thalamic-Reticular Connections

Each TCR neuron selects 13 RTN neurons randomly within an area of radius 0.5 mm. Each RTN cell innervates 25 TCR neurons within an area of radius 0.5 mm and 13 local RTN neurons within an area of radius 0.5 mm. GJ have been observed between RTN neurons in mice and rats without evidence of chemical synapses [78]. Other studies have found chemical synapses among RTN neurons [79–81]. This model includes chemical and electrical synapses in RTN neurons. Hughes & Crunelli [82] have observed only GJ between TCR excitatory neurons. Since none of the studies has confirmed the existence of any chemical synapses among TCR neurons [75, 83], only GJ among TCR neurons are included in the model. According to [82], not all thalamic neurons have GJ, consequently, only 1000 neurons of each thalamic population (TCR and RTN populations) have GJ in this modelling study.

Thalamocortical Connections

Each TCR neuron sends a long-range axon to a distal location on the cortical sphere. It's axonal collaterals span an area of 0.8 mm [75, 83]; thus innervating 40 cortical neurons.

Corticothalamic Fibers

The number of corticothalamic fibres is one order of magnitude larger than the number of thalamocortical axons [84, 85]. Each corticothalamic axon selects 40 RTN and 40 TCR neurons selected within an area of 0.5 mm.

Axonal Conduction Delays

Thalamocortical connections have 1 ms axonal conduction delay [86] while corticothalamic connections have a 5 ms axonal conduction delay [87]. In vivo, axonal conduction from cortex to thalamus is much slower than in the reverse direction [88]. A schematic diagram for the thalamocortical network model is presented in Fig. 9.2 below.

Model Dynamics

The model consists of different types of neurons, synaptic transmission with AMPA, GABA, and NMDA kinetics, short-term plasticity and a distribution of axonal conduction delays [68].

Neuronal Dynamics

Spiking dynamics of neurons are simulated based on Izhikevich's model of spiking neurons [65], which can reproduce the firing patterns of known types of hippocam-

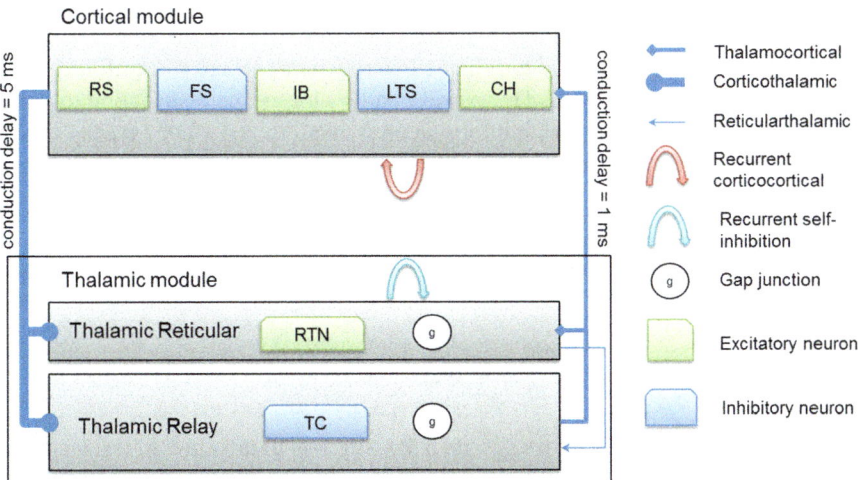

Fig. 9.2 A schematic diagram for the connectivity of the thalamocortical network model. Note that each cortical neuron selects its postsynaptic targets as described in Sect. "Cortical Model Structure". The number of corticothalamic fibers is ten times more than thalamocortical fibers (therefore, represented by a *bold line*). Abbreviations: regular spiking (*RS*), fast spiking (*FS*), intrinsically bursting (*IB*), low-threshold spiking (*LTS*) and chattering (*CH*)

pal, cortical, and thalamic neurons. A spiking neuron can be expressed in the form of ordinary differential equations (ODEs) as shown in Eqs. (9.1)–(9.3).

$$\frac{dV}{dt} = 0.04 \cdot V^2 + 5 \cdot V + 140 - u - I_{syn} \quad (9.1)$$

$$\frac{du}{dt} = a \cdot (b \cdot V - u) \quad (9.2)$$

with the auxiliary after-spike resetting as follows

$$if\ V \geq 30\ mV\ then\ V \leftarrow c,\ u \leftarrow u + d \quad (9.3)$$

where the dimensionless variables V and u represent the membrane potential and the recovery variable of the neuron, respectively. The variable I_{syn} denotes the total received cortico-cortical synaptic input as described below. The recovery variable u provides negative feedback to V, and it corresponds to the inactivation of Na+ ionic currents and activation of K+ ionic currents. Dimensionless parameters a, b, c and d can be tuned to simulate the dynamics of inhibitory and excitatory neurons. The parameter a describes the rate of recovery, the parameter b represents the sensitivity of the recovery variable u to the subthreshold fluctuations of the membrane potential V, the parameter c represents the after-spike reset value of variable V and the parameter d describes the after-spike reset of variable u [65].

The parameters of the cortical excitatory neurons are $(a, b) = (0.02, 0.2)$ and $(c, d) = (-65, 8) + (15, -6) \cdot r^2$ to achieve heterogeneity, where r is uniformly distributed on the interval [0, 1]. The expression (r^2) biases the distribution towards RS neurons. Each cortical inhibitory neuron has $(a, b) = (0.02, 0.25) + (0.08, -0.05) \cdot r$ and $(c, d) = (-65, 2)$ [68]. Dynamics of thalamic neurons are expressed in the form [89]

$$\frac{dV}{dt} = (k \cdot (V - v_r) \cdot (V - v_t) - u - Isyn - Igap) \quad (9.4)$$

$$\frac{du}{dt} = a \cdot (b \cdot (V - v_r) - u) \quad (9.5)$$

$$if\ V \geq v_{peak},\ then\ V \leftarrow c,\ u \leftarrow u + d \quad (9.6)$$

where the parameters for TCR are $C = 200$, $k = 1.6$, $v_r = -60$, $v_t = -50$, $a = 0.01$, $b = 0$ (if $v > -65$ and $b = 15$ otherwise), $c = -60$ and $d = 10$. The parameters for RTN are $C = 40$, $k = 0.25$, $v_r = -65$, $v_t = -45$, $a = 0.015$, $b = 2$ (if $v > -65$ and $b = 10$ otherwise), $c = -55$, $d = 50$ and $v_{peak} = 0$.

Synaptic Dynamics

Input

In addition to the input synaptic current, each neuron receives a noisy input of magnitude (15 pA) generated by a Poisson point process with 100 Hz mean firing rate as modelled in a previous study [50].

Synaptic Weights

The values of corticocortical excitatory synaptic weights are within the range [0, 0.5] as in [68]. The values of other types of synaptic weights are chosen such that the power spectra dynamics of the model has its peak at 10 Hz (as recorded in the wakeful relaxed state with closed eyes). The distribution of synaptic weights follows a Gaussian function with mean μ and standard deviation σ as described in Table 9.1.

Short-Term Plasticity

Short-term depression and facilitation are implemented using the synapse model in [90]:

$$\dot{R} = (1-R) \cdot L - R \cdot w \cdot \delta \cdot (t - t_n) \tag{9.7}$$

$$\dot{w} = \frac{(U-w)}{F} + U \cdot (1-w) \cdot \delta \cdot (t - t_n) \tag{9.8}$$

where R and w represent 'depression' and 'facilitation' variables, respectively. Excitatory synapses have $U=0.5$, $F=1000$ and $D=800$. Inhibitory synapses have $U=0.2$, $F=20$ and $D=700$. The parameters U, F and D were measured in [90, 91]. The expression δ is the Dirac function. The fractional amount of neurotransmitter available at time t is determined by $R(t) \cdot w(t)$. When the postsynaptic neuron

Table 9.1 The distribution of synaptic weights

Synaptic weights			
Type of synapse	Range [min, max]	Mean μ	Standard deviation σ
Corticocortical	[0, 0.5]	0.25	0.085
Thalamocortical	[0, 1]	0.5	0.180
Thalamicreticular	[0, 1]	0.5	0.180
Corticothalamic	[0, 4]	2	0.680
Corticoreticular	[0, 4]	2	0.680

receives a spike at time t_n (after axonal delay; i.e., when a presynaptic spike arrives at the synapse), the variable R decreases by $R \cdot w$ while the variable w increases by $U \cdot (1 - w)$.

Synaptic Kinetics

The total synaptic current of neuron, i, is calculated as

$$I_{syn} = g_{AMPA}(v-0) + \frac{g_{NMDA}(v-0)\left(\left[\frac{v+80}{60}\right]^2\right)}{1+\left[\frac{v+80}{60}\right]^2} + g_{GABAA}(v+70) + g_{GABA_B}(v+90) \tag{9.9}$$

where v is the postsynaptic membrane potential, and the subscript indicates the receptor type. Each conductance updates by first-order linear kinetics ($\dot{g} = -g/\tau$) with $\tau = 5$, 150, 6 and 150 ms for the simulated AMPA, NMDA, GABA$_A$ and GABA$_B$ receptors, respectively [68]. The ratio of NMDA to AMPA receptors is 1 for all excitatory neurons. Firing of a presynaptic excitatory neuron j increases g_{AMPA} and g_{NMDA} by $s_{ij} \cdot R_j \cdot w_j$ where s_{ij} is the strength of the synapse from neuron j to neuron i, R_j is the short-term depression variable and w_j is the short-term facilitation variable. The ratio of GABA$_B$ to GABA$_A$ receptors is 1 for all inhibitory neurons. Each firing of an inhibitory presynaptic neuron increases g_{GABA_A} and g_{GABA_B} by $R_j \cdot w_j$. The gap junction current is calculated according to the following formula

$$I_{gap} = \sum_{i \in \text{neighbours}} g \cdot (v - v_i) \tag{9.10}$$

where g (conductance) has a value of 2 and each thalamic neuron is electrically coupled to five neighbouring neurons of the same type.

Synaptic Degeneration and Compensation

Study 1– Cortical Network

We simulate ten random patterns of the model generated with different seeds (therefore representing ten individuals). All simulated networks preserve the ratios and distances described in Sect. "Cortical Model Structure". For each network pattern, we simulate different degrees of synaptic loss (SL) with values between 10 and 70% (accounting for different stages of neurodegeneration in AD). The GFS and spectral analysis is based on the average of these simulations. In the global compensation model, each network is simulated for 1 min model time with physiological values of all parameters. After the network reaches a steady-state, synaptic loss is introduced through a random deletion of a fraction, SL, of excitatory connections. Then, the firing rate of excitatory neurons is calculated every 5 s by averaging over all excitatory spikes in the preceding 5 s interval, as illustrated in Fig. 9.3. The remaining synaptic weights between excitatory neurons are then increased in these time-points by $\Delta p = \varepsilon \cdot (g^* - g) \cdot s$, where ε is a rate parameter ($\varepsilon = 0.1$), g^* is the target firing rate (the firing rate of the network during the steady-state before synaptic loss), and g is the current average firing rate.

This phenomenological model of degeneration and compensation is adapted from the established study by Fröhlich et al. [50]. The overall model time for each network is 260 s. We apply the compensation rule from 80 s until time-point 220 s (model time). As suggested in [50], this synaptic scaling rule is chosen because, firstly, it is computationally impossible to simulate the real biological timescale (hours to days) for homeostatic plasticity with this model, and second, the effect of scaling the synaptic weights on the firing activity is faster than the homeostatic regulation of synaptic weights.

In local compensation, each excitatory neuron maintains its activity in a local manner in response to synaptic loss (it employs its own compensation factor). The remaining synaptic weights of each excitatory neuron are increased by $\Delta p_i = \varepsilon \cdot (g_i^* - g_i) \cdot s_i$, where ε is a rate parameter ($\varepsilon = 0.05$), g_i^* is the target firing rate of the neuron (the firing rate of the neuron during the steady-state before synaptic loss), and g_i is the current average firing rate of the neuron (after synaptic loss). The total model time in the local compensation model is 1000 s. When applying

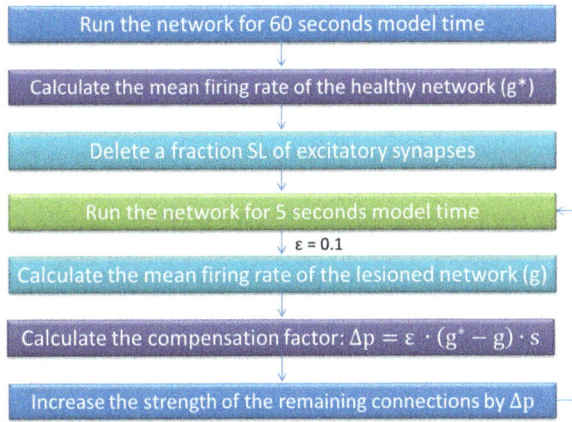

Fig. 9.3 A summary of the process of simulating synaptic degeneration and global compensation. The same model is utilized for local compensation. However, the compensation factor is estimated locally for each excitatory neuron based on the mean firing rate of the neuron during intervals of 40 s model time, settlement period is 80 s, the target firing rate of each excitatory neuron is calculated by averaging over all its spikes in the preceding 40 s interval (second 40 to second 80), synaptic loss is performed after second 80 and ε is 0.05

combined compensation mechanisms (CCM) simultaneously, we apply local and global compensations in parallel. The rate parameter (ε) is 0.05 and 0.025 for global and local compensation rules respectively. The local target firing rate of each excitatory neuron and the global target network firing rate are calculated by averaging over all spikes in the interval from second 40 to second 80. Synaptic loss is performed after second 80 and the total model time is 1000 s. The local and global compensation factors are estimated based on the mean firing rates during intervals of 40 s model time. The number of network patterns is 5 random patterns generated with different seeds (i.e., representing five individuals).

The choice of modelling synaptic loss in this study is supported by the finding that synapse loss occurs before neuronal death [92–94] and the early reduction in synapse density is not proportional to the degree of neuronal loss [92, 95]. Furthermore, the loss of dendritic spines (excitatory postsynaptic contacts) is suggested to occur long before or even in the absence of neurodegeneration [93, 94, 96]. Indeed, the degree of synaptic loss is the most correlated factor with the severity of AD and cognitive decline [92, 97], even more than neurodegeneration or NFT components [98]. Hence, the main focus of this study is on the investigation of synaptic loss as well as the global and local synaptic reactions. Synaptic loss is considered a major correlate of cognitive impairment in aging as synapses are essential for interneuronal communication and are the smallest anatomical structures that appear to change easily [46].

Experimental studies have shown that the loss occurs in excitatory synapses ; [96, 98–101], in particular, in the early stages of AD [101]. These synapses are attacked by neuropathological forms of Aβ protein [92, 99]. Deficits in numerous neurotransmitters (including corticotropin-releasing factor, somatostatin, GABA, and serotonin) have been observed to accrue as the disease progresses. However, the early symptoms appear to correlate with dysfunction of cholinergic and glutamatergic synapses [101]. These findings provide supportive evidence to justify the implementation choice of varying the number of excitatory (glutamatergic) synapses to investigate AD in this modelling study.

Study 2– Thalamocortical Network

The thalamocortical model was simulated for 80 s model time with ten different baseline (normal) setups generated with different seeds (therefore representing ten individuals) to collect the baseline data for analytical purposes. Similarly, there are ten random patterns of the model for each type of connectivity loss. For each network pattern, different degrees of synaptic loss (SL) were simulated with values between 10 and 60% (representing different stages of neurodegeneration). The spectral analysis is based on the average of these simulations.

Each network is simulated for 10 s model time with physiological values of all parameters. Then, synaptic loss is performed by a random deletion of a fraction, SL, of connections followed by the synaptic compensation mechanism as described below

- Corticocortical connectivity loss

Synaptic loss is implemented in this case by deleting excitatory synapses among cortical neurons on the cortical surface. The compensation rule is applied from 80 s until 200 s (model time). The firing rate of excitatory neurons is calculated every 5 s by averaging over all excitatory spikes in the preceding 5 s interval. The remaining synaptic weights between excitatory neurons are then increased in these time-points by the following formula

$$\Delta p = \varepsilon \cdot (g^* - g) \cdot S \qquad (9.11)$$

where $\varepsilon = 0.1$. The total model time is 300 s.

- Thalamocortical connectivity loss

Cortical neurons receive input from thalamic neurons as described in Sect. "Corticothalamic Fibers". This case examines the effect of losing such input (synapses) on the spectral output of the network. This case incorporates the presented compensation mechanism in case 1 (above) where corticocortical synaptic weights are scaled up to compensate for the loss of thalamic input signals that is induced by thalamocortical synaptic loss.

- Corticothalamic connectivity loss

Several studies have reported the fundamental role of the cortex in generating slow oscillations and alpha waves [77, 102]. Interestingly, it has been found that slow oscillations [103] and alpha waves [104] can be maintained in the presence of lesions in the thalamus. On the other hand, slow oscillations in the thalamus have been supressed in decorticated cats [105]. Such studies pointed out the central role of the cortex in generating oscillations and driving the activity of the thalamus as observed in the rastergram analysis in [77] and the simulated LFPs in the current study (see Fig. 9.17).

This case explores the dynamics of the network after abnormal reduction of the distribution of cortical efferents (input) to the TCR neurons in the thalamic network. It is mentioned earlier (in Sect. "Corticothalamic Fibers" above) that TCR afferents are received from cortical and RTN neurons. To compensate for the loss of cortical input, the study has scaled down the inhibitory input from RTN neurons according to Eq. (9.11) with ε has an initial value of 0.05 and evolves autonomously such that it is increased by 0.00125 if g is less than g^* and decreased by 0.00125 otherwise. The computation is allowed to run for a long period (500 s) to allow the model to stabilise. When using a constant parameter value (as in the above two cases), a significant depolarized phase followed by highly hyperpolarized intervals is observed (similar to sleep oscillations) [77]. The aim is to maintain the asynchronous firing pattern of the system (as in the wakeful state). From a neurobiological point of view, an *in vivo* study has observed a reduced RTN inhibition to TCR neurons in response to corticothalamic synaptic degeneration (initiated by cortical neuronal death) and consequently, leading to enhanced TCR activity and recovered thalamocortical activity [106].

- Corticoreticular connectivity loss

A recent thalamocortical modelling study based on neural mass models has speculated that reduced RTN afferents contribute to abnormal oscillatory activity in AD [1]. Using a microscopic model with more details such as GJ, this study considers this finding and includes a synaptic reaction mechanism that scales down the inhibition among RTN neurons to recover their output activity. The model employs the above mentioned technique in estimating ε with an increase (or decrease) in magnitude of 0.0025.

Data Analysis

Power Spectra Analysis

The power spectra analysis is based on the relative band power, an analytic tool recommended in dementia studies [29]. The averaged spike trains over all excitatory neurons are smoothed using a Gaussian kernel with standard deviation (SD = 20 ms) as modelled in [50]. To analyse the relative power spectra, the Fast Fourier Transform (FFT) is applied on the smoothed spike trains during an 18-second window (after recovering the target firing rate for simulations of compensation mechanisms) in each simulation as follows: (1) the 18-second window is divided equally into six smaller windows of 3-s width each; (2) the FFT is applied separately on the spike trains within each small window; (3) the average power spectra over all six windows is then calculated; and (4) the relative power spectra are found by dividing the power spectrum at each frequency band by the mean of the total power spectra as in [1]. We consider the frequency bands from 1 to 100 Hz when calculating the total power spectra. One-way repeated measures ANOVA is used to analyse the statistical significance of the difference in relative band power within each frequency band between the physiological (baseline) case and cases with different degrees of synaptic loss (p-values less than 0.01 indicate a significant difference).

Global Field Synchronization Analysis

The study utilizes the GFS tool to measure the correlation in neural activity across the network at a given frequency band. GFS quantifies the phase coupling between the recorded EEG signals over several locations on the scalp [9, 21, 25, 107]. Each signal is transformed into the frequency domain and is represented as a two-dimensional vector in a complex plane where the phase of the signal is represented by the direction of the vector. Then, GFS tests the locations of the endpoints (in the complex plane) of the vectors representing all EEG signals. GFS value ranges from zero to one. Enhanced functional connectivity results in greater values of GFS whereas decreased functional connectivity has lower GFS values indicating the absence of

Fig. 9.4 The configuration of the six LFP locations are indicated on the diagram. The great-circle (*orthodromic*) distance between location 1 and location 2 is 25.1327 mm

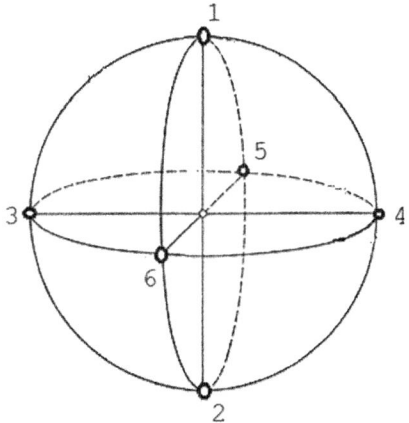

common phase [9, 21, 25, 107]. GFS is not correlated with the total power of the EEG data [21]. The software used for computing GFS is freely available in [108].

For each network pattern, this study analyses the correlation in neural activity over six Local Field Potential (LFP) signals obtained from six locations on the cortical sphere as illustrated in Fig. 9.4. Each LFP signal is a 5-s epoch and is obtained from the activity of neurons chosen within a circle of radius 8 mm. LFP is estimated by averaging the spike trains smoothed with a Gaussian kernel (SD of 20 ms) as in [50]. For each network pattern, the GFS measure at a particular frequency band is computed under the following conditions: (1) baseline, (2) synaptic degeneration without compensation; (3) synaptic degeneration with local compensation; (4) synaptic degeneration with global compensation and (5) synaptic degeneration with combined compensation mechanisms (CCM).

For all modelled cases, the Sum of Squared Differences (SSD) is computed over all degrees of synaptic loss (10–70%) at each frequency band to quantify the relative difference from the baseline value (in a single scalar value). SSD has the following simple formula:

$$SSD = \frac{1}{n}\sum_{i=1}^{n}(y_i - y_{base})^2 \quad (9.12)$$

where y_i is the baseline value at a given frequency band, y_i is the mean relative power (or GFS) at a particular degree of synaptic loss and n is the number of simulated cases of synaptic loss ($n=7$).

Event Related Synchronisation and Desynchronisation

EEG and modelling studies have quantified frequency alterations in the ongoing oscillatory signal in response to a stimulus (event) based on the event related desynchronisation/synchronisation (ERD/S) measure [109–111]. ERD refers to diminished power density in certain EEG waves after the internal or external stimulation, whereas ERS is observed if the event causes an enhancement in the power amplitude of an EEG frequency band. The measure first appeared in [112] and has been extensively utilised in BCI studies such as [113–115].

Study 2 employs an ERD/ERS tool[1] to analyse the impact of synaptic loss (the event) and compensation on the network oscillatory activity. ERD/ERS estimation is based on a previous study [111] that is accompanied with an online freely available MATLAB® routine [116]. The approach uses a Short time Fourier transform (STFT or spectrogram) to compute the time-frequency power of different waves and bootstrapping with pseudo-t statistics to mark the significant increase (ERS) or decrease (ERD) of the frequency band power in a particular band.

The simulated LFP are estimated by averaging spike trains smoothed with a Gaussian kernel (SD of 20 ms). Time-frequency spectrogram is a common EEG (and signal processing) analytical method used to visualise changes on the spectral power of frequency bands as a function of time. In the study 2 (TC network) the spectograms of the simulated LFP signals based on the MATLAB® (MathWorks) function *spectrogram()* provided with a LFP signal of length 6 s and a Hamming window of size 2 s. For presentation purposes, the spectrogram plots have been filtered with the MATLAB® (MathWorks) function *filter2()*. Dynamics of neurons are simulated using the first-order Euler method with 0.5 ms time step to avoid numerical instabilities. The synaptic dynamics are simulated with 1 ms time step [68].

Computer Simulations

Study 1

The anatomy of the network is implemented in MATLAB and saved to ASCII files. Then a C implementation with MPI is used to load the network (ASCII files) and simulate the model dynamics. The models run on a HPC facility that consists of 31 Dell R401 computing servers, each with two physical Intel Xeon CPUs with six Cores running at 2.66 MHz. For study 1 the cortical model takes about 3 min to simulate 1 s model time on 40 cores. The total number of simulations for local and global compensation mechanisms is 140. As described above, each network pattern has 14 simulations, 7 simulations for global compensation (each representing a

---

[1] The event in real EEG experiments corresponds to a motor task that lasts for a short time (few seconds). Such events stimulate certain populations of neurons and results in an attenuation or potentiation in the power of certain frequency bands. In this study, the event corresponds to a massive loss of synapses. By utilising the ERD/ERS measure, the study aims to examine the effects of synaptic loss and compensation on the oscillatory activity of the network. Synaptic compensation is implemented by increasing the weights of the remaining synapses. Synaptic weights and the intrinsic membrane properties of the neurons are responsible for the dynamics of the EEG signal (Pfurtscheller & Lopes da Silva 1999). This justifies the choice of utilising an ERD/ERS analysis in this study.

certain degree of synaptic loss) plus 7 simulations for local compensation. Half of the network patterns are involved in simulations of the CCM case (35 simulations). As in [68], neuron dynamics are simulated using the first-order Euler method with 0.5 ms time step to avoid numerical instabilities. The synaptic dynamics are simulated with 1 ms time step [68]. Data analysis is performed with custom-written MATLAB scripts and the GFS tool.

Study 2

The models were simulated on the HPC facility described above. It takes about 20 s to simulate 1 s model time on 40 cores. The anatomy of the network was implemented in MATLAB whereas the dynamics were simulated by using C programing language with MPI.

Results

Study 1– Cortical Network

Investigating the interplay between synaptic loss and compensation, and abnormal EEG power spectra based on a computational modelling approach is the focus of this study. Compensation mechanisms (i.e., homeostatic plasticity) play a critical role in controlling the activity levels of neurons and neural circuits. Excitatory synaptic loss leads to a decrease in the activity levels of neural circuits. Synaptic compensation increases the efficacy of the remaining excitatory synapses to maintain the activity of the network (compensate for synaptic loss).

Here, we model two compensatory mechanisms independently; a global compensation mechanism and a local compensation mechanism. In addition, we develop a third model of synaptic compensation that combines both local and global compensations in parallel. We also analyse the GFS and output power spectra of the lesioned network before and after the onset of synaptic compensation. Homeostatic plasticity is developed over the course of hours to days and possibly over much longer durations. However, it is simulated in this study by updating the remaining excitatory synapses at equally spaced time intervals. The underlying approximation of this separation of timescales is well justified by the fact that the effect of changes in synaptic conductance on firing rates is immediate, whereas homeostatic scaling triggered by changes in activity levels occurs on a much slower timescale [50].

With global compensation, the scaling factor estimate is based on the mean firing rate of excitatory neurons during a 5-s interval. The compensation factor estimates in local and CCM compensation mechanisms are based on the mean firing rate of the excitatory neurons (and the network in case of the global compensation factor in the CCM model) within a 40 s interval to gain an accurate estimate of the activity level of every excitatory neuron.

Rhythms and Asynchronous Firing

The network exhibits an asynchronous firing activity before synaptic degeneration and after synaptic loss with/without synaptic compensation mechanisms. It is suggested that stimulating the neurons with $a \geq 1$ Hz noisy input results in asynchronous firing behaviour, while low frequency input (<1 Hz) causes synchronized spiking activity [68]. This synchronized activity can be observed in networks with a small number of neurons such as in cortical slabs [77, 117] or cultures [118, 119].

The output power spectrum for all conditions is displayed in Fig. 9.5a. Figure 9.5a illustrates the network exhibited cortical-like asynchronous dynamics. The network produces collective rhythmic behaviour in the frequency range corresponding to that of the mammalian cortex in the awake state. However, the power spectra dynamics of the lesioned network (dashed line) demonstrate a significant loss compared to the other cases where compensation mechanisms are applied. After applying compensatory mechanisms, the network recovers its activity. The recovered dynamics with global compensation and CCM have shown a shift to lower frequency bands. This shift is in agreement with the analysed EEG signals from animal models of AD (Fig. 9.5b) and AD patients (Fig. 9.6) [4, 13]. In the following section, this study utilizes the relative band power [1, 29] and the GFS measure [21] to evaluate changes in oscillations across various degrees of synaptic loss as well as to explore the impacts of compensation mechanisms.

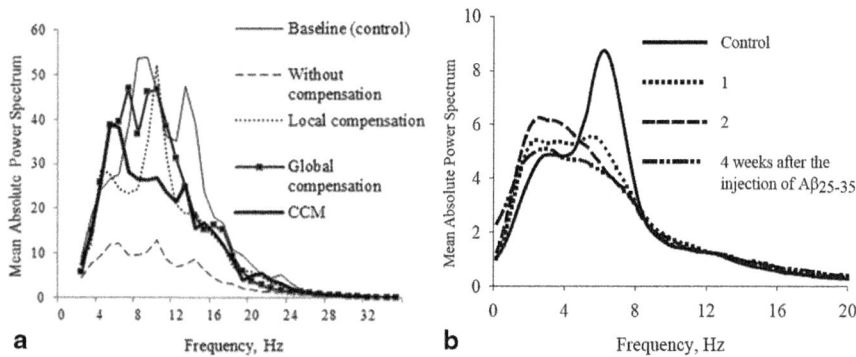

Fig. 9.5 (**a**) The mean power spectra of ten network simulations in each case (five network simulations for the CCM case). The degree of synaptic loss is 50%. With compensation mechanisms, the power spectra are computed after the networks recover their baseline firing rate. The power spectral density vector for each network pattern is computed using Welch's method with 2 s hamming window and 50% overlap between windows. This is a 2 s sliding window over a 30 s time interval. The power spectrum for each network pattern is the average of all the sliding windows and (**b**) the mean power spectra of cortical EEG signal for the adult rats in control and after 1, 2 and 4 weeks of the injection of Aβ25–35 in animal models of AD. (Reprinted from Ref. [4] with kind permission from Springer Science and Business Media B.V. Copyright © 2009. Springer Science and Business Media B.V.)

Neural Correlates of Synaptic Loss and Compensation Mechanisms

Initially, we show the typical changes in the frequency bands in different stages of AD as presented in Fig. 9.7 [69]. Theta band power (6–7.5 Hz) increases in the mild stage of AD (score 3 on the GDS) and enhances further as the disease progresses to the moderate stage (scoring 4 and 5 on the GDS). An initial increase in delta band power is observed in moderate AD (scoring 4 and 5 on the GDS). A sharp increase in the activity of slow frequency bands (2–7.5 Hz) occurs in the sever stage of AD (score 6 on the GDS). The power spectra of faster frequency bands (8–22.5 Hz) is disrupted at this stage [69].

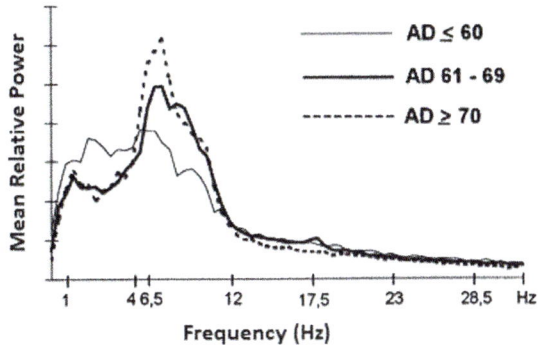

Fig. 9.6 Mean relative power spectra of three cohorts of AD patients categorized according to the age of AD onset. (Reprinted from Ref. [13] with kind permission from Elsevier. Copyright © 1999, Elsevier)

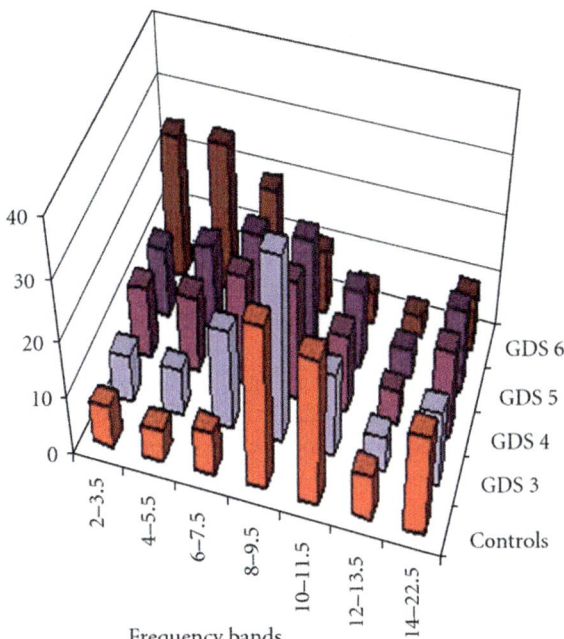

Fig. 9.7 Illustration of the association between AD severity and 7 EEG frequency bands (2–3.5; 4–5.5; 6–7.5; 8–9.5; 10–11.5; 12–13.5; 14–22.5 Hz). Five groups are included: a group of normal controls and four clinical classes of severity; Global Deterioration Scale (*GDS*) ranges from three to six. (Adapted with permission from Ref. [69]

The results of the current modelling study show that synaptic loss underlies the power spectra increase of delta and theta frequency bands with a parallel decrease in beta and alpha frequency powers, as illustrated in Fig. 9.7a, b, c, d.

Figure 9.8a shows the delta band power response to synaptic loss and the relationship to the type of compensation. It is clear that the relative power exhibits an abnormal increase with the increasing degree of synaptic loss (in the absence of the synaptic scaling rules). In contrast, global compensation controls the relative power magnitude up to 60% synaptic loss (then, the relative power increases significantly). In local and CCM compensation mechanisms, the output relative power tends to increase (after 10% synaptic loss) with the increasing degree of synaptic loss up to 50 and 60% synaptic degeneration, respectively. As mentioned in Sect. "Introduction", increased delta band power is one of the EEG hallmarks in AD patients (in later stages of the disease). The relative theta band power [120] increases in earlier stages of synaptic loss (10%) as shown in Fig. 9.8b. Overall, theta band relative power demonstrated a progressive increase with an increased rate of synaptic loss. However, this abnormal increase is less pronounced with local compensation compared to the other cases.

Figure 9.8b illustrates that local compensation can preserve the baseline level even at 40% synaptic loss. Similarly, the CCM case maintains the normal value of theta band power at 40% but thereafter causes the greatest power increase at the next stage. It can be seen also from Fig. 9.8b that local compensation fails to maintain the baseline value of theta relative power after 50% synaptic loss. Therefore, even with local compensation the increase in theta relative power cannot be prevented and is ineffective after this level of synaptic loss. The results of global compensation show the highest increase in relative theta band power after 50% synaptic loss. Before this point, the two cases, synaptic loss with global compensation and synaptic loss without compensation, have almost the same values of relative theta band power. As mentioned in Sect. "Introduction", increased theta band power has been detected in the early stages of AD [2, 120].

Figure 9.8c shows that relative alpha band power is decreased with increasing levels of synaptic loss. Interestingly, with local compensation and CCM, the relative power is also maintained until 50% synaptic loss. Within the beta frequency band, compensation processes have reduced the abnormal decrease in relative beta band power (compared to synaptic loss without compensation as presented in Fig. 9.8d). Decreased alpha and beta band powers have also been reported in the early stages of AD onset.

The detailed analysis in Fig. 9.8 can be generalized via computing the Sum of Squared Differences (SSD) across all degrees of synaptic loss (10–70%) at each frequency band. This global method results in a single index value that reflects the efficiency of the modelled compensation mechanisms in preserving the relative power at a particular frequency band. It can be observed from Table 9.1 that delta and beta relative powers are maximally preserved with global compensation whereas theta and alpha relative powers are maximally preserved with local compensation. For all conditions, the SSD values for the CCM case are in the range between the SSD values of local and global compensation mechanisms. Considering the

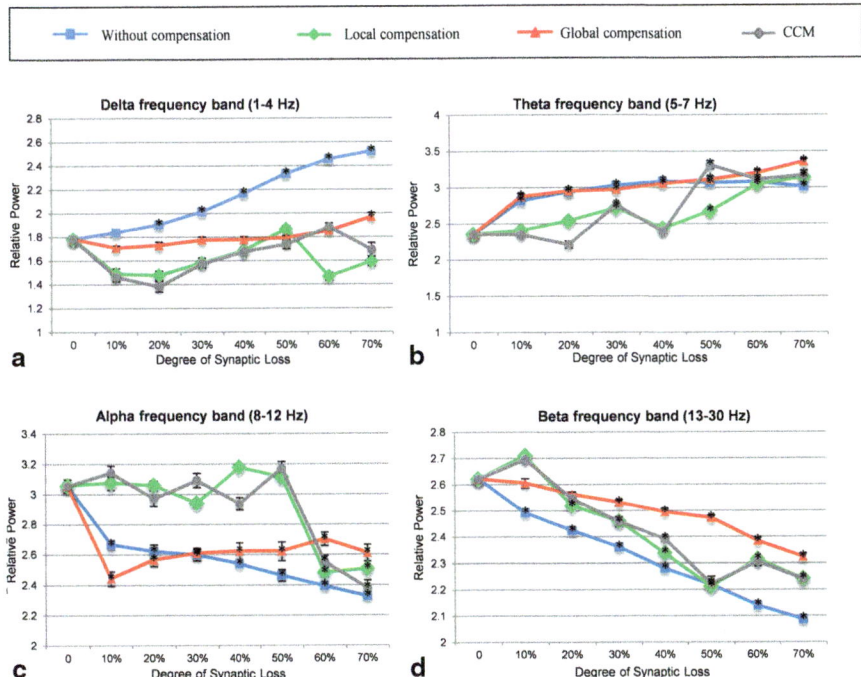

Fig. 9.8 Comparative power spectra analysis of the effects of synaptic degeneration without compensation (*blue line*), synaptic degeneration with local compensation (*green line*), synaptic degeneration with global compensation (*red line*) and synaptic degeneration with CCM (*grey line*) on the relative power of (**a**) delta, (**b**) theta, (**c**) alpha and (**d**) beta frequency band powers. *Asterisk marks* indicate cases where the difference between baseline and loss without/with compensation is significant ($p<0.01$). The demonstrated power spectra represent mean values which are determined from ten simulations with different random patterns of synaptic loss for each degree of synaptic loss. Error bars indicate standard error of the mean (*SEM*). In (**a**) and (**b**) there is clear increase in power in slow bands (delta and theta) and (**c**) and (**d**) show a significant decrease in fast bands (alpha and beta) relative to baseline with increasing synaptic loss (*blue line*). The *blue line* shows varying degree maintaining consistency with baseline when different types of compensation mechanisms are applied

mean SSD value over all frequency bands (the bottom row in Table 9.2), it can be concluded that local compensation can best maintain the spectral power dynamics in the range (1–30 Hz) as compared to other compensation mechanisms.

Perturbations on the frequency bands can also be analyzed using the Global Field Synchronization (GFS) measure [21]. Figure 9.9a shows that the mean GFS for the delta band before applying compensation increases remarkably, for local compensation increases are observed at 10, 30 and 50 % whereas for global compensation the minimal GFS increase is observed compared to the other cases. A marked decrease in mean GFS is observed at 30 and 50 % for the CCM case preceded by an early increase at 10 % synaptic loss. Increased GFS values in the delta band has been reported in an EEG study that involves 419 healthy and AD

Table 9.2 SSD values reflect the global preservation of relative power for a given frequency band. The optimal values (closest to the baseline) at each frequency band are marked with bold font. Rows represent frequency bands and columns represent the modelled cases. The bottom row presents the mean SSD value over all frequency bands for each case

	Without compensation	Local compensation	Global compensation	CCM compensations
Delta frequency band (1–4 Hz)	0.217346	0.053018	*0.006962*	0.047723
Theta frequency band (5–7 Hz)	0.426037	*0.194941*	0.536856	0.333622
Alpha frequency band (8–12 Hz)	0.305937	*0.094795*	0.217464	0.107403
Beta frequency band (13–30 Hz)	0.131522	0.075926	*0.028224*	0.070483
Mean	0.270211	*0.10467*	0.197377	0.139808

particpants [21]. However, a more recent EEG study (22 AD patients and 23 age-matched healthy controls) reported a minor decrease in GFS values for AD pateints at delta band [25].

The mean GFS for the theta frequency band without compensation persistently increases. Local compensation can maintain the mean GFS for theta band at moderate (40%) and late (70%) stages of synaptic loss. However, it triggers abnormal increases in the early stages as presented in Fig. 9.9b. CCM and global compensation mechanisms break down at 40 and 50% synaptic loss (though for CCM an early decrease is observed), respectively. Koenig et al. [21] have not found significant GFS alterations at theta frequency in AD pateints. On one hand, a nonsignificant GFS increase in the theta frequency band has been reported for EEG datasets aquired in New York (264 healthy and AD subjects), whilst GFS analysis of another EEG database from Stockholm (155 healthy and AD particpants) has shown a nonsignificant GFS decrease in the theta frequency band which is consistent with the EEG study in [25].

The mean GFS values at alpha frequency band are shown in Fig. 9.9c. The mean GFS with local compensation is closer to the baseline value compared to other compensation mechanisms. Local compensation and global compensation mechanisms break down at 40 and 50% synaptic loss, respectively. The greatest differences are generated with CCM which can also be observed in the beta frequency band (Fig. 9.9d). The GFS values in Fig. 9.9d are relatively reduced with the increasing degree of synaptic loss. Local compensation and global compensation mechanisms can maintain the GFS dynamics until 60 and 70% synaptic loss, respectively. Decreased GFS values in alpha and beta frequency bands have been reported in EEG studies in AD [21, 25]. Table 9.2 shows that GFS values at delta and beta frequency bands are maximally preserved with global compensation whereas local compensation can best maintain the GFS values in theta and alpha frequency bands. From the mean SSD value over all frequency bands (the bottom row in Table 9.2), it can be concluded that local compensation can best maintain the GFS values in the range (1–30 Hz) compared to other compensation

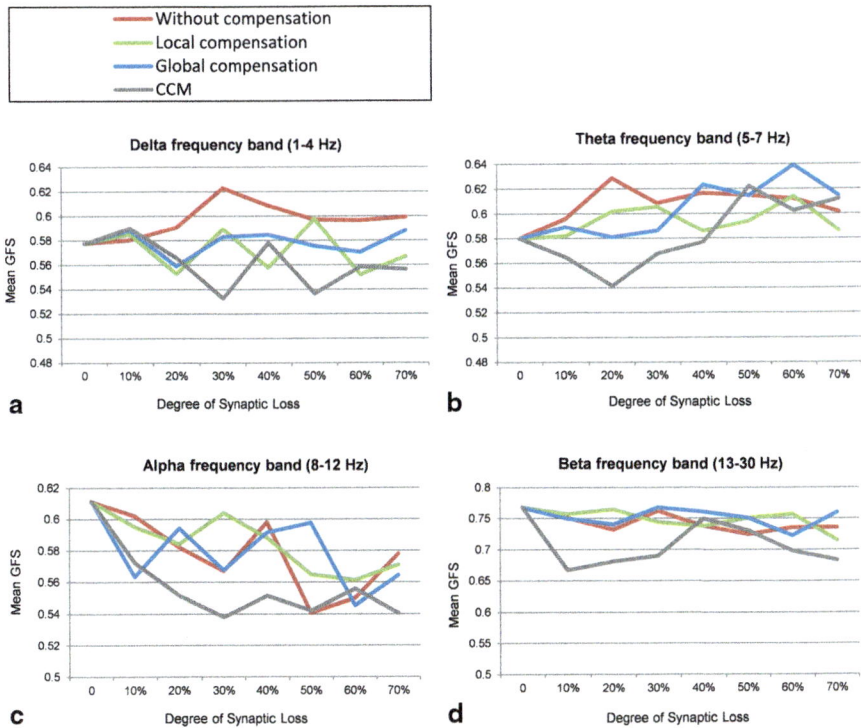

Fig. 9.9 Mean GFS values for (**a**) delta, (**b**) theta, (**c**) alpha and (**d**) beta frequency bands under four conditions: synaptic degeneration without compensation (*red lines*), synaptic degeneration with local compensation (*green lines*), synaptic degeneration with global compensation (*blue lines*) and synaptic degeneration with CCM (*grey lines*)

mechanisms. These observations are relatively correlated with the frequency band findings shown in Table 9.2.

The Effect of Compensation Mechanisms on the Organization of Spiking Activity Patterns Across the Network

Global compensation changes the distribution of firing rates across excitatory neurons. Global compensation stimulates silent excitatory neurons (0 Hz) and increases the fractions of low activity excitatory neurons (1–3 Hz) and active excitatory neurons (4–10 Hz). Furthermore, global compensation could not preserve the performance of higher-activity neurons (11 Hz or more); the fraction of high-activity excitatory neurons decrease as a function of synaptic loss. The correlation coefficient (R^2) of the linear regression lines (Fig. 9.10) in global compensation shows a linear relationship between the rate of synaptic loss and the fraction of excitatory neurons. The significant linear changes can also be ob-

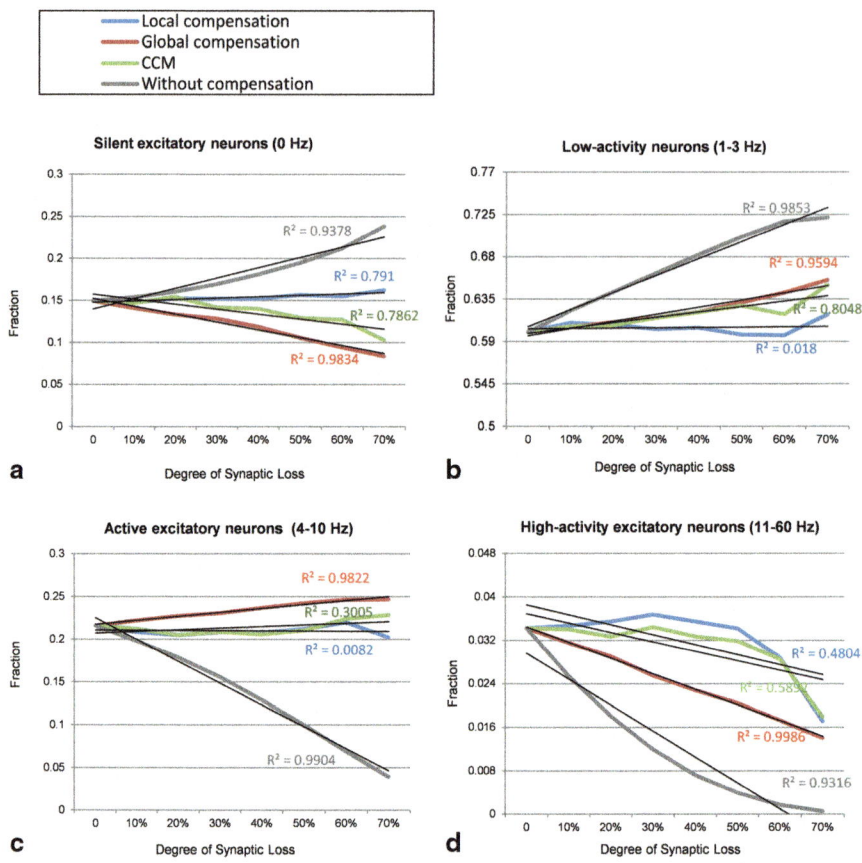

Fig. 9.10 Distribution of firing rates of (**a**) silent excitatory neurons (0 Hz), (**b**) low-activity excitatory neurons (1–3 Hz), (**c**) active excitatory neurons (4–10 Hz) and (**d**) high-activity excitatory neurons (11–60 Hz) without applying compensation (*grey line*) as well as after recovering the activity level with local compensation (*blue line*), global compensation (*red line*) and CCM (*green line*). *Straight lines* represent linear regression lines. The correlation coefficient (R^2) of the linear regression lines are shown in the figures

served before applying synaptic compensation. The results in Fig. 9.9a, b, c, d reveal that global synaptic scaling reorganizes network dynamics while recovering the activity level of the network (Fig. 9.11). On the contrary, local compensation mechanism maintains firing rate distributions close to their baseline value. The output histograms of the CCM case (the green lines in Fig. 9.10) range between the outputs of local compensation and global compensation mechanisms (blue and red lines, respectively).

We therefore complemented this comparative analysis of compensation mechanisms by quantifying the temporal structure of low-activity (0–3 Hz), active (4–10 Hz) and high-activity (11–60 Hz) excitatory neurons after recovering/maintaining the baseline activity level. A more detailed analysis of the 50 % synaptic loss

9 Modelling Cortical and Thalamocortical Synaptic Loss and Compensation ...

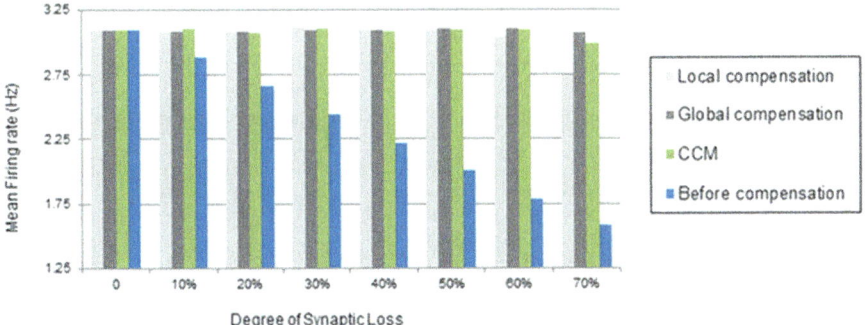

Fig. 9.11 The mean firing rate of the network during synaptic loss (0–70%) without applying compensation (*blue bars*) as well as after recovering the activity level with local compensation (*light grey bars*), global compensation (*dark grey bars*) and CCM (*green bars*)

case is conducted to estimate the frequency shift of the above neuronal types. The frequency shift Δf for time point t_2 is defined as the arithmetic difference between the firing rate at time point t_2 (after compensation) and the initial firing rate before synaptic loss. If the activity of neurons is higher at t_2 than before synaptic loss, then Δf is positive. Negative values of Δf indicate that the neurons have not recovered their baseline activity levels.

As mentioned previously (see Fig. 9a and b), global synaptic scaling stimulates silent excitatory neurons to recover the activity level of the network. This observation is in agreement with the results in Fig. 9.12a where we can see that the frequency-shift distributions (in global compensation) are more skewed toward positive values. Specifically, we find that: (1) the fraction of neurons with 1, 2 and 3 Hz frequency-shifts are higher than the cases with negative frequency-shifts (−1, −2 and −3 Hz, respectively); (2) the fraction of neurons with −4 Hz or less frequency-shifts is zero; and (3) the fraction of neurons with frequency-shifts between 8 and 16 Hz (data not shown) have positive values but they are much lower than the cases between 1 and 7 Hz. The aforementioned frequency shift histogram explains the decrease in the fraction of silent excitatory neurons and the increase in the fraction of low-firing and active excitatory neurons. Local compensation, on the contrary, recovers the activity levels of the network (and individual excitatory neurons) with minor changes on the network dynamics. As demonstrated in Fig. 9.12a (green line), 35 % of silent excitatory neurons (0–3 Hz) have not changed their firing frequency, 23.9 % of silent excitatory neurons have increased their firing by 1 Hz, 23.0 % of silent excitatory neurons have decreased their firing by 1 Hz (this would compensate for the 1 Hz increase), 7.86 % of silent excitatory neurons have increased their firing by 2 Hz and 6.2 % of silent excitatory neurons have decreased their firing by 2 Hz (this would almost compensate for the 2 Hz increase). Overall, the arithmetic summation of these fractions is 95.96 %. This explains why local synaptic scaling preserves the network dynamics while recovering the system activity level. It is worth mentioning that low-activity (0–3 Hz) excitatory neurons constitute 75 % of excitatory neurons in the network while active (4–10 Hz) and

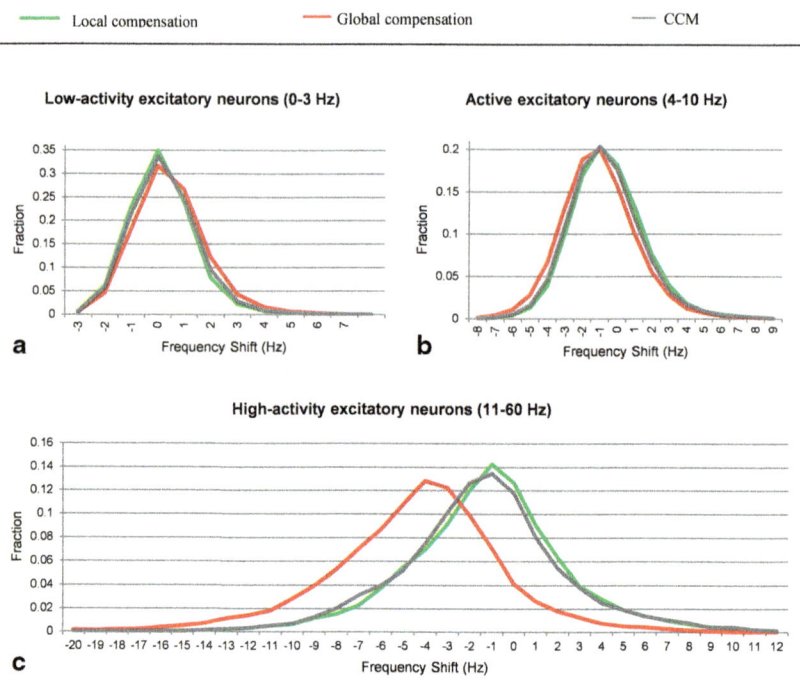

Fig. 9.12 Distribution of frequency-shifts for (**a**) low-activity excitatory neurons (1–3 Hz), (**b**) active excitatory neurons (4–10 Hz) and (**c**) high-activity excitatory neurons (11–60 Hz), after recovering the activity level with local compensation (*green line*), global compensation (*red line*) and CCM (*grey line*)

high-activity (11–60 Hz) excitatory neurons constitute 21.6 and 3.4% of excitatory neurons, respectively, in the baseline case (see Fig. 9a and b).

In Fig. 9.12b, it is shown that with global compensation, the frequency-shift distribution for active excitatory neurons exhibits a higher move towards negative values than the case with local compensation. A comparison of the fractions at both sides of the histogram reveals that the arithmetic differences between the fractions for each opposite pair of data points (e.g. 1 versus -1 Hz) in local compensation is lower than global compensation. This negative shift underlies the increase in the fraction of low-activity cells (in addition to other factors, as mentioned above). We also find that: (1) the minimum negative frequency-shift for global and local compensation is -10 and -9 Hz, respectively; (2) the maximum positive frequency-shift for global and local compensation is 17 and 20 Hz, respectively; and (3) the integrated fraction for -8 Hz or less frequency-shifts is 0.119% in global compensation and 0.0371% in local compensation, while the integrated fraction for 8 Hz or more frequency-shifts is 0.234% in global compensation and 0.292% in local compensation. The frequency-shift distribution in global compensation has also shown a distinct move to the left band in the case of high-activity neurons (see Fig. 9.12c) in agreement with the demonstrated results in Fig. 9.10d which reflect a failure in maintaining the firing rate of high-activity excitatory neurons with global compensation. The output of the histograms of the CCM case range between the outputs of local compensation and global compensation mechanisms in Fig. 9.11a and b. The histograms in Fig. 10c show that local compensation and the CCM case have similar distribution of frequency-shifts in line with the findings in Fig. 9.10d.

Study 2– Thalamacortical Model

Dynamics of the Thalamacortical Model in the Baseline Condition

The network exhibits an asynchronous firing activity in the baseline condition as shown in Fig. 9.13. The output power spectrum of the cortical network in Fig. 9.14a illustrates that the network oscillates at 10 Hz. The spectral analysis of the thalamic system shows that the power peak appears at 9 Hz, as shown in Fig. 9.5b.

EEG recordings from healthy and MCI subjects in the wakeful relaxed state with eyes closed show a power peak at 10 Hz for young healthy adults, 9.5 Hz for elderly healthy subjects and 9 Hz (or less) in MCI subjects [17] as demonstrated in Fig. 9.15 (the configuration of the electrodes is provided in Fig. 9.16).

The simulated LFPs of thalamic and cortical networks in Fig. 9.17 shows correlated activity between the two networks. The corticothalamic feedback together with the gap junctions (among TCR neurons) underlies the observed alpha oscillatory activity in the thalamic network [82]. It has been found that the strong

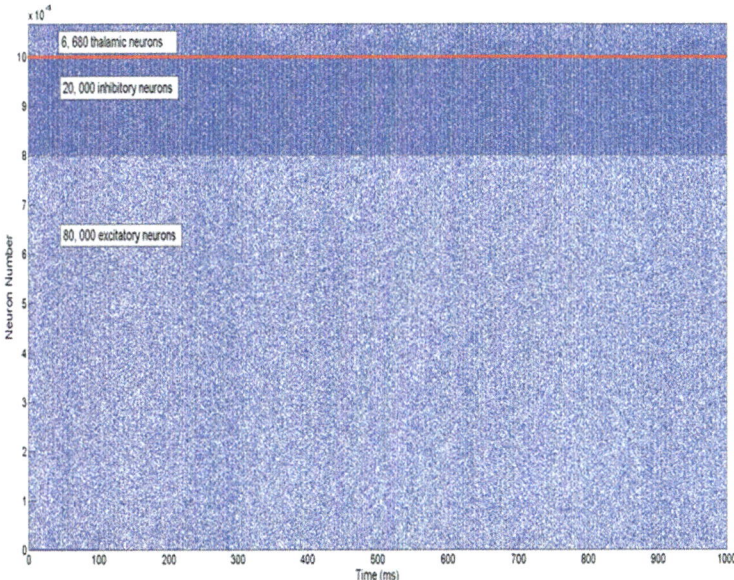

Fig. 9.13 A rastergram illustrates the asynchronous firing activity of the neurons

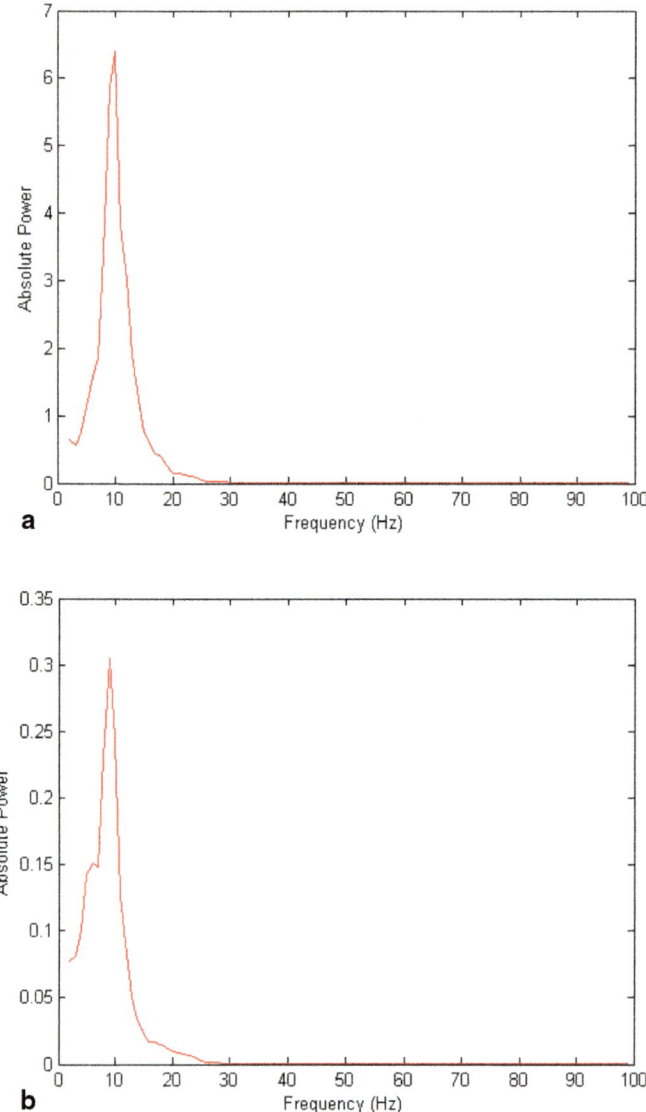

Fig. 9.14 Results of spectral power analysis of the thalamic and cortical modules. Averaged power spectra of (**a**) the cortical network and (**b**) the thalamic network. The number of network patterns (trials) is 10. The power spectral density vector for each network pattern is computed using Welch's method with 2 s hamming window and 50% overlap. This 2 s window is a sliding window over 60 s time interval. The power spectrum for each network pattern is the average of all the sliding windows

activation of the metabotropic glutamate receptor (mGluR), mGluR1a, that is located postsynaptically to corticothalamic fibers, results in thalamic alpha waves while a weak activation leads to thalamic theta waves (that is associated with decreased arousal state).

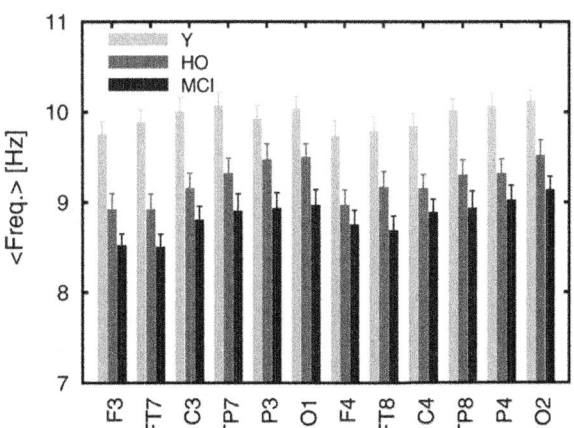

Fig. 9.15 Averaged frequency of the power spectrum peak of the EEG signals recorded from 12 electrodes. The graph includes three categories: young (Y), healthy old (HO) subjects, and MCI patients. (Reprinted from Ref. [17] with kind permission from Elsevier. Copyright © 2010, Elsevier)

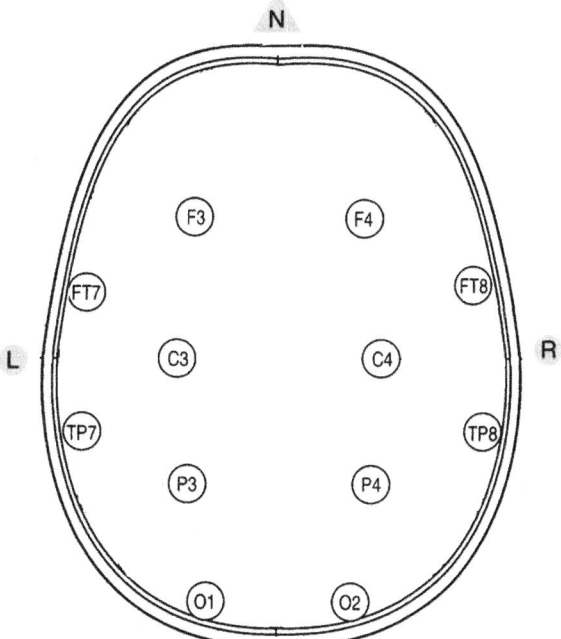

Fig. 9.16 Configuration of the electrodes' positions. (Adapted and modified from Ref. [121]. Frontal (F), Central (C), Parietal (P), Occipital (O), and Temporal (T))

The thalamus is situated in the middle of the human brain. Therefore, the definitive contribution of human thalamus to the recorded EEG signals at the scalp surface (non-invasively) cannot be identified. Using the invasive intracranial EEG (iEEG) recording from brains of dogs, a synchronised occurrence of alpha waves in visual cortical and thalamic regions as well as correlated LFP signals recorded from both systems have been observed [82]. This section has revealed the rhythmic activity in both, the thalamic and the cortical modules in the baseline condition. Thalamic alpha activity appears simultaneously with cortical alpha activity as seen

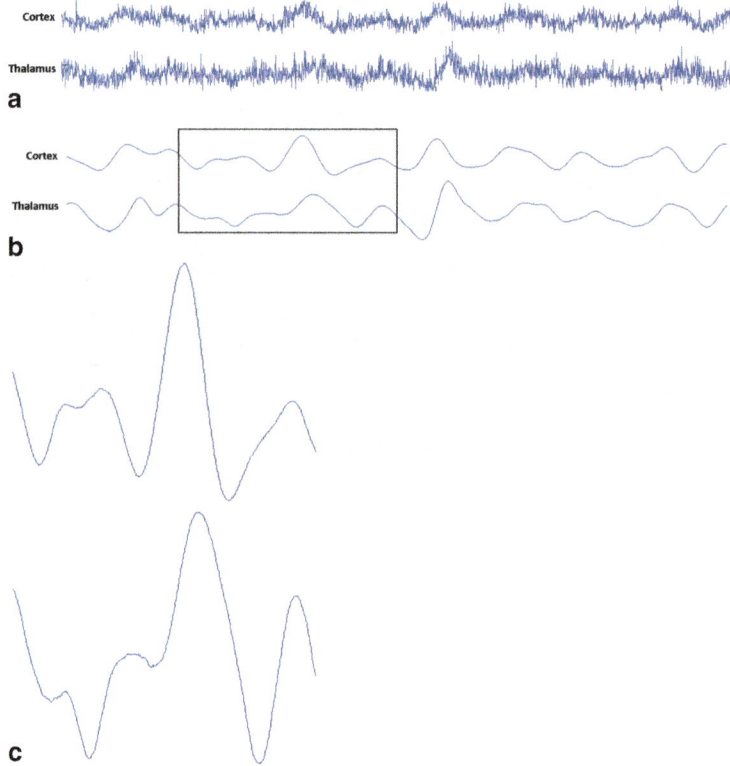

Fig. 9.17 The simulated LFP of thalamic and cortical networks. (**a**) Raw LFP signals of length 1800 ms, (**b**) smoothed LFP signals of length 1800 ms (calculated using a Gaussian kernel with SD = 20 ms, see [50] and (**c**) enlarged section of the smoothed LFP signal (length equals 600 ms)

in Fig. 9.17. The results from power spectra analysis and LFP signals from the network model are in line with previous experimental studies [17, 82] and confirm the plausibility of the model. The follow outlines how the verified model is used to investigate the influence of various cases of connectivity loss on the power spectral dynamics of the modelled network.

Spectra and ERD/S Plots for Various Cases of Connectivity Loss and Compensation

The thalamocortical model described above has been used to study the influence of different types of structural disconnections among cortical, thalamic and reticular areas. This section presents an analysis of the behaviour of thalamic and cortical networks in the thalamocortical model with various degrees of connectivity loss based on visualisation and analytical EEG methods, namely the spectrogram and ERD/S analysis. Variations of synaptic degeneration are assumed to reflect different stages of AD where 10 and 60 % synaptic loss correspond to early and later stages of AD, respectively.

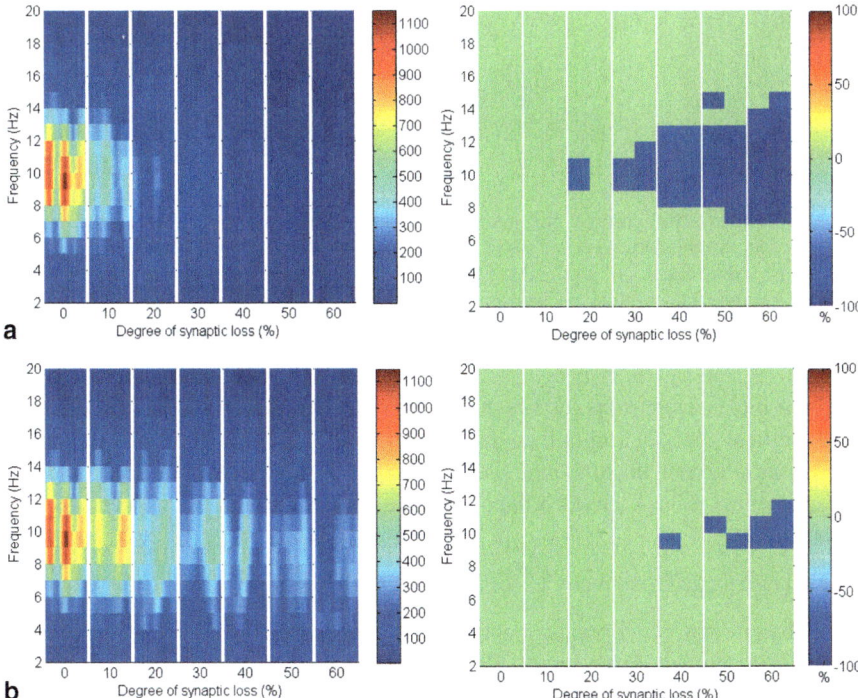

Fig. 9.18 The effect of corticocortical connectivity loss on cortical oscillations. Spectrograms (*left column*) and the corresponding ERD/S diagrams (*right column*) of the cortical oscillatory activity at various degrees of corticocortical connectivity loss. (**a**) Before synaptic compensation and (**b**) after applying compensation

- Corticocortical connectivity loss

The effect of modelling corticocortical synaptic loss is shown in Figs. 9.18 and 9.10. The plots reveal that a significant power spectra decrease ($P<0.05$) occurs in cortical alpha band, specifically, in middle alpha frequency band (9–11 Hz) as demonstrated in Fig. 9.18a.

Upper alpha band (12 Hz) power is affected when the decrease approaches (and exceeds) 30 % synaptic loss. It is also observed that delta (2–4 Hz) and lower theta (5–6 Hz) bands power is not affected even with a massive loss of synapses whereas upper theta band (7 Hz) power is significantly affected after 50 % synaptic loss. In Fig. 9.19a, a similar effect can be seen with respect to the thalamic module in a later (30 %) rather than an earlier (20 %) stage compared to the modelled cortex, a finding which suggests that the abnormal thalamic activity is induced by a disruption in corticothalamic drive. Interestingly, the cortical and thalamic ERD/S outputs (after synaptic compensation) in Figs. 9.18b and 9.19b respectively have confirmed the correlated spectral changes among both modules as can be seen at 50 % synaptic loss.

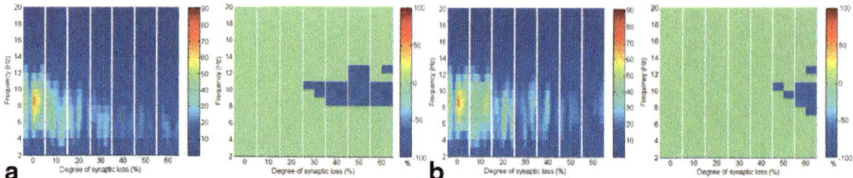

Fig. 9.19 The effect of corticocortical connectivity loss on thalamic oscillations. Spectrograms (*left column*) and the corresponding ERD/S diagrams (*right column*) of the thalamic oscillatory activity at various degrees of corticocortical connectivity loss. (**a**) Before synaptic compensation and (**b**) after the compensation mechanism is applied. For presentation purposes, the spectrogram plots have been filtered with the Matlab (MathWorks) function *filter2()*

In the most severe stage (60 % synaptic loss), compensation mechanism recovers more cortical spectral output than thalamic output. This is not surprising, since the compensation mechanism is modelled by increasing the weights of corticocortical synapses, thus it has a direct influence on the cortex and indirect influence on the thalamus.

- Thalamocortical connectivity loss

Next, the effect of abnormal dismantling of thalamocortical contacts on the cortical surface is examined. From Fig. 9.20a and b, it can be seen that the cortical spectral activity is not affected by this deafferentation. The same behaviour is observed in the thalamic module in Fig. 9.11c and d.

- Corticothalamic connectivity loss

The synaptic connectivity of the cortical afferents pathways to the thalamic neurons controls oscillatory activity in the thalamic module as confirmed by experimental [122, 123] and theoretical studies [124]. Decortication (i.e., removal of the cortical slice) results in synchrony loss in the thalamus [122–124].

The findings of this modelling study show that thalamic activity is diminished at an early stage starting with a clear shift to low frequency waves at 10 % connectivity loss followed by a marked loss of activity in the next stages as presented in Fig. 9.22a. On the contrary, cortical oscillations are not significantly affected by this loss, as demonstrated in Fig. 9.21a.

Applying compensation mechanism for the thalamic network assists in (partial) recovery of the network activity as displayed in the ERD plot in Fig. 9.22b. The compensation mechanism in this case is modelled by scaling down the inhibition from RTN neurons. Taking corticothalamic connectivity loss before compensation (Fig. 9.22a) together with the case after reducing RTN to TCR inhibition (Fig. 9.22b), the data indicates that the spiking activity of the inhibitory RTN neurons is another important factor contributing significantly to the breakdown of thalamic activity. Scaling down this factor helps excitatory TCR neurons to recover (partially) it's spiking activity. The reader can also see that the decrease ratio (darkness of the blue colour) in Fig. 9.22b is less than that in Fig. 9.22a.

9 Modelling Cortical and Thalamocortical Synaptic Loss and Compensation ... 257

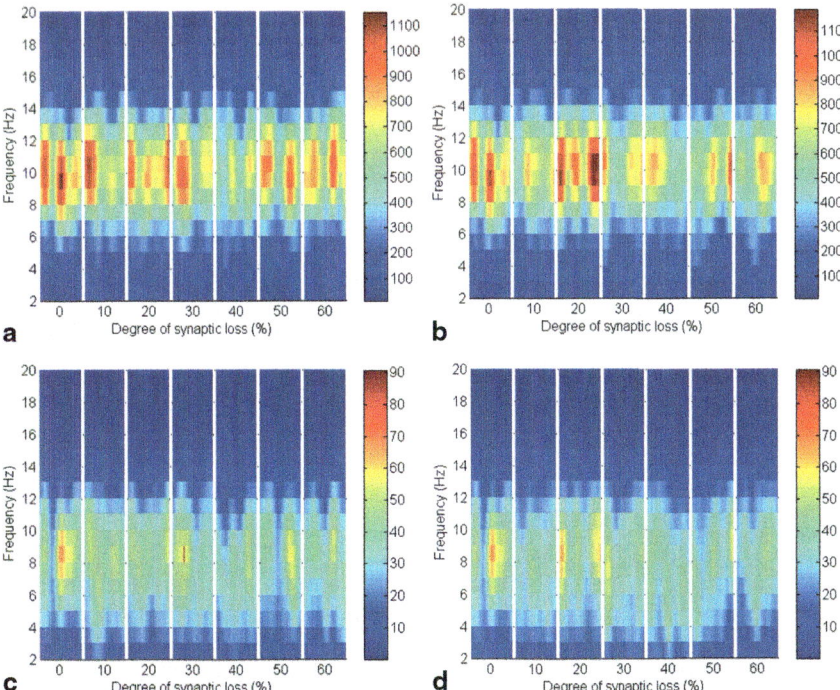

Fig. 9.20 The effect of thalamocortical connectivity loss on cortical and thalamic oscillations. Spectrograms of the cortical (**a** and **b**) and thalamic (**c** and **d**) LFPs after thalamocortical connectivity loss. The *left*- and right-hand side figures correspond to the case synaptic loss before compensation and the case synaptic loss after compensation, respectively. Significant changes have not been observed (all regions have a *green colour* as in the first bar of the ERD/S diagrams in Figs. 9.18b and 9.19b)

- Corticoreticular connectivity loss

As mentioned earlier, RTN neurons play a crucial role in regulating thalamic oscillatory activity. Any impairment in this circuit (due to neuronal death or loss of afferent input from other areas) is expected to have a direct influence on the thalamus. This section presents the results of partial deafferentation of RTN neurons modelled by a loss of cortical contacts on the RTN surface. The results in Fig. 9.23a and b show that cortical oscillations are not significantly affected by this loss. On the other hand, the thalamic oscillatory activity is disturbed as seen in Fig. 9.24a despite the significant decrease in RTN—TCR inhibition and the increase in TCR firing rate. Interestingly, the spectral dynamics of the thalamic network is fully recovered when decreasing the mutual inhibition in the RTN population. This helps RTN neurons depolarize recovering their firing rate and regulates the thalamic oscillatory activity as shown in Fig. 9.24b.

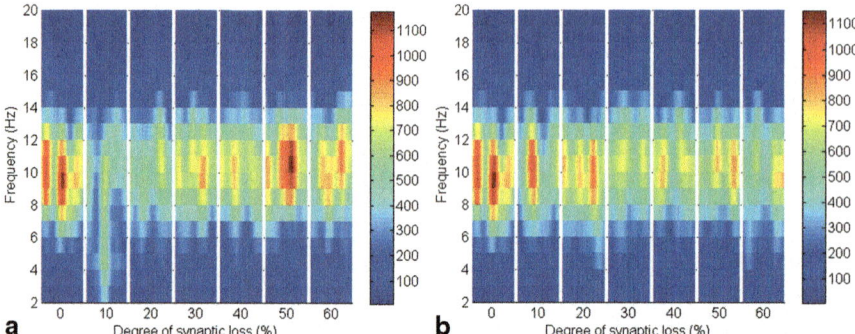

Fig. 9.21 Responses of the cortical module to corticothalamic disconnection. (**a**) Before synaptic scaling and (**b**) after applying synaptic scaling. Significant changes have not been observed in the cortical spectrograms

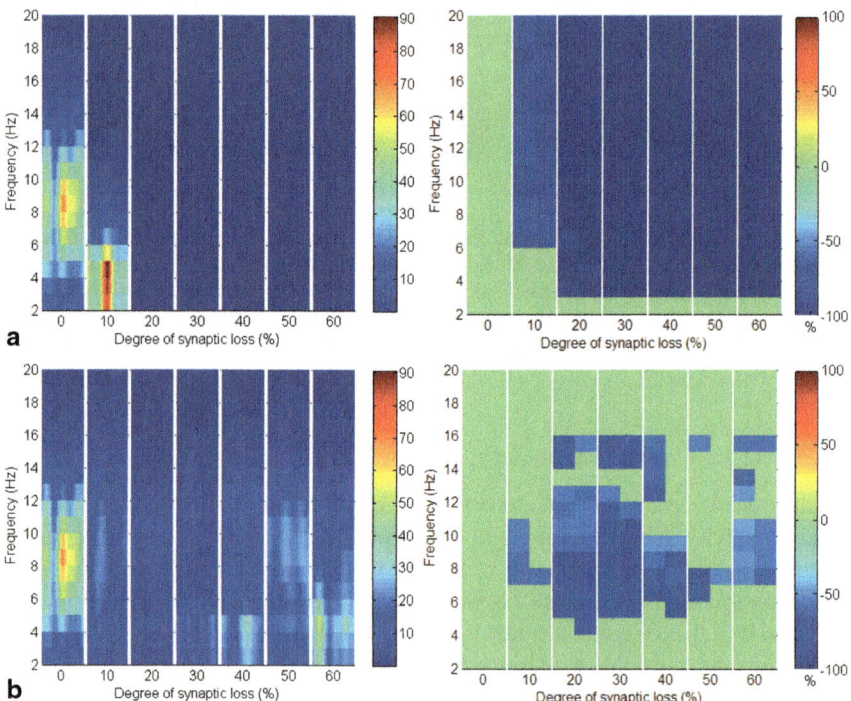

Fig. 9.22 Responses of the thalamic network to corticothalamic disconnection. (**a**) Before synaptic scaling and (**b**) after applying synaptic scaling

9 Modelling Cortical and Thalamocortical Synaptic Loss and Compensation … 259

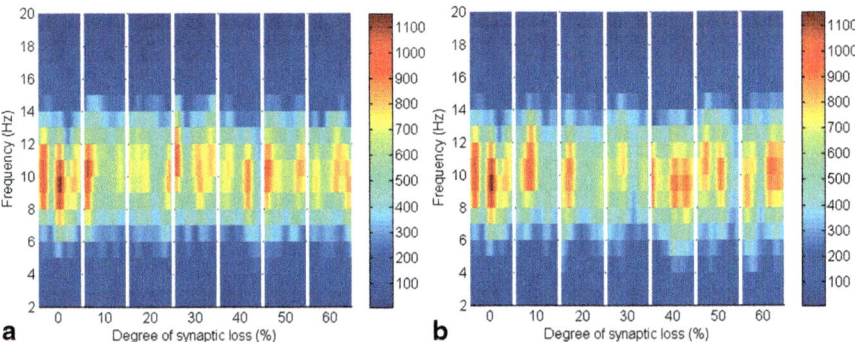

Fig. 9.23 The effect of corticoreticular connectivity loss on the cortical network

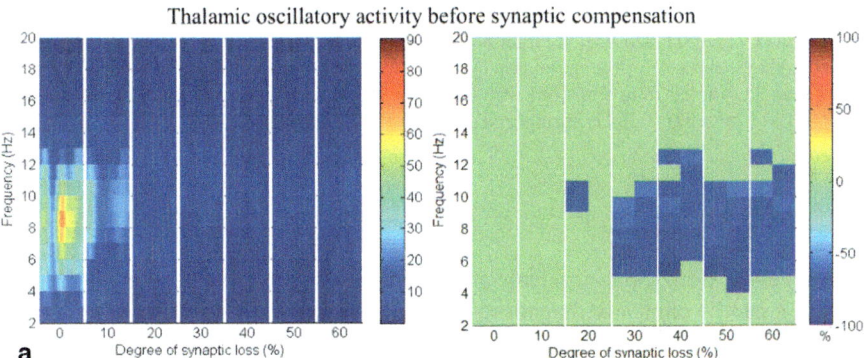

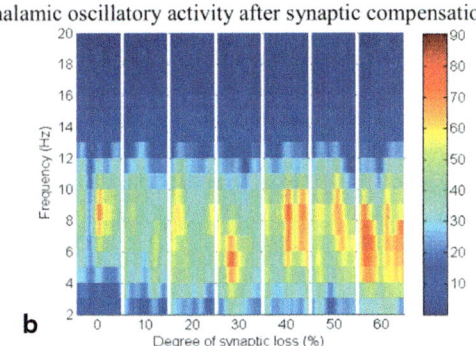

Fig. 9.24 The effect of corticoreticular connectivity loss on the thalamic network. (**a**) Thalamic oscillatory activity before synaptic scaling and (**b**) after applying compensation mechanism (significant changes have not been observed after synaptic compensation)

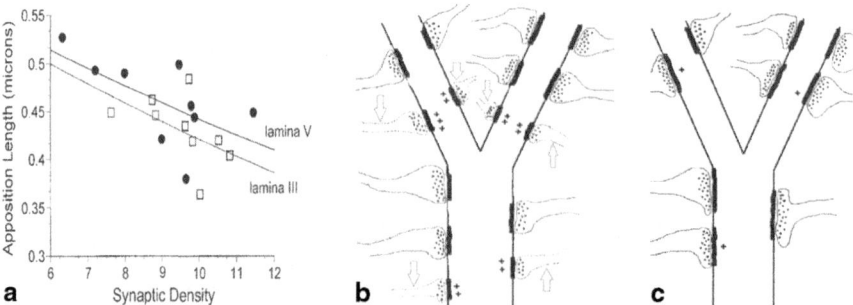

Fig. 9.25 Findings of the experimental study [66]. (**a**) The interplay between the synaptic density (number of synapses) and the apposition length (synaptic size) in lamina III and V of the frontal cortex (Brodmann area 9) in AD patients (Points indicate subjects; open symbols represent lamina III and filled symbols represent lamina V). Regression analysis show that the size of the remaining synapses increased when the synaptic density declines [66] in turn adapted the data from Ref. [67], (**b**) a schematic diagram of speculated neuronal-based responses to synaptic loss (*red arrows*) in early AD where the denervated target neuron sends requests (+) for increased synaptic input; In later AD (**c**) the size of the remaining synapses were increased whereas the signals (+) were decreased. (Reprinted from Ref. [66] with kind permission from Springer Science and Business Media B.V. Copyright © 2003. Springer Science and Business Media B.V.)

Discussion

AD develops slowly, progressively and silently over several years or even decades before it can be clinically diagnosed [125]. Synaptic loss is the earliest pathological change in AD as well as the most correlated with cognitive decline [125]. In addition to synaptic loss, an experimental study observed an increase in the size of the remaining synapses as a function of synaptic density in lamina III and V of the frontal cortex in AD patients, as illustrated in Fig. 9.25a. This behaviour has also been observed in the superior and middle temporal gyrus [66, 126]. This reaction mechanism is able to counteract synaptic loss by maintaining the mean total synaptic contact area per unit volume of the cortex. This has been further explained by the proposed model in [66] which suggests that the denervated target neuron informs (via signalling) the remaining afferents about its need for synaptic input, as illustrated in Fig. 9.24b and c.

Study 1

In this computational modelling study of study 1, we combined synaptic loss with compensation mechanisms in our investigations of the underlying neuropathological process of abnormal cortical oscillations in AD. The work is based on the model proposed by [68]. To simulate the severity of AD, we simulated the models with different fractions of random synaptic loss. Our choice of random synaptic loss is based on the phenomenological approximation that cortical disorders result in a random degeneration in cortical networks [50]. We investigated local as well as global compensation mechanisms. Detection and control of activity levels can occur at the network level, neuronal level or synaptic level [44, 50, 58]. The performance of the simulated compensation mechanisms in this study (described in Sect. "Thalamic-Reticular Connections") depends on the rate parameter (ε) and the target and current firing rates of the network. Experimental and computational studies have shown that synaptic homeostasis changes as a function of age [61, 127]. Thus, various values of the rate parameter (ε) may correspond to different ages (or resistance toward cortical damage) of the patients [61].

As mentioned earlier, EEG abnormality in AD is associated with increased theta (in early stages) and delta (in late stages) band powers and decreased alpha and beta band powers in early stages [1, 2]. This work is mainly concerned with evaluating the GFS and power spectra output of the simulated networks after incorporating more realistic neural processes. We have explored the impact on the network in the absence of compensation mechanisms during synaptic loss by analyzing the GFS and power spectra of the lesioned network before activating the synaptic scaling rule (the case 'synaptic loss without compensation' in Sect. "Spectra and ERD/S Plots for Various Cases of Connectivity Loss and Compensation"). Based on the GFS and power spectra analysis of the output spike trains in the lesioned network (without compensation), we observed that the structural impairment in the excitatory network has (1) increased relative delta and theta band powers and GFS as a function of synaptic loss; (2) diminished relative alpha and beta band powers and GFS (as a function of synaptic loss); and (3) changed relative theta band power at an earlier stage than delta band power changes (as a function of synaptic loss). All these outputs are compatible with EEG studies on AD [2, 16, 29, 128, 129].

The same power spectra and GFS analysis is repeated after the lesioned network has recovered its baseline activity level via the global scaling rule. Overall, the results show that global compensation has reduced the abnormal power spectra and GFS changes across all frequency bands, in particular, delta and beta frequency bands. However, it triggers network reorganization when stimulating silent excitatory neurons. A relevant computational study suggested that global homeostatic synaptic scaling can cause secondary pathological effects in severe deafferentation cases (80–100%) via modulating the distribution of firing activities across the network and triggering the occurrence of periodic EEG patterns at slow frequency bands [50]. The presented study observes abnormal increase of relative theta power with a parallel decrease in relative alpha power in the global compensation model. In the CCM model, it is shown that the power spectra dynamics can be recovered compared to the case 'without compensation' whereas the model fails at maintaining the baseline GFS values.

Local compensation mechanism has maintained the network dynamics. The GFS and power spectra dynamics at theta and alpha frequency bands can be best recovered with local compensation. Increasing the rate parameter (ε) may enhance the stability of the network and the neuron numbers (neuronal death results when the synaptic input decreases below a certain threshold) as suggested by experimental and computational studies [61, 66]. Indeed, it has been speculated that AD neuropathology is characterized by a failure of local synaptic scaling mechanisms when the synaptic input decreases below a certain threshold [61]. The finding of disturbed neuronal-based synaptic homeostatic plasticity mechanisms in AD have been derived from experimental results of deficits in the structural plasticity of dendrites and axons [125] and computational modelling studies [61, 130] which may point to a deregulation of a synaptic turnover mechanism [125]. According to our GFS and power spectral analysis, deactivating local compensation mechanisms will result in rapid decline (cognitive deficit) of the network dynamics at theta and alpha bands. Therefore, methods which can enhance local compensation could play a major role in the stimulation of neural processes and cognitive functions that are associated with these frequency bands. Similarly, some neural (cognitive) processes which are associated with delta and beta bands can be maintained by stimulating global compensation. Such observations imply that various cortical areas employ several types of synaptic compensation mechanisms. As compensating for synaptic loss is speculated to differ from one cortical area to another, the study suggests that activating an inappropriate compensation mechanism in a particular area may fail to recover the network dynamics and/or may induce secondary pathological changes in the network. This speculation is supported by the observation that local compensation fails at recovering/maintaining the baseline delta and beta oscillations whilst theta and alpha oscillations are least preserved with global compensation.

Compensation mechanisms have shown positive outcomes in therapeutic solutions. An experimental therapeutic study has already found that stimulating synaptogenesis (with quercetin and bilobalide) can aid in AD treatment and prevention [131]. Quercertin vitamin is derived from plants and plays a neuroprotective role in the brain [132]. Bilobalide can be found in Ginkgo leaves. It can reduce the production of amyloid β-peptide (Aβ) and soluble amyloid precursor protein β-peptide (sAPPβ) in the brain [133].

Limited knowledge is available about neuropathological processes in AD. The findings in this work provide valuable insights for experimentalists to conduct laboratory experiments on compensation mechanisms during several stages of AD and to investigate the impacts on the EEG dynamics. Our conclusions are inherently restricted by the biological details embedded in the computational model (as for any computational modelling study). Although the model lacks cortical layers, it preserves fundamental features of a mammalian cortex [68]. In particular, the model includes neurons firing in a variety of known types of spiking and bursting patterns. The neurons have receptors with AMPA, NMDA, $GABA_A$ and $GABA_B$ kinetics and interconnected with 8.5 million synapses. Synaptic dynamics include short-term synaptic plasticity [68]. As mentioned in Sect. "Cortical Model Structure", the model preserves important ratios and relative distances of the cortex with biologically plausible parameters for the modelling of neuronal and synaptic dynamics [68, 71]. The spectral analysis of spike trains illustrate that the model can oscillate in the frequency bands associated with the awake state (Figs. 9.4 and 9.7). To the best of the authors' knowledge, this is the first study to explore compensation mechanisms using a model with these characteristics.

The study has provided novel evidence linking abnormal brain oscillations to compensation mechanisms in synaptic loss. Another novelty is in the design of the local compensation mechanism and the parallel model of combined compensation mechanisms (CCM) as well as the comparison of the power spectra dynamics among four groups (in addition to the baseline group): synaptic loss (1) without compensation, (2) with global compensation, (3) with local compensation and (4) with combined compensation mechanisms (CCM).

Study 2

Study 2 involved modelling structural synaptic impairment of fundamental connectivity types based on a novel thalamocortical network model consisting of different types of cortical excitatory and inhibitory neurons recurrently connected to thalamic and reticular thalamic regions with the ratios and distances found in the mammalian thalamocortical system. The model includes short- and long-range corticocortical couplings, synaptic transmission with AMPA, GABA, NMDA and GJ kinetics, short-term plasticity and a distribution of axonal conduction delays. The model was implemented using C programming language with MPI and was run on HPC facility. The current study presents several novel findings: (1) synaptic compensation plays a significant role in preserving the dynamics of the network (after degeneration), (2) thalamic atrophy can be a secondary pathology to cortical shrinkage and (3) a deficit in the activity of the RTN population (disinhibition) causes a disruption in the oscillatory activity of the TCR population.

The study builds on the presented cortical models in our previous two studies [134] et al. 2014[63] by including the thalamus, an important subcortical structure involved in language, executive functioning, attention and memory functions [135, 136]. Such cognitive functions are deteriorated in AD [136, 137]. The investigated models incorporate a synaptic compensation mechanism to maintain the firing rate of the lesioned network [138]. Experimental [30, 43, 46, 47] and neuroimaging [139, 140] studies have observed coexistent impaired and compensatory processes in the neuronal networks.

In this modelling study, the death (impairment) of synaptic contacts on the cortical surface (either corticocortical or thalamocortical) can be compensated by increasing the weights of corticocortical or thalamocortical excitatory synapses. Both possibilities are discussed subsequently in the context of biological plausibility. According to functional Magnetic Resonance Imaging (fMRI)-based neuroimaging studies [140, 141], the results obtained from an AD group (compared to a healthy group) have shown decreased coupling between the thalamus and a number of cortical areas, namely, temporal, frontal and occipital lobes. In contrast to reduced thalamocortical connectivity, coupling within and between cortical prefrontal and frontal areas are higher in the AD group than that in the healthy group [141]. A neurochemical study has observed an increase in the size of the remaining synapses as a function of synaptic density in lamina III and V of the frontal cortex in AD patients [66,

126]. This behaviour has also been observed in the superior and middle temporal gyrus [66, 126]. Considering this experimental evidence, the demonstrated model implements a compensation process that increases the weights of corticocortical synapses in response to corticocortical (case 1) and thalamocortical (case 2) connectivity loss. In other words, adopting a compensation approach that increases the weights of thalamocortical (on the cortical surface) or corticothalamic synapses (on the thalamic surface) should lead to increased thalamocortical coupling; this may not be a biologically plausible choice as it contradicts the aforementioned experimental observations.

Wang et al. [140] have detected increased coupling between the right and left thalamus in the fMRI datasets from an AD cohort. This behaviour coexists with the decreased thalamocortical connectivity. Another in vivo study has observed a reduced RTN inhibition to TCR neurons in response to corticothalamic synaptic degeneration (initiated by cortical neuronal death) and subsequently, leading to enhanced TCR activity and recovered thalamocortical activity [106]. Based on these experimental findings, the compensation for corticothalamic synaptic loss (case 3) is implemented by down-scaling the RTN—TCR inhibition rather than increasing the excitatory corticothalamic synaptic weights on the thalamic surface. In the fourth case (corticoreticular disconnection), compensation is performed by down-scaling the mutual RTN—RTN inhibition.

The cerebral cortex includes the majority of neurons in the brain. It forms about 85 % of the brain's weight [142]. The presented results show that only the structural impairment of the cortical module (case 1) causes aberrant oscillatory behaviour of the whole network (see Figs. 9.18 and 9.19). Firstly, it is observed that the spectral outputs of both the cortical and the thalamic modules are significantly affected. Secondly, the patterns of changes in thalamic spectral activity are correlated with those in the cortical model. These observations are not detected in the other investigated case studies (cases 2–4). Corticothalamic and corticoreticular connectivity loss impact the thalamic network but not the modelled cortex. It can be seen in case study 1 that the power density within the range (9–11 Hz) is affected at an early stage; this is consistent with EEG and modelling studies in AD and MCI [1, 17, 29]. The spectrogram in the baseline condition in Fig. 9.18a shows that the power content is in the range (6–14 Hz). Consequently, the significant changes (decreases) are observed within this band. Compensation mechanisms delay and slow down the spectral changes in the network. These changes occur in spite of recovering the firing rate activity via synaptic scaling.

Jong et al. [27] have observed reduced volumes of thalamus, neocortex, putamen and hippocampus in AD based on a structural MRI study. However, the issue of whether thalamic atrophy is a principal or secondary pathology to cortical shrinkage have not been answered [27]. The majority of studies have focused on measuring the volume shrinkage of cortical lobes and the hippocampus. Less emphasis, however, is placed on thalamus despite its critical role in cognitive functions and thalamocortical oscillations [143]. Recent studies reported atrophy in thalamus and other cortical regions in MCI [143] and AD [144] patients compared to controls. Thalamic shrinkage is also observed in presymptomatic familial AD (FAD) [145] but not in healthy subjects with high risk at developing late-onset AD (LAD) [146]. A comparative study observed a faster thalamic volumetric decrease in early-onset AD (EOD) than LAD [147]. Again, such studies stress that cortical shrinkage is presented in addition to thalamic atrophy. However, the temporal sequence of thalamic and cortical atrophy remains unclear.

In Fig. 9.22a, it is can be seen that the thalamic activity is abolished after partial corticothalamic denervation whereas the cortical behaviour is not significantly affected by thalamocortical synaptic degeneration (see Fig. 9.20). The latter finding is consistent with a recent study based on a neural mass model [1]. The outputs shown in Figs. 9.18 and 9.19 suggest spectral alterations begin in the cortical part, then appear at a later stage in the thalamus with similar patterns. Moreover, the simulated LFPs of thalamic and cortical networks in Fig. 9.17 demonstrate a driven thalamic activity by the cortex. Considering all these observations together, it can be speculated that corticothalamic denervation can be a crucial factor in the observed thalamic shrinkage during AD. Therefore, these findings suggest that thalamic atrophy is a secondary pathology.

Despite the spectrogram (Fig. 9.20) and ERD/S (data not shown) plots not showing a significant spectral changes triggered by anomalies in the thalamocortical connectivity type, it is proposed to investigate this factor with a larger-scale and more detailed thalamocortical model rather than drawing extrapolative conclusions.

RTN regulation of the thalamic activity is visible in Figs. 9.22a and b and 9.24. Preserving the basal RTN inhibition level to TCR neurons (in the presence of corticothalamic synaptic loss) contributes significantly to thalamic activity 'shut-down' while scaling down this inhibition improves thalamic functionality. Interestingly, a significant blocking of reticulathalamic inhibition (as a result of corticoreticular deafferentation) stimulates TCR neurons but disrupts the thalamic oscillatory activity as demonstrated in Fig. 9.22a. Reversing the disinhibition by scaling down the mutual inhibition within the RTN population does recover thalamic oscillations as in Fig. 9.24b. Bhattacharya et al. [1] have investigated a hypothesized pathological influence of RTN on the oscillatory activity within alpha frequency band and reported a correlative result of this factor. The key role played by RTN neurons in modulating thalamic activity has been confirmed by biological studies [148, 149].

Considered as a limitation of the models, the results have not replicated the observed power increase in slow frequency bands in EEG studies on AD patients (represented by ERS). This observation can be associated with the reduced anatomical details of the model. Moreover, the spectral analysis of the model has shown a narrow peak with alpha band power. This means that the model oscillates mainly in alpha frequency band. A recent study proposes a thalamocortical network model oscillating only in alpha frequency band [1]. The model has been mainly utilized to investigate the changes within alpha frequency band in AD [1]. Power increase in slow frequency bands has not been shown by the model [1]. However, the model has replicated the power increase in lower alpha band (8–10 Hz) as observed in EEG studies in AD, referred to as alpha slowing [1]. Notably, [1] demonstrated that alpha band power decrease is better correlated with EEG studies in AD than alpha slowing. As mentioned earlier, another limitation is in the model's anatomy;

its cortical network lacks the columnar structure, the cortical shape and coordinates and cortical layers. The thalamic region in the human brain is divided into a number of nuclei depending on their association with other cortical areas. Despite this limitation, the developed model can oscillate within alpha frequency band. It is a future goal to extend the current framework so that it captures such realistic details.

Model Validation

Validation of the model is essential to assure that it provides accurate predictions. Valid models of brain disorders should initially reproduce the dynamics of the healthy brain before investigating the abnormal states. Brain dynamics can be mainly observed with EEG (or Intracranial EEG as in many animal studies), fMRI, Functional transcranial Doppler sonography (fTCD) and neurophysiological tests (for perceptual processing and learning).

The network connectivity as well as neuron and synapse models and parameters in this study were derived from other modelling and experimental studies as described in Sect. "Materials and Methods". Those modelling studies have validated the neuron and synapse models against experimental data. The network models in this study produce collective rhythmic behaviour in the frequency range corresponding to that of the healthy mammalian cortex in the awake state as seen in Figs. 9.5 and 9.14. This assures the validity of the models at the baseline condition. Note that throughout this study the trajectories of the frequency power were compared visually. This can be thought of as a qualitative validation. The presented work can be extended by performing quantitative validation with statistical approaches to assess how well the model fits the EEG data.

Given the EEG changes in AD (described in Sect. "Introduction") the spectral output of the models shows a qualitative correlation as seen in Figs. 9.5 and 9.6. Again future extensions of this work can apply fitting techniques to validate the AD models and to achieve higher impacts of the modelling studies. The modelled network lesions and compensation mechanisms were motivated by biological studies and were not chosen arbitrarily. In this study there are range of observation based on the findings which can be corroborated with evidence from other modelling studies, electrophysiological and neuroimaging studies, validating that that the models presented here are biologically plausible models of AD. A summary of the key findings that support model validation are outlined in the conclusion section that follows.

Conclusion and Future Work

Study 1 has explored the impact of local (neuronal-based) and global (network-based) synaptic compensation mechanisms on the dynamics of lesioned networks based on a large-scale cortical model with anatomy and dynamics inspired by the cerebral cortex. We have shown that local compensation has relatively maintained the GFS and power spectra dynamics, firing rate and frequency shift distributions across the network until 50 % synaptic loss. The baseline activity at theta and alpha bands is optimally recovered by local compensation whereas the baseline level of delta and beta bands is maintained best by global compensation. The findings provide insights for experimentalists to conduct comprehensive lab experiments on both compensation mechanisms. Results suggest that therapeutics targeting compensation mechanisms may be efficient for treating AD.

Study 2 involved a large-scale thalamocortical network model which oscillates within the alpha frequency band as recorded in the wakeful relaxed state with closed eyes. The model consists of different types of cortical excitatory and inhibitory neurons recurrently connected to thalamic and reticular thalamic neurons with the ratios and distances found in the mammalian thalamocortical system. Synaptic dynamics include AMPA, GABA, NMDA and GJ kinetics, short-term plasticity and a distribution of axonal conduction delays. Compared to other computational models of AD [150–152], this model can be considered as the most detailed microscopic model used to investigate the neural causes of abnormal oscillations in AD. A number of neural mass models of AD [1, 17, 153–155] can also be considered biological plausible as they incorporate important biological details such as connectivity data and cortical maps from neuroimaging studies. This is planned as a future extension of this work.

The results have shown that the dynamics of the network are significantly influenced by corticocortical synaptic loss. The thalamic activity is driven by the cortical module. However, RTN inhibitory efferent to the TCR neurons is another key regulator. Including the compensation mechanisms in the study assists in exploring the role of more complicated changes rather than the role of a single factor or parameter. ERD diagrams show that alpha band power decreases in mild stages. It has been previously suggested that AD is associated with a decrease in alpha band power [1, 29]. In summary, this work provides two novel findings: (1) it speculates that thalamic atrophy is a secondary pathology to cortical shrinkage since thalamic oscillations are sharply disrupted after corticocortical and corticothalamic denervations and (2) RTN inhibition to TCR neurons plays a key role in regulating thalamic oscillations; disinhibition disrupts thalamic oscillatory activity even though TCR neurons are more depolarized after being released from RTN inhibition.

The utilised model in this paper lacks multi-compartmental neuron models. The number of neurons is greatly reduced compared to real thalamocortical systems. The cortical and thalamic anatomy is significantly simplified. The geometry and boundaries of cortical regions and thalamic nuclei are not incorporated in the model. Incorporating such details is proposed as a future extension of this work.

Additionally, a neurobiological study has shown that astrocytes act as a sensor and regulator of neuronal activity [51]. Astrocytes have compensated for the prolonged blocking of neuronal activity by producing TNF-α that binds to the neuronal TNF-α receptors causing an elevation in AMPA receptors accompanied with a subsequent decrease in GABA receptors [51]. Accordingly, we speculate that including astrocytes in modelling studies might contribute further to the understanding of homeostatic mechanisms in the brain in physiological and pathologic states.

Acknowledgment This work is supported by the Northern Ireland Department for Education and Learning under the Strengthening the All Island Research Base Programme.

References

1. Bhattacharya BS, Coyle D, Maguire L. A thalamo-cortico-thalamic neural mass model to study alpha rhythms in Alzheimer's disease. Neural Netw. 2011;24:631–45 (Elsevier Ltd.).
2. Jeong J. EEG dynamics in patients with Alzheimer's disease, Clinical Neurophysiol. 2004;115(7):1490–505.
3. Lizio R, Vecchio F, Frisoni GB, Ferri R, Rodriguez G, Babiloni C. Electroencephalographic rhythms in Alzheimer's disease. Int J Alzheimers Dis. 2011;2011:927573.
4. Mugantseva EA, Podolski IY. Animal model of Alzheimer's disease: characteristics of EEG and memory. Cent Eur J Biol. 2009;4(4):507–14.
5. Besthorn C, Zerfass R, Geiger-Kabisch C, Sattel H, Daniel S, Schreiter-Gasser U, Förstl H. Discrimination of Alzheimer's disease and normal aging by EEG data. Electroencephalogr Clin Neurophysiol. 1997;103:241–8.
6. Coben L, Danziger W, Berg L. Frequency analysis of the resting awake EEG in mild senile dementia of Alzheimer type. Electroencephalogr Clin Neurophysiol. 1983;55:372–80.
7. Coben L, Danziger W, Storandt M. A longitudinal EEG study of mild senile dementia of Alzheimer type: changes at 1 year and at 2.5 years, Electroencephalogr Clin Neurophysiol. 1985;61:101–12.
8. Hier DB, Mangone CA, Ganellen R, Warach JD, Van Egeren R, Perlik SJ, Gorelick PB. Quantitative measurement of delta activity in Alzheimer's disease. Clini EEG. 1991;22:178–82.
9. Dauwels J, Vialatte F, Musha T, Cichocki A. A comparative study of synchrony measures for the early diagnosis of Alzheimer's disease based on EEG. NeuroImage. 2010;49(1):668–93 (Elsevier Inc.). http://www.ncbi.nlm.nih.gov/pubmed/19573607. Accessed 6 Feb 2013.
10. De Waal H, Stam CJ, de Haan W, van Straaten ECW, Scheltens P, van der Flier WM. Young Alzheimer patients show distinct regional changes of oscillatory brain dynamics. Neurobiol Aging. 2012;33(5):1008.e25–31.
11. Baker M, Akrofi K, Schiffer R, Boyle MWO. EEG patterns in mild cognitive impairment (mci) patients. Open Neuroimag J. 2008;2:52–5.
12. Van der Hiele K, Vein AA, Reijntjes RHAM, Westendorp RGJ, Bollen ELEM van Buchem MA, van Dijk JG, Middelkoop HAM. EEG correlates in the spectrum of cognitive decline. Clin Neurophysiol. 2007;118:1931–9.
13. Pucci E, Belardinelli N, Cacchiò G, Signorino M, Angeleri F. EEG power spectrum differences in early and late onset forms of Alzheimer's disease. Clin Neurophysiol. 1999;110(4):621–31.
14. Dauwels J, Vialatte F-B, Cichocki A. On the early diagnosis of Alzheimer's disease from EEG signals: a mini-review. In: Wang R, Gu F, editors. Advances in cognitive neurodynamics (II) SE—106. Netherlands: Springer; 2011. pp. 709–16.
15. Jelles B, Scheltens P, van der Flier WM, Jonkman EJ, da Silva FHL, Stam CJ. Global dynamical analysis of the EEG in Alzheimer's disease: frequency-specific changes of functional interactions. Clin Neurophysiol. 2008;119:837–41.
16. Jelic V, Johansson SE, Almkvist O, Shigeta M, Julin P, Nordberg A, Winblad B, Wahlund LO. Quantitative electroencephalography in mild cognitive impairment: longitudinal changes and possible prediction of Alzheimer's disease. Neurobiol Aging. 2000;21:533–40.
17. Pons AJ, Cantero JL, Atienza M, Garcia-Ojalvo J. Relating structural and functional anomalous connectivity in the aging brain via neural mass modeling. NeuroImage. 2010;52(3):848–61 (Elsevier Inc.). http://www.ncbi.nlm.nih.gov/pubmed/20056154. Accessed 29 Jan 2013.

18. Jones SR, Pinto DJ, Kaper TJ, Kopell N. Alpha-frequency rhythms desynchronize over long cortical distances: a modeling study. J Comput Neurosci. 2000;9(3):271–91.
19. De Waal H, Stam CJ, Blankenstein MA, Pijnenburg YAL, Scheltens P, van der Flier WM. EEG abnormalities in early and late onset Alzheimer's disease: understanding heterogeneity. J Neurol Neurosurg Psychiatry. 2011;82(1):67–71.
20. Babilon C, Lizio R, Vecchio F, Frisoni GB, Pievani M, Geroldi C, Claudia F, Ferri R, Lanuzza B, Rossini PM. Reactivity of cortical alpha rhythms to eye opening in mild cognitive impairment and Alzheimer's disease: an EEG study. J Alzheimers Dis. 2010;22(4):1047–64.
21. Koenig T, Prichep L, Dierks T, Hubl D, Wahlund LO, John ER, Jelic V. Decreased EEG synchronization in Alzheimer's disease and mild cognitive impairment. Neurobiol Aging. 2005;26:165–71.
22. Delbeuck X, Van der Linden M, Collette F. Alzheimer's disease as a disconnection syndrome? Neuropsychol Rev. 2003;13:79–92.
23. De Haan W, Pijnenburg YAL, Strijers RLM, van der Made Y, van der Flier WM, Scheltens P, Stam CJ. Functional neural network analysis in frontotemporal dementia and Alzheimer's disease using EEG and graph theory. BMC Neurosci. 2009;10:101. http://www.pubmedcentral.nih.gov/articlerender.fcgi?artid=2736175tool=pmcentrezrendertype=abstract. Accessed: 1 Feb 2013.
24. Li X, Coyle D, Maguire L, Watson DR, McGinnity TM. Gray matter concentration and effective connectivity changes in Alzheimer's disease: a longitudinal structural MRI study. Neuroradiology. 2011;53(10):733–48. http://www.ncbi.nlm.nih.gov/pubmed/21113707. Accessed: 31 Jan 2013.
25. Park Y-M, Che H-J, Im C-H, Jung H-T, Bae S-M, Lee S-H. Decreased EEG synchronization and its correlation with symptom severity in Alzheimer's disease. Neurosci Res. 2008;62(2):112–7. http://www.ncbi.nlm.nih.gov/pubmed/18672010. Accessed: 6 Feb 2013.
26. Jones EG. Thalamic circuitry and thalamocortical synchrony. Philos Trans R Soc Lond B Biol Sci. 2002;357(1428):1659–73.
27. De Jong LW, van der Hiele K, Veer IM, Houwing JJ, Westendorp RGJ, Bollen ELEM, de Bruin PW, Middelkoop HAM, van Buchem MA, van der Grond J. Strongly reduced volumes of putamen and thalamus in Alzheimer's disease: an MRI study. Brain. 2008;131(12):3277–85. http://www.pubmedcentral.nih.gov/articlerender.fcgi?artid=2639208tool=pmcentrezrendertype=abstract. Accessed 6 Feb 2013.
28. Sherman SM. A wake-up call from the thalamus. Nature Neurosci. 2001;4:344–6.
29. Moretti DV, Babiloni C, Binetti G, Cassetta E, Dal Forno G, Ferreric F, Ferri R, Lanuzza B, Miniussi C, Nobili F, Rodriguez G, Salinari S, Rossini, PM. Individual analysis of EEG frequency and band power in mild Alzheimer's disease. Clin Neurophysiol. 2004;115:299–308
30. Savioz A, Leuba G, Vallet PG, Walzer C. Contribution of neural networks to Alzheimer disease's progression. Brain Res Bull. 2009;80(4–5):309–14. http://www.ncbi.nlm.nih.gov/pubmed/19539730. Accessed 6 Feb 2013.
31. Minati L, Edginton T, Bruzzone MG, Giaccone G. Current concepts in Alzheimer's disease: a multidisciplinary review. Am J Alzheimers Dis Dement. 2009;24:95–121.
32. Spires-Jones TL, Stoothoff WH, de Calignon A, Jones PB, Hyman BT. Tau pathophysiology in neurodegeneration: a tangled issue. Trends Neurosci. 2009;32:150–9.
33. Brion JP. Neurofibrillary tangles and Alzheimer's disease. Eur Neurol. 1998;40:130–40.
34. Bear M, Connors B, Paradiso M, Bear MF, Connors BW, Paradiso MA. Neuroscience: exploring the brain. Philadelphia: Lippincott Williams & Wilkins; 2002.
35. Boche D, Nicoll JAR. Are we getting to grips with Alzheimer's disease at last? Brain. 2010;133:1297–9.
36. Davies C, Mann D, Sumpter P, Yates P. A quantitative morphometric analysis of the neuronal and synaptic content of the frontal and temporal cortex in patients with Alzheimer's disease. J Neurol Sci. 1987;78:151–64.
37. Bell RD, Winkler EA, Singh I, Sagare AP, Deane R, Wu Z, Holtzman DM, Betsholtz C, Armulik A, Sallstrom J, Berk BC, Zlokovic BV. Apolipoprotein E controls cerebrovascular integrity via cyclophilin A. Nature. 2012;485:512–16.

38. Zlokovic BV. Neurovascular pathways to neurodegeneration in Alzheimer's disease and other disorders. Nat Rev Neurosci. 2011;12(12):723–38 (Nature Publishing Group). http://www.ncbi.nlm.nih.gov/pubmed/22048062. Accessed 29 Jan 2013.
39. Carmeliet P, De Strooper B. Alzheimer's disease: a breach in the blood-brain barrier. Nature. 2012;485:451–2.
40. Pimentel-Coelho PM, Rivest S. The early contribution of cerebrovascular factors to the pathogenesis of Alzheimer's disease. Eur J Neurosci. 2012;35:1917–37.
41. Zlokovic BV. The blood-brain barrier in health and chronic neurodegenerative disorders. Neuron. 2008;57:178–201.
42. Armulik A, Genové G, Mäe M, Nisancioglu MH, Wallgard E, Niaudet C, He L, Norlin J, Lindblom P, Strittmatter K, Johansson BR, Betsholtz C. Pericytes regulate the blood-brain barrier. Nature. 2010;468:557–61.
43. Small DH. Mechanisms of synaptic homeostasis in Alzheimer's disease. Curr Alzheimer Res. 2004;1:27–32.
44. Turrigiano G. Homeostatic synaptic plasticity: local and global mechanisms for stabilizing neuronal function. Cold Spring Harb Perspect Biol. 2012;4:a005736.
45. Turrigiano G. Too many cooks? Intrinsic and synaptic homeostatic mechanisms in cortical circuit refinement. Annual Rev Neurosci. 2011;34:89–103.
46. Uylings HBM, De Brabander JM. Neuronal changes in normal human aging and Alzheimer's disease. Brain Cognit. 2002;49:268–76 (Elsevier).
47. Jin K, Peel AL, Mao XO, Xie L, Cottrell BA, Henshall DC, Greenberg DA. Increased hippocampal neurogenesis in Alzheimer's disease. Proc Natl Acad Sci U S A. 2004;101:343–7.
48. Leuba G, Savioz A, Vernay A, Carnal B, Kraftsik R, Tardif E, Riederer I, Riederer BM. Differential changes in synaptic proteins in the Alzheimer frontal cortex with marked increase in PSD-95 postsynaptic protein. J Alzheimer's Dis. 2008;15:139–51.
49. Holtmaat A, Svoboda K. Experience-dependent structural synaptic plasticity in the mammalian brain. Nat Rev Neurosci. 2009;10:647–58 (Nature Publishing Group).
50. Fröhlich F, Bazhenov M, Sejnowski TJ. Pathological effect of homeostatic synaptic scaling on network dynamics in diseases of the cortex. J Neurosci. 2008;28(7):1709–20.
51. Stellwagen D, Malenka RC. Synaptic scaling mediated by glial TNF-alpha. Nature. 2006;440:1054–9.
52. Auld DS, Robitaille R. Glial cells and neurotransmission: an inclusive view of synaptic function. Neuron. 2003;40(2):389–400.
53. Aguado F, Espinosa-Parrilla JF, Carmona MA, Soriano E. Neuronal activity regulates correlated network properties of spontaneous calcium transients in astrocytes in situ. J Neurosci. 2002;22(21):9430–44.
54. Diamond JS. Deriving the glutamate clearance time course from transporter currents in CA1 hippocampal astrocytes: transmitter uptake gets faster during development. J Neurosci. 2005;25(11):2906–16.
55. Lin, S-C, Bergles, DE. Synaptic signaling between neurons and glia. Glia. 2004;47(3):290–8.
56. Wigley R, Hamilton N, Nishiyama A, Kirchhoff F, Butt AM. Morphological and physiological interactions of NG2-glia with astrocytes and neurons. J Anat. 2007;210(6):661–70.
57. Nedergaard M, Rodríguez JJ, Verkhratsky A. Glial calcium and diseases of the nervous system. Cell Calcium. 2010;47(2):140–9.
58. Vlachos A, Becker D, Jedlicka P, Winkels R, Roeper J, Deller T. Entorhinal denervation induces homeostatic synaptic scaling of excitatory postsynapses of dentate granule cells in mouse organotypic slice cultures. PloS ONE. 2012;7:e32883.
59. Topolnik L, Steriade M, Timofeev I. Partial cortical deafferentation promotes development of paroxysmal activity. Cereb Cortex. 2003;13(8):883–93.
60. Horn D, Ruppin E, Usher M, Herrmann M. Neural network modeling of memory deterioration in Alzheimer's disease. Neural Comput. 1993;5:736–49 (Cambridge: MIT Press).
61. Horn D, Levy N, Ruppin E. Neuronal-based synaptic compensation: a computational study in Alzheimer's disease. Neural Comput. 1996;8:1227–43.

62. Rowan M. Effects of compensation, connectivity and tau in a computational model of Alzheimer's disease. The 2011 International joint conference on neural networks, IEEE. San Jose. 2011. pp. 543–50.
63. Abuhassan K, Coyle D, Maguire LP. Investigating the neural correlates of pathological cortical networks in Alzheimer's disease using heterogeneous neuronal models. IEEE Trans Biomed Eng. 2012;59:890–6.
64. Jensen O, Goel P, Kopell N, Pohja M, Hari R, Ermentrout B. On the human sensorimotor-cortex beta rhythm: sources and modeling. NeuroImage. 2005;26:347–55.
65. Izhikevich EM. Simple model of spiking neurons. IEEE Trans Neural Netw. 2003;14(6):1569–72. http://www.ncbi.nlm.nih.gov/pubmed/18244602. Accessed 20 Jan 2012.
66. Scheff SW. Reactive synaptogenesis in aging and Alzheimer's disease: lessons learned in the Cotman laboratory. Neurochem Res. 2003;28:1625–30.
67. Scheff SW, DeKosky ST, Price DA. Quantitative assessment of cortical synaptic density in Alzheimer's disease. Neurobiol Aging. 1990;11:29–37.
68. Izhikevich EM, Gally JA, Edelman GM. Spike-timing dynamics of neuronal groups. Cereb Cortex. 2004;14:933–44.
69. Rodriguez G, Arnaldi D, Picco A. Brain functional network in Alzheimer's disease: diagnostic markers for diagnosis and monitoring. Int J Alzheimers Dis. 2011;2011:481903.
70. Izhikevich EM. Polychronization: computation with spikes. Neural Comput. 2006;18(2):245–82. http://www.ncbi.nlm.nih.gov/pubmed/16378515. Accessed 20 Jan 2012.
71. Braitenberg V, Schüz A. Cortex: statistics and geometry of neuronal connectivity. Berlin: Springer; 1998.
72. Swadlow HA. Efferent neurons and suspected interneurons in motor cortex of the awake rabbit: axonal properties, sensory receptive fields, and subthreshold synaptic inputs. J Neurophysiol. 1994;71:437–53.
73. Waxman SG, Bennett MV. Relative conduction velocities of small myelinated and non-myelinated fibres in the central nervous system. Nat New Biol. 1972;238:217–19.
74. Binzegger T, Douglas RJ, Martin KAC. A quantitative map of the circuit of cat primary visual cortex. J Neurosci. 2004;24(39):8441–53. http://www.ncbi.nlm.nih.gov/pubmed/15456817. Accessed 29 Jan 2013.
75. Izhikevich EM, Edelman GM. Large-scale model of mammalian thalamocortical systems. Proc Natl Acad Sci U S A. 2008;105(9):3593–8. http://www.pubmedcentral.nih.gov/articlerender.fcgi?artid=2265160tool=pmcentrezrendertype=abstract. Accessed 20 Jan 2012.
76. Traub RD, Contreras D, Cunningham MO, Murray H, LeBeau FEN, Roopun A, Bibbig A, Wilent WB, Higley MJ, Whittington MA. Single-column thalamocortical network model exhibiting gamma oscillations, sleep spindles, and epileptogenic bursts. J Neurophysiol. 2005;93:2194–232.
77. Bazhenov M, Timofeev I, Steriade M, Sejnowski T. Model of thalamocortical slow-wave sleep oscillations and transitions to activated states. J Neurosci. 2002;22:8691–704.
78. Landisman CE, Long MA, Beierlein M, Deans MR, Paul DL, Connors BW. Electrical synapses in the thalamic reticular nucleus. J Neurosci. 2002;22:1002–9.
79. Benarroch EE. GABAB receptors: structure, functions, and clinical implications. Neurology. 2012;78(8):578–84.
80. Zhang SJ, Huguenard JR, Prince DA. GABAA receptor-mediated Cl- currents in rat thalamic reticular and relay neurons. J Neurophysiol. 1997;78:2280–6.
81. Sohal VS, Huntsman MM, Huguenard JR. Reciprocal inhibitory connections regulate the spatiotemporal properties of intrathalamic oscillations. J Neurosci. 2000;20:1735–45.
82. Hughes SW, Crunelli V. Thalamic mechanisms of EEG alpha rhythms and their pathological implications. Neuroscientist. 2005;11:357–72.
83. Jones EG. The thalamus. 1st edn. New York: Plenum Press; 1985.
84. Castro-Alamancos MA, Calcagnotto ME. Presynaptic long-term potentiation in corticothalamic synapses. J Neurosci. 1999;19:9090–7.
85. Kirkcaldie M. Neocortex. In: Watson C, Paxinos G, Puelles L, editors. The mouse nervous system. London: Academic Press; 2012. pp. 52–94.

86. Agmon A, Connors BW. Correlation between intrinsic firing patterns and thalamocortical synaptic responses of neurons in mouse barrel cortex. J Neurosci. 1992;12:319–29.
87. Gentet LJ, Ulrich D. Electrophysiological characterization of synaptic connections between layer VI cortical cells and neurons of the nucleus reticularis thalami in juvenile rats. Eur J Neurosci. 2004;19:625–33.
88. Steriade M, Jones EG, Llinás RR. Thalamic oscillations and signaling. New York: Wiley; 1990.
89. Izhikevich EM. Dynamical systems in neuroscience: the geometry of excitability and bursting. In: Sejnowski TJ, Poggio TA, editors. Dynamical systems. Cambridge; The MIT Press; 2007 (25:441).
90. Markram H, Wang Y, Tsodyks M. Differential signaling via the same axon of neocortical pyramidal neurons. PNAS. 1998;95(9):5323–8.
91. Gupta A. Organizing principles for a diversity of gabaergic interneurons and synapses in the neocortex. Science. 2000;287:273–8.
92. Shankar GM, Walsh DM. Alzheimer's disease: synaptic dysfunction and Abeta. Mol Neurodegen. 2009;4(1):48.
93. Yoshiyama Y, Higuchi M, Zhang B, Huang S-M, Iwata N, Saido TC, Maeda J, Suhara T, Trojanowski JQ, Lee VM-Y. Synapse loss and microglial activation precede tangles in a P301S tauopathy mouse model. Neuron. 2007;53:337–51.
94. Yu W, Lu B. Synapses and dendritic spines as pathogenic targets in Alzheimer's disease. Neural Plast. 2012;2012:24715 (Department of Pathology, Stanford University School of Medicine, Stanford CA).
95. Bertoni-Freddari C, Fattoretti P, Casoli T, Caselli U, Meier-Ruge W. Deterioration threshold of synaptic morphology in aging and senile dementia of Alzheimer's type. Analayt Quant Cytol Histol. 1996;18:209–13.
96. Knobloch M, Mansuy IM. Dendritic spine loss and synaptic alterations in Alzheimer's disease. Mol Neurobiol. 2008;37:73–82.
97. Terry RD, Masliah E, Salmon DP, Butters N, DeTeresa R, Hill R, Hansen LA, Katzman R. Physical basis of cognitive alterations in Alzheimer's disease: synapse loss is the major correlate of cognitive impairment. Ann Neurol. 1991;30:572–80.
98. Penzes P, Cahill ME, Jones KA, VanLeeuwen J-E, Woolfrey KM. Dendritic spine pathology in neuropsychiatric disorders. Nat Neurosci. 2011;14:285–93.
99. Koffie RM, Hyman BT, Spires-Jones TL. 2011, Alzheimer's disease: synapses gone cold. Mol Neurodegen. 2011;6:63.
100. Sheffler-Collins SI, Dalva MB. EphBs: an integral link between synaptic function and synaptopathies. Trends Neurosci. 2012;35:293–304.
101. Selkoe DJ. Alzheimer's disease is a synaptic failure. Science. 2002;298:789–91.
102. Manshanden I, De Munck JC, Simon NR, Lopes da Silva FH. Source localization of MEG sleep spindles and the relation to sources of alpha band rhythms. Clin Neurophysiol. 2002;113(12):1937–47.
103. Steriade M, Nuñez A, Amzica F. Intracellular analysis of relations between the slow (<1 Hz) neocortical oscillation and other sleep rhythms of the electroencephalogram. J Neuroscience. 1993;13(8):3266–83.
104. Yazawa S, Kawasaki S, Kanemaru A, Kuratsuwa Y, Yabuoshi R, Ohi T. Bilateral paramedian thalamo-midbrain infarction showing electroencephalographic alpha activity. Intern Med. 2001;40(5):443–8.
105. Timofeev I, Steriade M. Low-frequency rhythms in the thalamus of intact-cortex and decorticated cats. J Neurophysiol. 1996;76(6):4152–68.
106. Paz JT, Christian CA, Parada I, Prince DA, Huguenard JR. Focal cortical infarcts alter intrinsic excitability and synaptic excitation in the reticular thalamic nucleus. J Neurosci. 2010;30:5465–79.
107. Koenig T, Lehmann D, Saito N, Kuginuki T, Kinoshita T, Koukkou M. Decreased functional connectivity of EEG theta-frequency activity Ève patients with schizophrenia†⁻: preliminary results. NeuroReport. 2001a;50:55–60.

108. Koenig T, Lehmann D, Saito N, Kuginuki T, Kinoshita T, Koukkou M. Decreased functional connectivity of EEG theta-frequency activity in first-episode, neuroleptic-naive patients with schizophrenia: preliminary results. Schizophr Res. 2001b;50:55–60.
109. Pfurtscheller G, Lopes da Silva FH. Event-related EEG/MEG synchronization and desynchronization: basic principles. Clin Neurophysiol. 1999;110(11):1842–7. http://www.ncbi.nlm.nih.gov/pubmed/10576479. Accessed 20 Jan 2012.
110. Bhattacharya BS, Coyle D, Maguire LP. Assessing alpha band event-related synchronisation/desynchronisation using a bio-inspired computational model. J Univ Comp Sci. 2012;18(13):1888–904.
111. Durka PJ, Zygierewicz J, Klekowicz H, Ginter J, Blinowska KJ. On the statistical significance of event-related EEG desynchronization and synchronization in the time-frequency plane. IEEE Trans Biomed Eng. 2004b;51(7):1167–75.
112. Pfurtscheller G, Aranibar A. Event-related cortical desynchronization detected by power measurements of scalp EEG. Electroencephalogr Clin Neurophysiol. 1977;42(6):817–26.
113. Prasad G, Herman P, Coyle D, McDonough S, Crosbie J. Applying a brain-computer interface to support motor imagery practice in people with stroke for upper limb recovery: a feasibility study. J Neuroeng Rehabil. 2010;7(1):60 (BioMed Central Ltd.). http://www.pubmedcentral.nih.gov/articlerender.fcgi?artid=3017056tool=pmcentrezrendertype=abstract. Accessed 25 Oct 2012.
114. Coyle D, Prasad G, McGinnity TM. A time-frequency approach to feature extraction for a brain-computer interface with a comparative analysis of performance measures. EURASIP J Adv Signal Process. 2005;2005(19):3141–51. http://asp.eurasipjournals.com/content/2005/19/861614. Accessed 20 Jan 2012.
115. Herman P, Prasad G, McGinnity TM, Coyle D. Comparative analysis of spectral approaches to feature extraction for EEG-based motor imagery classification. IEEE Trans Neural Syst Rehabil Eng. 2008;16(4):317–26. http://www.ncbi.nlm.nih.gov/pubmed/18701380. Accessed 20 Jan 2012.
116. Durka PJ, Zygierewicz J, Klekowicz H, Ginter J, Blinowska KJ. On the statistical significance of event-related EEG desynchronization and synchronization in the time-frequency plane. IEEE Trans. Biomed. Eng. 2004b;51:1167–75. 10.1109/TBME.2004.827341.
117. Timofeev I, Grenier F, Bazhenov M, Senjnowski T, Steriade M. Origin of slow cortical oscillations in deafferented cortical slabs. Cereb Cortex. 2000;10:1185–99.
118. Latham PE, Richmond BJ, Nelson PG, Nirenberg S. Intrinsic dynamics in neuronal networks. I. Theory. J Neurophysiol. 2000;83:808–27.
119. Latham PE, Richmond BJ, Nirenberg S, Nelson PG. Intrinsic dynamics in neuronal networks. II. Experiment. J Neurophysiol. 2000;83:828–35.
120. Prichep LS, John ER, Ferris SH, Rausch L, Fang Z, Cancro R, Torossian C, Reisberg B. Prediction of longitudinal cognitive decline in normal elderly with subjective complaints using electrophysiological imaging. Neurobiol Aging. 2006;27:471–81.
121. Nuwer MR, Comi G, Emerson R, Fuglsang-Frederiksen A, Guérit JM, Hinrichs H, Ikeda A, Luccas FJ, Rappelsberger P. IFCN standards for digital recording of clinical EEG. International Federation of Clinical Neurophysiology. Electroencephalogr Clin Neurophysiol. 1999;52:11–4.
122. Contreras D, Destexhe A, Sejnowski TJ, Steriade M. Control of spatiotemporal coherence of a thalamic oscillation by corticothalamic feedback. Science. 1996;274(5288):771–4.
123. Destexhe A. Modelling corticothalamic feedback and the gating of the thalamus by the cerebral cortex. J Physiol. 2000;94(5–6):391–410.
124. Mayer J, Schuster HG, Claussen JC, Mölle M. Corticothalamic projections control synchronization in locally coupled bistable thalamic oscillators. Phys Rev Lett. 2007;99(6):068102.
125. Arendt T. Alzheimer's disease as a disorder of mechanisms underlying structural brain self-organization. Neuroscience. 2001;102:723–65.
126. Scheff SW, Price DA. Synapse loss in the temporal lobe in Alzheimer's disease. Annals of Neurology. 1993;33:190–9.

127. Kirov SA, Goddard CA, Harris KM. Age-dependence in the homeostatic upregulation of hippocampal dendritic spine number during blocked synaptic transmission. Neuropharmacol. 2004;47:640–8.
128. Cantero JL, Atienza M, Gomez-Herrero G, Cruz-Vadell A, Gil-Neciga E, Rodriguez-Romero R, Garcia-Solis D. Functional integrity of thalamocortical circuits differentiates normal aging from mild cognitive impairment. Human Brain Map. 2009;30(12):3944–57. http://www.ncbi.nlm.nih.gov/pubmed/19449329. Accessed 6 Feb 2013.
129. Moretti DV, Pievani M, Fracassi C, Geroldi C, Calabria M, De Carli CS, Rossini PM Frisoni GB. Brain vascular damage of cholinergic pathways and EEG markers in mild cognitive impairment. J Alzheimers Dis. 2008;15:357–72.
130. Hasselmo M. Runaway synaptic modification in models of cortex: Implications for Alzheimer's disease. Neural Netw. 1994;7:13–40.
131. Tchantchou F, Lacor PN, Cao Z, Lao L, Hou Y, Cui C, Klein WL, Luo Y. Stimulation of neurogenesis and synaptogenesis by bilobalide and quercetin via common final pathway in hippocampal neurons. J Alzheimers Dis. 2009;18(4):787–98.
132. Dok-Go H, Lee KH, Kim HJ, Lee EH, Lee J, Song YS, Lee Y-H, Jin C, Lee YS, Cho J. Neuroprotective effects of antioxidative flavonoids, quercetin, (+)-dihydroquercetin and quercetin 3-methyl ether, isolated from Opuntia ficus-indica var. saboten. Brain Res. 2003;965:130–6.
133. Shi C, Zheng D, Wu F, Liu, J Xu, J. The phosphatidyl inositol 3 kinase-glycogen synthase kinase 3β pathway mediates bilobalide-induced reduction in amyloid β-peptide. Neurochem Res. 2012;37:298–306.
134. Abuhassan K, Coyle D, Belatreche A, Maguire L. Compensating for synaptic loss in Alzheimer's disease. J Comput Neurosci. 2014;36(1):19–37. doi:10.1007/s10827–013–0462–8.
135. Johnson MD, Ojemann GA. The role of the human thalamus in language and memory: evidence from electrophysiological studies. Brain Cognit. 2000;42(2):218–30.
136. Van der Werf YD, Scheltens P, Lindeboom J, Witter MP, Uylings HBM, Jolles J. Deficits of memory, executive functioning and attention following infarction in the thalamus; a study of 22 cases with localised lesions. Neuropsychologia. 2003;41(10):1330–44.
137. Zhao X, Liu Y, Wang X, Liu B, Xi Q, Guo Q, Jiang H, Jiang T, Wang P. Disrupted small-world brain networks in moderate Alzheimer's disease: a resting-state FMRI study. PloS ONE. 2012;7(3):e33540. doi:10.1371/journal.pone.0033540.
138. Abuhassan K, Coyle D, Maguire L. Compensating for thalamocortical synaptic loss in Alzheimer's disease. Front Comp Neurosci. 2014;8:65. doi:10.3389/fncom.2014.00065
139. Qi Z, Wu X, Wang Z, Zhang N, Dong H, Yao L, Li K. Impairment and compensation coexist in amnestic MCI default mode network. NeuroImage. 2010;50:48–55.
140. Wang Z, Jia X, Liang P, Qi Z, Yang Y, Zhou W, Li K. Changes in thalamus connectivity in mild cognitive impairment: evidence from resting state fMRI. Eur J Radiol. 2012;81:277–85.
141. Supekar K, Menon V, Rubin D, Musen M, Greicius MD. Network analysis of intrinsic functional brain connectivity in Alzheimer's disease. PLoS Comput Biol. 2008:4. doi:10.1371/journal.pcbi.1000100.
142. Woolfolk A. Educational psychology. 11th edn. Bacon: Pearson/Allyn; 2011
143. Pedro T, Weiler M, Yasuda CL, D'Abreu A, Damasceno BP, Cendes F, Balthazar MLF. Volumetric brain changes in thalamus, corpus callosum and medial temporal structures: mild Alzheimer's disease compared with amnestic mild cognitive impairment. Dement Geriat Cognit Disord. 2012;34(3–4):149–55.
144. Zarei M, Patenaude B, Damoiseaux J, Morgese C, Smith S, Matthews PM, Barkhof F, Rombouts SARB, Sanz-Arigita E, Jenkinson M. Combining shape and connectivity analysis: an MRI study of thalamic degeneration in Alzheimer's disease. NeuroImage. 2010;49(1):1–8.
145. Ryan NS, Keihaninejad S, Shakespeare TJ, Lehmann M, Crutch SJ, Malone IB, Thornton JS, Mancini L, Hyare H, Yousry T, Ridgway GR, Zhang H, Modat M, Alexander DC, Rossor MN, Ourselin S, Fox NC. Magnetic resonance imaging evidence for presymptomatic

change in thalamus and caudate in familial Alzheimer's disease. Brain. 2013. doi:10.1093/brain/awt065.
146. O'Dwyer L, Lamberton F, Matura S, Tanner C, Scheibe M, Miller J, Rujescu D, Prvulovic D, Hampel H. Reduced hippocampal volume in healthy young ApoE4 carriers: an MRI study. PloS ONE. 2012;7(11). doi:10.1371/journal.pone.0048895.
147. Cho H, Seo SW, Kim J-H, Kim C, Ye BS, Kim GH, Noh Y, Kim HJ, Yoon CW, Seong J-K, Kim C-H, Kang SJ, Chin J, Kim ST, Lee K-H, Na DL. Changes in subcortical structures in early- versus late-onset Alzheimer's disease. Neurobiol Aging. 2013;34(7):1740–7.
148. Zikopoulos B, Barbas H. Prefrontal projections to the thalamic reticular nucleus form a unique circuit for attentional mechanisms. J Neurosci. 2006;26(28):7348–61.
149. Steriade M. Sleep, epilepsy and thalamic reticular inhibitory neurons. Trends Neurosci. 2005;28(6):317–24.
150. Zou X, Coyle D, Wong-Lin K, Maguire L. Beta-amyloid induced changes in A-type K^+ current can alter hippocampo-septal network dynamics. J Comput Neurosci. 2012;32(3):465–77. http://www.ncbi.nlm.nih.gov/pubmed/21938438. Accessed 4 Jul 2013.
151. Zou X, Coyle D, Wong-lin K, Maguire L n.d. Computational study of hippocampal-septal theta rhythm changes due to beta-amyloid-altered ionic channels. PLoS ONE. 2011;6(6):e21579.
152. Menschik ED Finkel LH. Neuromodulatory control of hippocampal function: towards a model of Alzheimer's disease. Art Intell Med. 1998;13:99–121.
153. Pons AJ, Cantero JL, Atienza M, García-Ojalvo J. Modeling synchronization loss in large-scale brain dynamics. Proceedings of the 18th international Conference On Artificial Neural Networks, Part II. : Springer; 2008. pp. 675–84
154. De Haan W, Mott K, van Straaten ECW, Scheltens P, Stam CJ. Activity dependent degeneration explains hub vulnerability in Alzheimer's disease. PLoS Comput Biol. 2012;8(8):e1002582. http://www.pubmedcentral.nih.gov/articlerender.fcgi?artid=3420961 tool=pmcentrezrendertype=abstract. Accessed 1 Feb 2013.
155. Sen Bhattacharya B, Cakir Y, Serap-Sengor N, Maguire L, Coyle D. Model-based bifurcation and power spectral analyses of thalamocortical alpha rhythm slowing in Alzheimer's disease. Neurocomput. 2013;115:11–22. http://linkinghub.elsevier.com/retrieve/pii/S0925231212008600. Accessed 29 May 2013.
156. Dauwels J, Vialatte F, Cichocki A. Diagnosis of Alzheimer's disease from EEG signals: where are we standing? Curr Alzheimer Res. 2010;7:487–505.

Chapter 10
Toward Networks from Spikes

Mark Hereld, Jyothsna Suresh, Mihailo Radojicic, Lorenzo L. Pesce, Janice Wang, Jeremy Marks and Wim van Drongelen

Abstract Computational models of brain tissue provide important insights for understanding pathological behavior within neuronal networks. Validating these models poses difficult challenges due to the number of neurons and synaptic connections in even the most modest samples. An important step toward validation is determining connectivity within the biological network so that simulations can be configured to match and then compared directly to the observed behaviors. Cultures of dissociated neurons on multi-electrode arrays provide a flexible experimental platform for the study of fundamental network behaviors. Extracting connectivity from this in vitro setup is challenging because we are able to measure only a relatively small number of neurons. Today, cultures are routinely grown on arrays of microelectrodes, each reporting the activity of several neurons. With these techniques we can distinguish at most hundreds of spiking neurons while cultures comprise thousands of neurons. Even as the number of electrodes increases with gains in technology, it is important to understand how much information about the network connectivity can be discovered with sparse spatial sampling. We describe an approach to searching for repeating patterns in parallel spike train data, the presence of which can inform inferences of causality and connectivity in the network.

Keywords Spike trains · Neural networks · Shift sets · Multi-electrode recordings

---

M. Hereld (✉)
Argonne National Laboratory, The University of Chicago, BLDG 240, RM 4139, 9700 S. Cass Ave., Argonne, IL 60439, USA
e-mail: hereld@anl.gov

J. Suresh · M. Radojicic · L. L. Pesce · J. Wang · J. Marks
The University of Chicago, Chicago, IL, USA

W. van Drongelen
e-mail: wvandron@peds.bsd.uchicago.edu

Introduction

Computational modeling can complement laboratory and clinical research in an effort to understand and treat neurological disorders and diseases [27]. Epilepsy, for example, has proven remarkably resistant to therapy in a large fraction of cases despite significant progress in pharmacological treatments. Simulations of network behavior in neocortex is providing interesting results that may shed light on causative factors [25, 26, 28]. Modeling individual neurons may be on a stronger footing than modeling of networks, mainly due to the extraordinary explosion of model configuration information required by the densely interconnected networks coupled with the relative lack of knowledge of exactly how those networks are connected in live tissue.

Consequently, there is a critical need to develop analysis techniques that can help extract information about the underlying network connectivity responsible for generating observed network behaviors. For this purpose, we are advocating an approach to hybrid experimentation that uses dissociated cultures of neurons to provide large parallel spike trains for comparison to simulation output. Unraveling network connections requires detection of the spike events generated by single neurons, and recognition of consistent spatiotemporal sequences of these events across the network. We hope to use analysis of the spike trains to provide connectivity information to be used in configuring our simulations. By iterating on this process of observation, analysis, configuration, simulation, and comparison we will be developing a validated simulation capability that can be used to perform independent experiments.

Although sophisticated algorithms for detection of spike events in extracellular recordings have been developed over the past decades, they are far from perfect. Since these algorithms for detection of activity from individual nerve cells are mostly based on waveform analysis, high spike densities and/or high activity levels in single neurons require ad hoc decision rules to classify single units in multi-unit records. Another principal challenge of this aim is that we are able to measure the electrical activity from a relatively small number of neurons. For example, subsampling impedes accurate characterization of power law distributions [18]. In our lab the cultures are grown on arrays of 60 electrodes. With current techniques we can detect activity from and distinguish at most a handful of spiking neurons while these cultures comprise more than 10K neurons. In clinical recordings with microelectrode arrays, spike detection procedures and problems associated with sparse sampling are an even larger problem, neuronal densities are orders of magnitude higher while electrode interdistances are in the same ball park (e.g., [30]). It is especially critical in clinical applications to know to what extent a limited sample of activity of single neurons provides information about the networks functional properties.

We are therefore trying to develop an understanding of how much information about connectivity can be extracted from a sparse multi-channel sampling of active neuronal networks. Toward this end we are developing strategies in support of a brute force search for precisely matched timing differences between spike events in detailed 30-min recordings from these cultures. In this we are attempting to identify candidates for membership in neuronal groups, such as Izhikevich's polychronous groups [11], that have been triggered at least twice within the experimental window, producing an ordered firing pattern each time. To enhance the power of spike train detection, we will take advantage of the multi-channel aspect of the array records by considering sequences that involve activity on several electrodes. The data used in the current analysis came from a related work by Suresh et al. [24] involving systematic study of the effects of synaptic connectivity on network activity by selectively disabling inhibitory and excitatory transmission.

Many approaches have been applied to the problem of inferring connectivity or at least causality to the observed repeated patterns in parallel spike trains: transfer entropy [5, 10, 14, 23, 29]; granger causality [2, 17]; directed transfer function [31]; and statistical analysis of repeats [6, 7]. It has been shown that pairwise correlations may be sufficient for determining functional relationships involving many neurons, reducing the need to directly consider higher order correlations and, consequently, lessening the potential computational burden [22]. This finding is leveraged by Diekman et al. [3] in a study that strives to assign connections between units (neurons) in a sparsely sampled network.

Our approach of restricting the search space by applying a sequence of constraints on the event set may be related to the comparison of model-free vs. model-based methods by Kobayashi and Kitano [12]. They found that knowledge about the topology improved performance of reconstructions from spike data. By hierarchically asserting constraints based on data driven concerns, we are effectively imposing assumptions about the properties (perhaps including topology through fork, join, and chain constraints) of the underlying connected network.

While repeating sequences (referred to variously as motifs, patterns, episodes) have been reported and ascribed to functional connectivity in some cases [9], it has been found that statistical analysis of repeats is fraught with problems. Because it is difficult to compute analytically the expected distributions of events under random conditions it is difficult to assign significance to findings of repeating patterns. Detailed analysis of jittering methods [1, 20] underscore the difficulties encountered when trying to infer connectivity from parallel spike trains, particularly when response is not tied to known excitation of the circuits. It has been shown that stochastic processes can very easily generate repeating patterns. Mokeichev et al. [16] emphasize the importance of careful consideration of statistics when conferring significance to purported patterns found in spike trains. They found that complex repeating motifs occurring in cortical recordings may be generated stochastically rather than indicating the existence of underlying network structure. Others, while cautionary, continue to find some evidence of interconnected neural subgroups and legitimately generated "long patterns that do not repeat exactly" [20].

It is therefore of paramount importance that an accurate estimate of the baseline be computed so that detected patterns can be compared for statistical significance. There are several techniques for generating such surrogate data, including: interval randomization [4, 8], spike jittering [1, 19, 20], and Poisson simulation [16].

The methods described in the following sections are both selective and efficient. Presuming the presence of repeating spatiotemporal patterns in the parallel spike trains enables an interesting perspective on the distribution of lags between event pairs, leading to the notion of *shift sets*. The consequences of this perspective are discussed in section "Shift Sets". Because the search space is so large, finding ways to eliminate vast swaths from computational consideration is important to developing efficient algorithms. Section "Distribution of Event Pair Delays" discusses the distribution of delay times in the measured collection of shift sets and its significant departure from the nominal spectrum generated by shuffling intraspike intervals. Using this distribution to target a reduced subset of shift sets for further study, we show in section "Distribution of Shift Set Lengths" how the distribution of shift set length is affected by this filter. And finally, we discuss the results and what they might mean in terms of network behavior and repeating patterns supported by functional sub-networks.

The spike trains analyzed in this work were collected from dissociated rat hippocampal cell cultures (embryonic 18 days) [21] grown on multi-electrode arrays (MEA) having 60 channels [24]. Multichannel recordings were performed with the MEA 2100 device (Multi Channel Systems, Reutlingen, Germany). Data was recorded at a sample rate of 25 kHz/channel and passed through a band filter (300–1500 Hz). Spike rasters depict all spike events detected as negative deflections that exceeded a threshold of five times the standard deviation of the filtered signal. Spikes were not sorted since no pattern detections would be missed by leaving them unsorted, although the false positive rate might rise somewhat as a consequence. Recordings of 15 or 30 min were taken over the course of several weeks as the cultures matured.

Shift Sets

We seek signatures of repeating patterns that will enable them to be spotted above the noise of the stochastically generated event pairs. One property of these signatures will be how much they concentrate the signal into bins that will enable them to stand out above the stochastic background. Another is how they lend themselves to pruning the data to lessen the computational burden of the search.

Pattern searches of the sort we are considering here typically begin by studying the spectrum of event pairs separated by a common delay interval. Borrowing a convenient notation [3, 13], we can write $A[t_i]B$ to indicate the set of all event pairs wherein an event on neuron A precedes an event on neuron B by time t_i. These sets are sometimes referred to as *frequent episodes*, a term taken from the literature on data mining of event streams in general [15]. We are looking for repeating patterns in the multi-electrode event stream.

If a spatiotemporal pattern of spike events of length M is repeated N times in a single trial, how can we decide how to look for it in a large multi-neuron spike train? With the polychronous group model as our guide we focus our search on instances

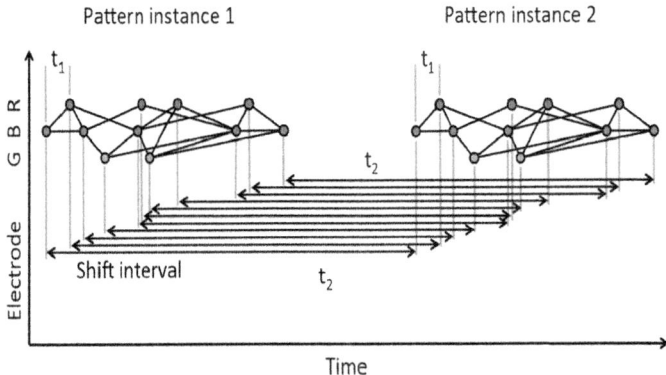

Fig. 10.1 A shift set. This example of repeated firing patterns illustrates the repeating pairwise intervals that they induce. Color of node (*red, blue, green*) corresponds to electrode (*neuron*) label on the *left* (R, B, G). Lines connecting events indicate causal connection, possibly synaptic connection. This diagram follows Izhikevich [11]

of fixed time differences between sets of events on electrode pairs. The schematic raster plot of a subset of the events in a trial (Fig. 10.1) shows why this might be appealing in this context. Imagine a set of events that are causally connected as in a polychronous group—initiating the firing sequence by exciting a small number of neurons leads to activation of the neurons in the group according to the pattern shown in Fig. 10.1. The pattern involves three electrodes (coded in the diagram as red, blue, and green) and includes 11 events in each instance. The pattern occurs twice in this example. Note that the time delay for corresponding events is constant (t_2 in this example), which we denote as the shift interval. We denote the collection of events in these two instances together with the shift interval as a *shift set*. We refer to the number of events in an instance of this firing pattern as the shift set length.

Using the color coding of the neuronal associations, the sequence reads: B-R-B-G-B-R-G-R-B-R-B. We can characterize two classes of repeated event pairs from this repeated pattern as follows.

Intra-pattern intervals: Taking the first two events in the sequence as an example, episodes $B[t_1]R$ will be found in the event stream as many times as the pattern is found, that is, N times. For this class of frequent episode, there are $\binom{M}{2} = \frac{M(M-1)}{2}$ pairs of neurons to consider within the pattern. Each of these pairs will contribute N instances of an interval to the spectrum of pairwise event time differences. That is, there will be $\binom{M}{2}$ peaks that are N instances tall. The total number of pairs that will contribute to the spectrum is $\binom{M}{2} \times N = \frac{NM(M-1)}{2}$, a measure of the overall impact of this feature on the spectrum of intervals.

Inter-pattern intervals: Likewise, considering the delay between start of each pattern instance, t_2, we note that episodes $X[t_2]X$, where X is any one of the recorded neurons, will be found as many times as there are events in the pattern. We call this list of event pairs a *shift set*. For N instances of the pattern, there are $\binom{N}{2}$ pattern instance pairs. Each of these pattern pairs, leading to a shift set, will contribute M

instances of the interval representing the offset between the onset of each instance. There will be $\binom{N}{2}$ peaks that are M tall. The total number of pairs participating is $\binom{N}{2} \times M = \frac{MN(N-1)}{2}$. A key distinguishing factor between the intra- and inter-pattern intervals is that the latter collects pairs where the participating neurons are the same in each pair, as in: $B[t_2]B, R[t_2]R, B[t_2]B, G[t_2]G, B[t_2]B, R[t_2]R$, etc.

One can add background noise and ask how the two perspectives will compare to one another. Conspicuous peaks in the spectrum will indicate either the presence of long patterns, where the number of event pairs in the peak will correlate to the length of the pattern, M; or, high repetition count of smaller patterns, large N. Interpreting such peaks would require different analysis approaches depending on which of these cases holds true.

Note that in addition to the two repeating sets of pairwise delays just described, the existence of an MN pattern in the spike train also contributes a great number of pairwise delays that derive from events taken from different instances of the pattern, but not the special subset of those comprising the shift sets. An example: the delay between the first event in one instance of the pattern, and the second event in another instance. Given the supposed N instances of the M-length pattern, there will be $\binom{MN}{2}$ total distinct event pairs that might contribute to the computed spectrum of delays. This is a number in excess of the two much smaller subsets computed above. For example, with $M = N = 10$ there will be 450 intra-pattern intervals and another 450 inter-pattern intervals. Compare this to 4050 delays that are neither inter- or intra-pattern. These will in general contribute to the noise floor against which we are trying to detect the patterns.

Distribution of Event Pair Delays

Spike rasters were obtained from a band filtered (300–1500 Hz) version of the broadband signals (Fig. 10.2). A typical raster plot from one trial is shown in Fig. 10.3a. Signals from nine electrodes were selected from the region of the MEA that was most active. Figure 10.3b depicts the same data plotted after shuffling the intervals in the event stream. This procedure preserves the overall number of events on each electrode, as well as the firing sequence across electrodes. Much of the bursting structure of the time series is preserved, but even by eye it is apparent that the distribution on bursting time scales has been affected by the interval shuffling procedure. Figure 10.3c and 10.3d show the corresponding fine-temporal distribution of spike events in three detailed views of typical bursts.

The event data from a trial can be described in terms of an event list, E_i, which is a set of tuples $\{T_i, N_i\}$ with T_i the time of the ith event, and N_i the channel number of the MEA. We compute all of the intervals between events in the list, giving a list I_{ij} of tuples $\{(T_i - T_j), N_i, N_j\}$ such that $(T_i - T_j) > 0$. We define a shift set as the set of all intervals I_{ij} such that $(T_i - T_j) = T_s$, some constant shift interval, and $N_i = N_j$

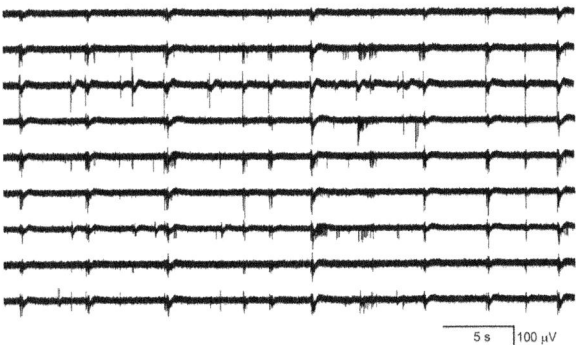

Fig. 10.2 A representative subset of broadband signals (bandwidth 1–3000 Hz) recorded with a multielectrode array setup. The activity patterns depict low frequency waveforms and synchronous as well as asynchronous multi-unit spiking activity

for each of the intervals in the set. This notation provides a little more detail than our earlier characterization in terms of episodes, $X[t_i]X$.

Note that the neuron set associated with the shift set is only a candidate as a polychronous group. This is because we don't have any information about the causal relationships between the events in the set. If a polychronous group is represented in the spike train by firing at least twice, then there will be a shift set that captures it's signature, perhaps many shift sets—$\binom{M}{2}$ if the group fires M times in the recording. The shift set is a list of neurons that fired two times separated by a specific delay time. A subset of that ordered list of neurons might belong to a polychronous group: (1) there may be neurons represented in the shift set that are serendipitously there because they happened to fire with the delay in question; (2) since we are not sampling all neurons, there may be more neurons in the full polychronous group that we are not even recording.

Given this definition, it is now possible to find all shift sets by sorting and filtering the aggregate interval list. For each value of T_s, there may be more than one entry in the interval list for which the above conditions (namely, I_{ij} such that $(T_i - T_j) = T_s$, some constant shift interval, and $N_i = N_j$). If so, we have a shift set whose length is given by the number of entries meeting this condition.

The distribution of shift set delays of the data shown in Fig. 10.4 has a number of important features. For delays longer than 1.5 s (middle of the plot), and in fact out to 80 s where the analysis ends (not shown), the observed (black dots) and surrogate data (blue line) are remarkably similar, fluctuating around a constant number of event pairs (around 350 for this 1.6 ms binned data) sharing the delay value. The delay range between approximately 850 ms and 1.5 s appears to be a transition zone. Below 850 ms we see that the principal difference in the distributions is evident. The observed distribution is much more peaked toward short delays than the shuffled surrogate data, as is evident in the 1 s burst details shown in Fig. 10.3c and 10.3d.

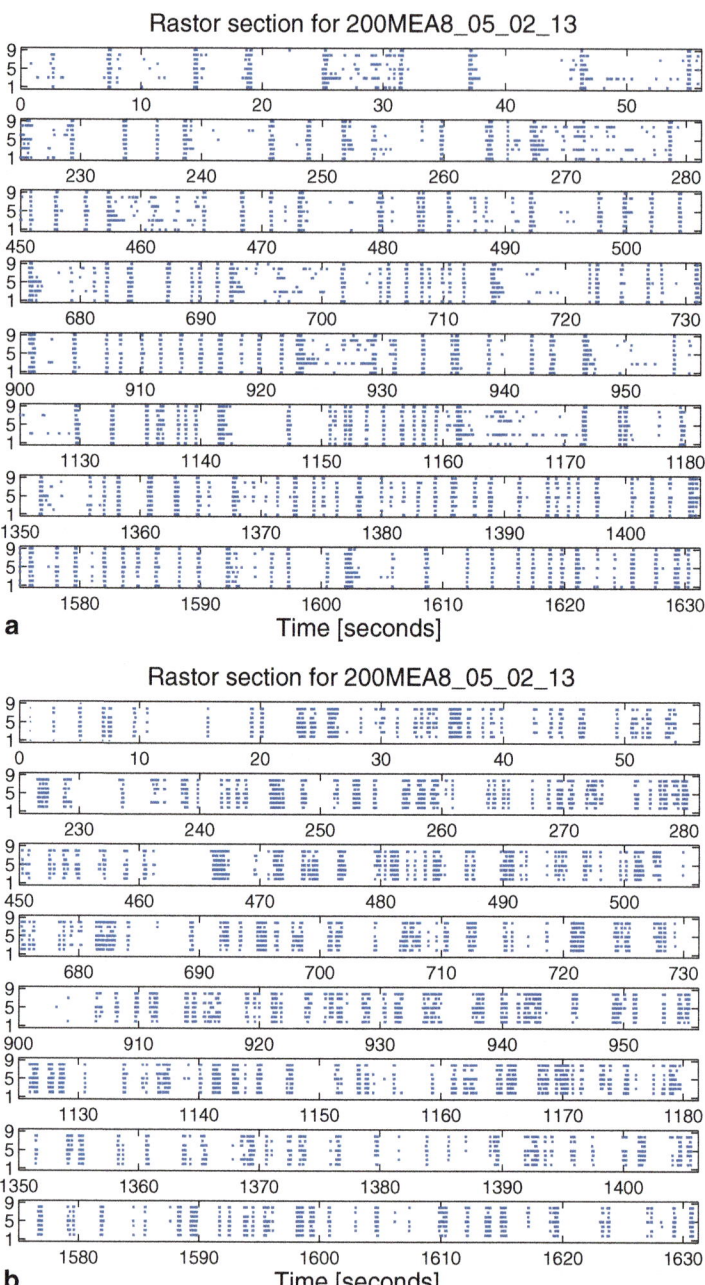

Fig. 10.3 Raster plot of spike train from a multi-electrode array recording: (**a**) typical raster of nine very active channels from a 30-min recording, and (**b**) the corresponding surrogate data generated by interval shuffling. They include eight slices of 56 s taken at the beginning of each 225 s segment of the 1800 s recording to show behavior across the entire recording. Also shown are three detailed examples of 1 s rasters of typical bursts showing observed spikes (**c**) and the surrogate data (**d**)

10 Toward Networks from Spikes

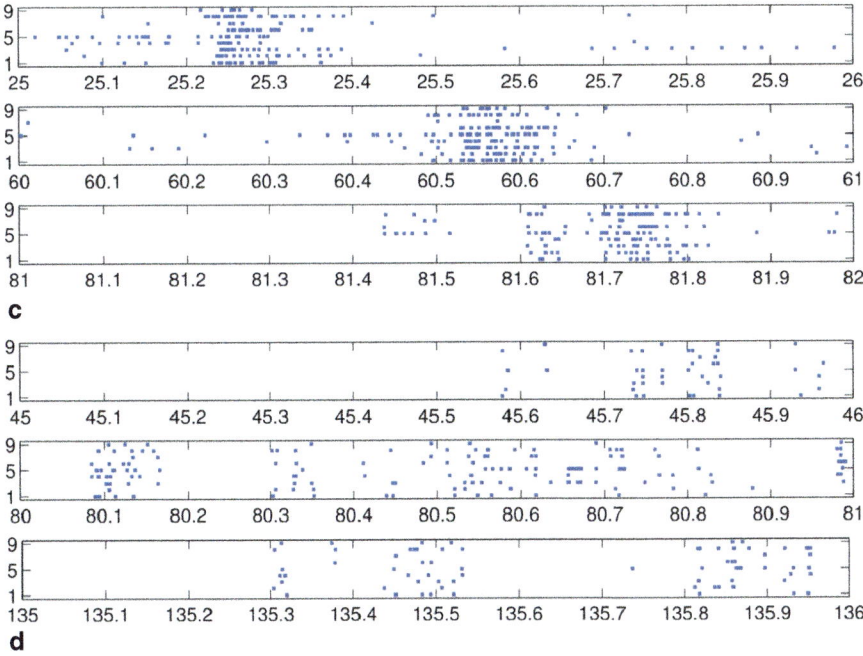

Fig. 10.3 (continued)

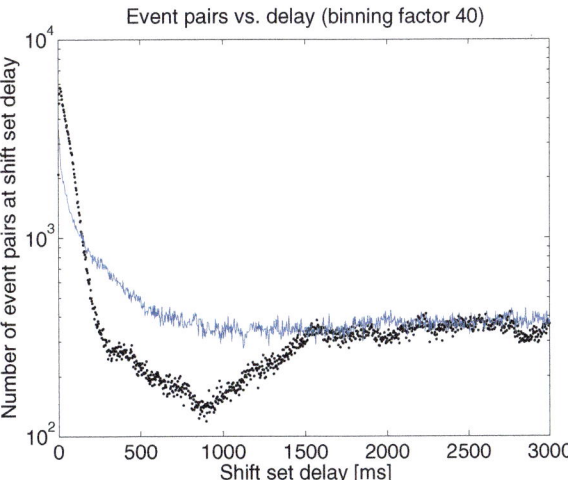

Fig. 10.4 The distribution of delays between event pairs on the same neuron, $X[t]X$, up to delay $t=3$ s. Observed data (*black dots*) and surrogate data (*blue line*) continue to overlap at a constant occurrence level out to 80 s, the maximum delay considered in the analysis

Distribution of Shift Set Lengths

The distribution of shift set lengths is shown in Fig. 10.5. In each of the plots, the measured data is shown with points while the surrogate data is shown as a line. Binning factors of 1, 4, and 20 are plotted from left to right.

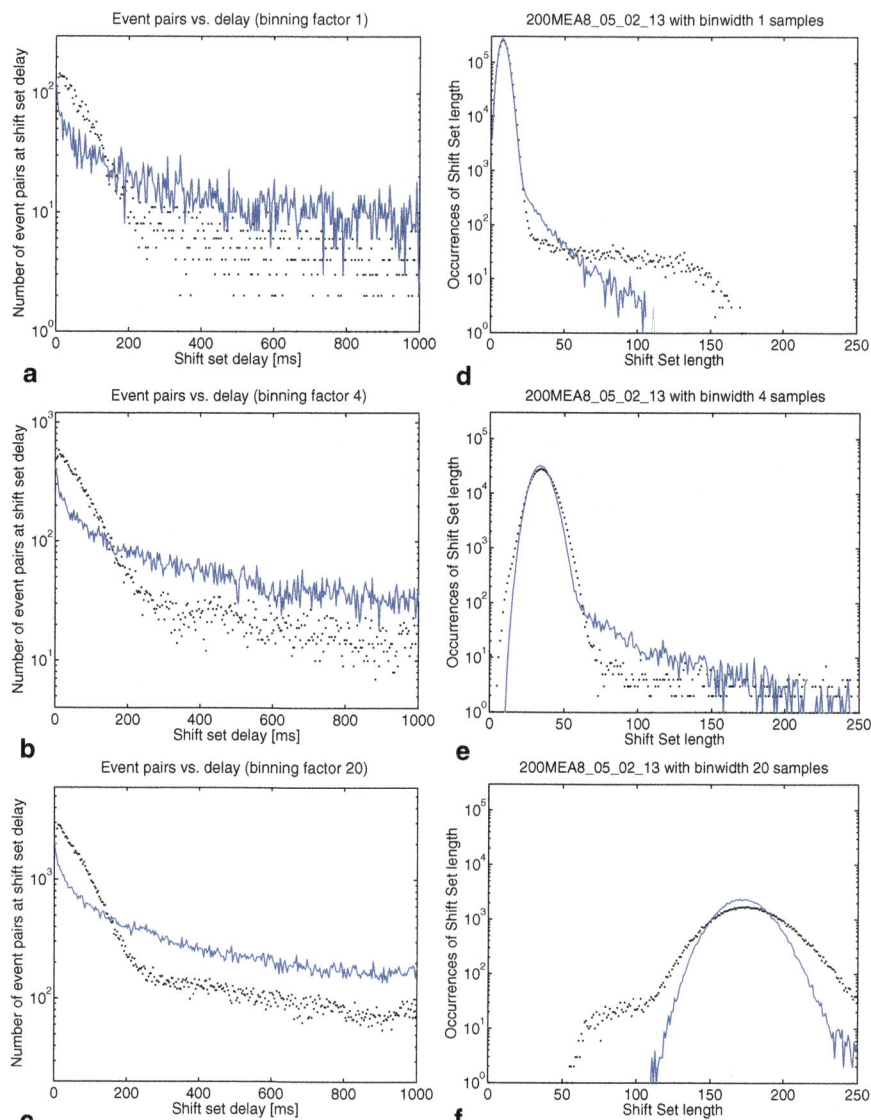

Fig. 10.5 Effects of binning on distributions drawn from the observed data and compared to surrogate data. Binning factors of 1, 4, and 20 are shown from *top to bottom*, corresponding to bin widths of 0.04, 0.16, and 0.8 ms. (**a–c**) The distribution of delays (as in Fig. 10.4) between event pairs up to 1 s; (**d–f**) the distribution of shift set lengths taking all delays up to 80 s; (**g–i**) includes only delays up to 850 ms

There are several noteworthy observations we would make about the distribution of shift set lengths and its relationship to the distribution of event pair delays (between a single neuron). Consider first the surrogate data (blue line) in all plots. Looking at the distribution of shift set lengths (Fig. 10.5d, 10.5e and 10.5f),

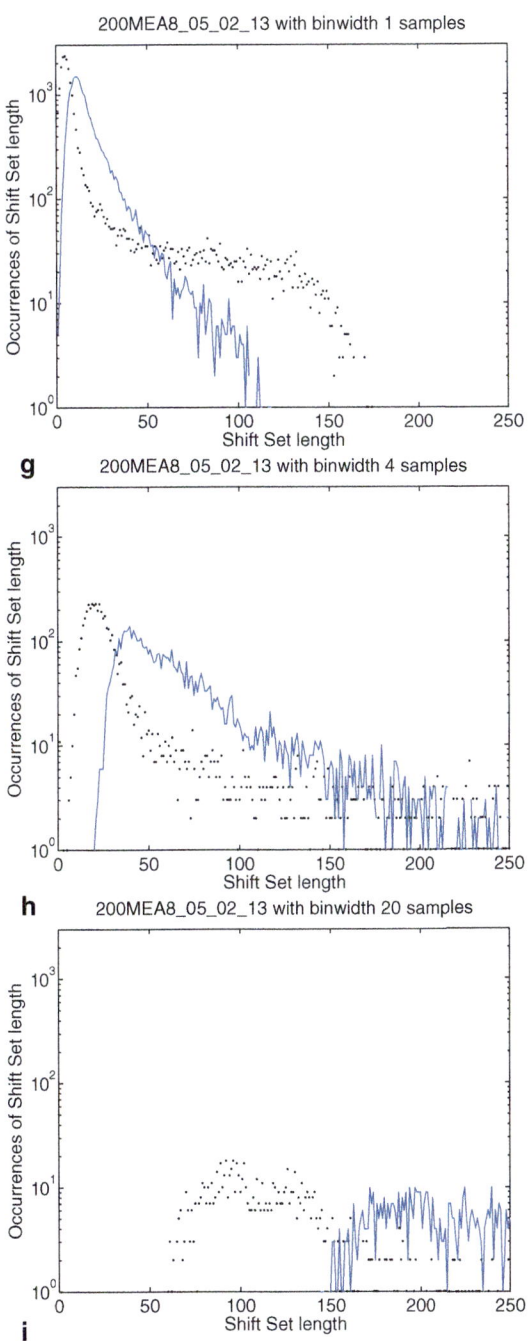

Fig. 10.5 (continued)

it appears to be comprised of two component subpopulations. Most of the expected shift set lengths are contained in a large peak that moves to the right, becoming shorter and wider with increasing binning factor. As noted in the previous section, the distribution of shift set delays asymptote to a constant value (Figs. 10.4 and 10.5a, 10.5b and 10.5c). The large peak in the length distribution contains all of the shift set delays from 1.5 s out to 80 s (we didn't analyze delays longer than this). The peak is centered on the asymptotic constant, and has a width set by the fluctuations in shift set length around this constant.

The long tail seen in the surrogate data dominates the fall-off on the right in Fig. 10.5d, 10.5e and 10.5f, only barely visible in (f) due to the choice of horizontal axis range. The distribution of this tail clearly derives from the spectrum of shift set lengths shortward of 1.5 s (most clearly seen in Fig. 10.4) where there are increasing numbers of shift sets with lengths larger than the asymptote.

Turning our attention to the observed data (black dots), we note first that the peak population corresponds closely to the surrogate data peak except that it is perhaps a little bit broader as is most clearly evident in Fig. 10.5f.

As has been noted, the shift sets with delay less than about 850 ms in this dataset have the most divergent spectrum when compared to the surrogate data. On the assumption that this subpopulation therefore carries the most relevant temporal structure, we can strategically focus our investigation on those shift sets. The length distributions shown in the bottom row (Fig. 10.5g, 10.5h and 10.5i) include only shift sets with delay 850 ms or less.

We can say something about the constituents in the main peak in the distribution of event pair delays. This subpopulation is evidently not present in either the observed or surrogate distributions shown in the bottom row of Fig. 10.5 as expected. The counts that occupy this constant regime of the delay distribution present a distribution that is approximately gaussian around that constant. This is the peak distribution, which derives from the timing of the bursts, and results in rather large occurrence counts because it includes counts from delays out to 80 s.

Next consider the distribution of the surrogate data in the third row (Fig. 10.5g, 10.5h and 10.5i). It appears to correspond to the long tail of the inclusive distribution shown in the second row (Fig. 10.5d, 10.5e and 10.5f). This tail distribution has arisen as an artifact of the interval shuffling applied to the measured data and is consistent with the rising counts in the delay distribution toward short delays, as noted.

Now, the highly selected subpopulation of the observed shift sets, shown in the third row (Fig. 10.5g, 10.5h and 10.5i), has two morphological components: one is strongly peaked at short length shift sets and the other is a shoulder or plateau at longer shift set lengths. The peak is evident in each of those panels, and can be seen most readily in the inclusive distribution as a bulge to the left of the main peaks in Fig. 10.5e and 10.5f. Looking at the total counts in Table 10.1 of the shift sets in this population (*delays below 850 ms*) appear to represent about 1% of the total counts (*data total counts*).

We can easily understand some of the effects of binning as illustrated in Fig. 10.5. In the top row (a–c) binning simply averages out fluctuations, tightening the vertical spread of the distribution around the local mean. In the middle row (d–f), the

Table 10.1 Key statistics drawn from the distribution of shift set lengths as illustrated in Fig. 10.5

Bin factor (ms)	1 (0.04)	2 (0.08)	4 (0.16)	10 (0.4)	20 (0.8)	40 (1.6)
Total counts	2,000,000	100,0000	50,0000	20,0000	10,0000	50,000
Observed peak[a]	254,000	85,600	28,000	5980	1720	518
At length	8	17	34	87	174	352
Surrogate peak[a]	264,000	93,300	31,800	7770	2300	807
At length	7	17	34	86	171	352
Delays below 850 ms	21,250	10,625	5312	2125	1062	531

Binning factor and total number of shift sets are tabulated in the top two rows. Properties of the observed distribution peak are in the second block of two rows. The third block shows the same for the surrogate data. The last row tabulates the number of shift sets remaining to be considered once the peak population has been removed

[a] Rounded to three significant figures

apparent peak broadening with increased bin width is an artifact of the linear scale of the x-axis and not due to binning at all; a log-log plot, instead of the semi-log plot shown, would preserve the apparent width of the distribution as it moves to the right. What is constant is the width as a fraction of the shift set length on the x-axis.

Binning also apparently shifts features to the right, toward larger shift set lengths, while decreasing their frequency of occurrence. Table 10.1 summarizes some of the statistics extracted from the data shown in Fig. 10.5 and includes data from a more complete set of binning factors. Keep in mind that the length distributions have not been normalized and that the total counts are reduced in proportion to the binning factor, as shown in the row labeled *total counts*. The position of the peak is shown in the row following *observed peak*. The rightward shift of the peak appears to track the binning factor very closely. The trend is not difficult to explain. Binning adjacent time samples $a_1 a_2$ will tend to increase the number of pairwise differences that are labeled with the same delay because in addition to $a_1[t_x]b_1$ and $a_2[t_x]b_2$, the combinations $a_1[t_x+1]b_2$ and $a_2[t_x-1]b_1$ would be included in the $A[t_x]B$ shift set, now longer but less temporally refined. This effect moves the features to the right in the length distribution. On the other hand, the decreasing frequency of occurrence of distribution features, as tabulated in the rows labeled *observed peak* and *surrogate peak*, do not appear to be inversely proportional to the binning factor. That might have been expected if the drop in peak were due solely to the drop in total counts. Instead, it may arise from the probability distribution for neighboring time sample pairs (in the above analysis) to include event pairs.

Conclusions and Next Steps

In this paper we have espoused an approach to validating simulations of brain tissue that uses long recordings of spike trains from dissociated cultures of neurons in a program aimed at understanding network connectivity. Armed with information about connectivity in such networks, we argue that it will be possible to configure our network models more realistically. The results from these tuned simulations can be compared to the observed spike trains from the biological networks. Repeating this process of observation, analysis, configuration, simulation, and comparison will enable iterative development of a validated model that could then be applied with a measure of confidence to other questions about network behavior both in the laboratory and in clinical applications.

We have introduced the concept of a *shift set* to describe a collection of event pairs that are computationally inexpensive to identify and, furthermore, map directly to candidate groups of neurons that may be producing complex spatiotemporal firing patterns. We derived combinatoric expressions for the number of occurrences of event pairs at delay values determined by repeated patterns of fixed length. These may be used to study the distribution of delays generated by ensembles of patterns in comparison to observed distributions.

We discussed the morphology of the distribution of shift set delays and of the distribution of shift set lengths, comparing results from analysis of observed spike trains with surrogate data generated by shuffling interspike intervals. In particular, for delays less than 850 ms we found a significant departure from the number of instances present in the surrogate data. This observation suggests that organized network behaviors are preferentially represented by shift sets in this range. Pruning the search space in this way, based on identifying subsets of the shift set collection that depart significantly from the shuffled distribution, will improve search efficiency by removing unlikely candidates ahead of time. The bulk of the distribution beyond delays of 1.5 s appear to be randomly fluctuating around a constant number of instances per delay bin. Binning in the time domain by a factor of 40 from the sample rate, leading to 1.6 ms bins, made estimating the breaks in the delay distribution much easier.

In a brute force search through the spike data we see significant departures from chance occurrence of repeated intervals in spikes from pairs of channels. These show up in excesses on both sides of the main peak of the shift set length histogram. It is hoped that with further analysis of this ilk, based on the shift set and relying on judicious pruning of the search space, we will be able to develop a set of candidate repeating patterns that prove significant as indicators of internal network connectivity.

The pieces of the analysis pipeline described in this work are only part of the story, and are presented here as they represent novel perspectives on the problem of finding informative spatiotemporal firing patterns in large parallel spike train data streams. Spike sorting can help to provide more selective tests of patterns purported to be generated by underlying sub-networks. This will come at the expense of additional combinatorial explosion of the search space. The methods for pruning the search space would benefit from a more solid statistical underpinning. Specifically, it would be valuable to attach a likelihood estimate for the observed departures from the surrogate distributions as has been done in similar contexts [3, 6, 16, 20]. Comparison of results derived from these methods with well-studied measures of causality [14, 17, 29] would further improve understanding of their significance.

Finally, we need to introduce simulation of artificial neural networks with known connectivity. Subjecting the generated spike trains to the proposed pipeline of analysis would enable us to understand better the limitations of the proposed analysis methodology and possibly suggest improvements.

Acknowledgments This work was supported in part by the Ralph and Marian Falk Medical Research Trust, the National Institutes of Health under award number R01NS084142, and the U.S. Department of Energy under contract DE-AC02–06CH11357. The content is solely the responsibility of the authors and does not necessarily represent the official views of the granting agencies.

References

1. Amarasingham A, Harrison MT, Hatsopoulos NG, Geman S. Conditional modeling and the jitter method of spike resampling. J Neurophysiol. 2012;107(2):517–31. doi:10.1152/jn.00633.2011.
2. Barnett L, Barrett AB, Seth AK. Granger causality and transfer entropy are equivalent for gaussian variables. Phys Rev Lett. 2009;103:238701. doi:10.1103/PhysRevLett.103.238701.
3. Diekman C, Dasgupta K, Nair V, Unnikrishnan KP. Discovering functional neuronal connectivity from serial patterns in spike train data. Neural Comput. 2014;26:1263–97. doi:10.1162/NECO_a_00598.
4. Gansel KS, Singer W. Detecting multineuronal temporal patterns in parallel spike trains. Front Neuroinform. 2012;doi:10.3389/fninf.2012.00018.
5. Garofalo M, Nieus T, Massobrio P, Martinoia S. Evaluation of the performance of information theory-based methods and cross-correlation to estimate the functional connectivity in cortical networks. PLoS ONE. 2009;4(8):e6482. doi:10.1371/journal.pone.0006482.
6. Grün S, Diesmann M, Aertsen A. Unitary events in multiple single-neuron spiking activity: I. Detection and significance. Neural Comput. 2002;14(1):43–80. doi:10.1162/089976602753284455.
7. Grün S, Diesmann M, Aertsen A. Unitary events in multiple single-neuron spiking activity: II. Nonstationary data. Neural Comput. 2002;14(1):81–119. doi:10.1162/089976602753284464.
8. Humphries MD. Spike-train communities: finding groups of similar spike trains. J Neurosci. 2011;31:2321–2336. doi:10.1523/JNEUROSCI.2853-10.2011.
9. Ikegaya Y, Aaron G, Cossart R, Aronov D, Lampl I, Ferster D, Yuste R. Synfire chains and cortical songs: temporal modules of cortical activity. Science. 2004;304(5670):559–64. doi:10.1126/science.1093173.
10. Ito S, Hansen ME, Heiland R, Lumsdaine A, Litke AM, Beggs JM. Extending transfer entropy improves identification of effective connectivity in a spiking cortical network model. PLoS ONE. 2011;6(11):e27431. doi:10.1371/journal.pone.0027431.
11. Izhikevich EM. Polychronization: computation with spikes. Neural Comput. 2006;282:245–82.
12. Kobayashi R, Kitano K. Impact of network topology on inference of synaptic connectivity from multi-neuronal spike data simulated by a large-scale cortical network model. J Comput Neurosci. 2013;35:109–24. doi:10.1007/s10827-013-0443-y.
13. Laxman S, Sastry PS, Unnikrishnan KP. A fast algorithm for finding frequent episodes in event streams. Proceedings of the 13th ACM SIGKDD international conference on Knowledge discovery and data mining—KDD '07. New York: ACM; 2007. p. 410. doi:10.1145/1281192.1281238.
14. Li Z, Li X. Estimating temporal causal interaction between spike trains with permutation and transfer entropy. PLoS ONE. 2013;8(8):e70894. doi:10.1371/journal.pone.0070894.

15. Mannila H, Toivonen H, Verkamo AI. Discovery of frequent episodes in event sequences. Data Min Knowl Discov. 1997;1(3):259–89. doi:10.1023/A:1009748302351.
16. Mokeichev A, Okun M, Barak O, Katz Y, Ben-Shahar O, Lampl I. Stochastic emergence of repeating cortical motifs in spontaneous membrane potential fluctuations in vivo. Neuron. 2007;53:413–25. doi:10.1016/j.neuron.2007.01.017.
17. Nedungadi AG, Rangarajan G, Jain N, Ding M. Analyzing multiple spike trains with non-parametric granger causality. J Comput Neurosci. 2009;27:55–64. doi:10.1007/s10827-008-0126-2.
18. Priesemann V, Munk MHJ, Wibral M. Subsampling effects in neuronal avalanche distributions recorded in vivo. BMC Neurosci. 2009;10:40. doi:10.1186/1471-2202-10-40.
19. Rolston JD, Wagenaar DA, Potter SM. Precisely timed spatiotemporal patterns of neural activity in dissociated cortical cultures. Neuroscience. 2007;148:294–303. doi:10.1016/j.neuroscience.2007.05.025.
20. Roxin A, Hakim V, Brunel N. The statistics of repeating patterns of cortical activity can be reproduced by a model network of stochastic binary neurons. J Neurosci. 2008;28:10734–45. doi:10.1523/JNEUROSCI.1016-08.2008.
21. Shelat PB, Plant LD, Wang JC, Lee E, Marks JD. The membrane-active tri-block copolymer Pluronic F-68 profoundly rescues rat hippocampal neurons from oxygen–glucose deprivation-induced death through early inhibition of apoptosis. J Neurosci. 2013;33(30):12287–99.
22. Schneidman E, Berry MJ, Segev R, Bialek W. Weak pairwise correlations imply strongly correlated network states in a neural population. Nature. 2006;440(7087):1007–12. doi:10.1038/nature04701.
23. Stetter O, Battaglia D, Soriano J, Geisel T. Model-free reconstruction of excitatory neuronal connectivity from calcium imaging signals. PLoS Comput Biol. 2012;8(8):e1002653. doi:10.1371/journal.pcbi.1002653.
24. Suresh J, Radojicic M, Pesce L, Bhansali A, Tryba AK, Wang J, Marks JD, Van Drongelen W. Role of excitatory and inhibitory synaptic transmission in shaping network burst dynamics in hippocampal neuronal cultures in-vitro. In preparation 2015.
25. Van Drongelen W, Koch H, Elsen FP, Lee HC, Mrejeru A, Doren E, Marcuccilli CJ, Hereld M, Stevens RL, Ramirez JM. Role of persistent sodium current in bursting activity of mouse neocortical networks in vitro. J Neurophysiol. 2006;96:2564–77. doi:10.1152/jn.00446.2006.
26. Van Drongelen W, Lee HC, Hereld M, Chen Z, Elsen FP, Stevens RL. Emergent epileptiform activity in neural networks with weak excitatory synapses. IEEE Trans Neural Syst Rehabil Eng. 2005;13:236–41. doi:10.1109/TNSRE.2005.847387.
27. Van Drongelen W, Lee HC, Hereld M, Jones D, Cohoon M, Elsen F, Papka ME, Stevens RL. Simulation of neocortical epileptiform activity using parallel computing. Neurocomputing. 2004;58–60:1203–9. doi:10.1016/j.neucom.2004.01.186.
28. Van Drongelen W, Lee HC, Stevens RL, Hereld M. Propagation of seizure-like activity in a model of neocortex. J Clin Neurophysiol. 2007;24:182–8. doi:10.1097/WNP.0b013e318039b4de.
29. Vicente R, Wibral M, Lindner M, Pipa G. Transfer entropy-a model-free measure of effective connectivity for the neurosciences. J Comput Neurosci. 2011;30:45–67. doi:10.1007/s10827-010-0262-3.
30. Weiss SA, Banks GP, McKhann GM, Goodman RR, Emerson RG, Trevelyan AJ, Schevon CA. Ictal high frequency oscillations distinguish two types of seizure territories in humans. Brain. 2013;136(Pt 12):3796–808. doi:10.1093/brain/awt276.
31. Wilke C, Van Drongelen W, Kohrman M, He B. Neocortical seizure foci localization by means of a directed transfer function method. Epilepsia. 2010;51(4):564–72. doi:10.1111/j.1528-1167.2009.02329.x.

Chapter 11
Epileptogenic Networks: Applying Network Analysis Techniques to Human Seizure Activity

Sofija V. Canavan, Tahra L. Eissa, Catherine Schevon, Guy M. McKhan, Robert R Goodman, Ronald G. Emerson and Wim van Drongelen

Abstract A key aspect of the treatment of patients with medically intractable focal epilepsy is surgical resection of the seizure focus. This requires the seizure onset zone to be localized to a specific area of brain tissue. Currently, this seizure locus is largely identified via expert visual assessment of the electroencephalogram (EEG) and electrocorticogram (ECoG). Therefore, localization could potentially be improved by (1) the incorporation of recording techniques with higher spatial and temporal resolution; and (2) by analyses that help to objectively and quantitatively identify the source of epileptic activity. Here, we discuss the application of signal processing and graph theory measures to human ECoG and depth electrode data. We demonstrate the application of a subset of these techniques, including cross correlation, coherence, and Granger causality, to analyze a recording of a seizure from a multi-electrode array (MEA) implanted in human cortex. The results are used to define epileptogenic networks, which can be validated against the visually identifiable development and propagation of the seizure. We then characterize each

S. V. Canavan (✉)
Department of Computational Neuroscience, University of Chicago, 5841 S. Maryland Avenue, O-132, MC 2121, Chicago, IL 60637, USA
e-mail: svcanavan@uchicago.edu

T. L. Eissa
Department of Neurobiology, University of Chicago, Chicago, IL 60637, USA

C. Schevon · R. G. Emerson
Department of Neurology, Hospital for Special Surgery, Weill Cornell Medical Center, New York, NY 10021, USA

G. M. McKhan
Department of Neurological Surgery, Columbia University Medical Center, New York, NY 10032, USA

R. R. Goodman
Department of Neurosurgery, St. Luke's-Roosevelt and Beth Israel Hospitals, New York, NY 10019, USA

W. van Drongelen
Department of Pediatrics, University of Chicago, Chicago, IL 60637, USA

Department of Neurology, University of Chicago, Chicago, IL 60637, USA

© Springer International Publishing Switzerland 2015
B. S. Bhattacharya, F. N. Chowdhury (eds.), *Validating Neuro-Computational Models of Neurological and Psychiatric Disorders,* Springer Series in Computational Neuroscience 14, DOI 10.1007/978-3-319-20037-8_11

network and quantify the degree of similarity between networks in order to directly compare analysis techniques.

Keywords Epilepsy · Cross correlation · Coherence · Granger causality · Graph theory

Abbreviations

MEA	Multi-electrode array
EEG	Electroencephalography
PET	Positron emission tomography
MEG	Magnetoencephalography
MRI	Magnetic resonance imaging
ECoG	Electrocorticography

Introduction

Epilepsy is a serious neurological disorder that affects 70 million people in the world [1, 2]. Despite the development of numerous antiepileptic medications, approximately one third of patients are refractory to medical treatment [3]. Among intractable patients who have focal epilepsy, one of the most efficacious remaining treatment options is to surgically resect the brain tissue thought to contain the seizure focus [4–6]. The rate of success of this treatment varies from approximately 30–70 %, depending on the location of the seizure focus [7], and relies on accurately localizing and removing the source of the seizure, while sparing the maximum possible amount of healthy tissue in order to minimize the risk of adverse effects.

Currently, focal localization is accomplished using a combination of techniques including long-term scalp electroencephalography (EEG), positron emission tomography (PET), magnetoencephalography (MEG), magnetic resonance imaging (MRI), and finally, electrocorticography (ECoG) with or without depth electrodes that must be surgically placed prior to resection of the seizure focus. Among these techniques, MRI and PET have the best spatial resolution, but do not have sufficient time resolution to directly measure the fast development and propagation of seizures. Instead, they are primarily implemented to detect the baseline structural and metabolic abnormalities that are sometimes associated with epileptogenic neural tissue, as in cases of cortical dysplasia. MEG has the required time resolution for measuring fast neural activity, but is not suitable for standard long term monitoring and does not reflect activity associated with radial dipoles. Therefore, the current standard of care for seizure localization relies most on expert visual evaluation of the EEG and ECoG.

This clinical gold standard of seizure localization has several recognized drawbacks, some of which may be addressed by the incorporation of signal processing techniques. A successful example of this in clinical practice is the application of the fast Fourier transform (FFT) to detect and quickly evaluate specific, often low-amplitude, frequencies that may be simultaneously occurring in segments of the

EEG/ECoG recording. More sophisticated techniques may also add valuable information. Many research efforts have employed such methods as cross correlation, coherence and Granger causality to analyze neural recordings obtained from ECoG arrays or depth electrodes implanted in epileptic patients. For example, early efforts used coherence to quantify the development of synchrony between medial temporal depth electrodes during seizures [8] and, with the addition of phase information, to characterize the direction of propagation of seizures between the amygdala and the hippocampus [9].

These techniques have been applied to intracranial recordings of both active seizures and interictal periods (between seizures), including specific interictal events that are thought to arise from the seizure focus. One early ECoG study showed local increases in coherence during the interictal period [10]. An analysis of interictal spikes with the adaptive directed transfer function (DTF), a frequency-domain analogue of Granger causality, was able to estimate the location of the seizure focus; these locations overlapped well with those determined by the epileptologists, and the eight patients examined had positive outcomes following resection of these locations [11]. In a similar analysis of ictal recordings, the results of spectral Granger causality were used to determine causal flow in ECoG and depth electrodes and found that high frequency (>80 Hz) Granger-causal connections appeared a few seconds before visible seizure onset, suggesting network activity involved in generating the seizure [12]. Another study applied a measure of phase synchronization, calculated from the Hilbert transform, to bilateral depth electrode recordings [13]. When the interictal periods were analyzed with this technique, higher levels of interictal synchrony were observed within structures ipsilateral to the seizure focus, as compared to the contralateral side. During seizures, a surprising pattern emerged wherein the mean phase synchronization decreased at the onset of the seizure and began to exhibit high variance relative to interictal levels, with peaks of high phase synchronization appearing sporadically throughout the seizure.

Efforts to validate signal processing techniques for use in intracranial recordings include the analysis of brain activity during the performance of tasks, such as simple motor tasks, for which the flow of information through specific brain regions is already known [14]. This approach has also been taken with the development of new analyses. One example is event-related causality, a new method based on the DTF: when applied to ECoG activity during a word repetition task, it showed increased flow of activity between language areas [15]. Another strategy for validating new methods includes the use of simulated data along with the ECoG, as was done for the development of nonlinear measures of functional connectivity [16].

The addition of techniques from graph theory promises to further clarify the results of signal processing analyses of epileptiform activity. A recent study constructed networks based on cross correlation analysis of ECoG and depth electrode data, and analyzed these networks with five measures drawn from graph theory [17]. This approach found that a few large dominant networks were formed at seizure onset and near seizure termination, but during the middle of the seizure, these "fractured" into smaller components. Another study used DTF to build networks from ECoG data and pinpointed network "hubs" using another graph theory measure, betweenness centrality [18]. This measure was also observed to decrease following seizure onset. These findings potentially reflect dynamics similar to those found in the phase synchronization study above [13], but the use of graph theory allows for additional depth of analysis.

So far, analyses of epilepsy in humans have focused on the activity of macroscale networks between relatively distant brain regions with thousands of neurons contributing to each signal. ECoG is well suited for recording this type of activity, and it lends itself to tracking global progression of seizure activity across large regions. However, in order to investigate human epilepsy at the mesoscale level of small local networks, alternative recording techniques must be found.

Recently, the application of multi-electrode (Utah) arrays (MEAs) was approved for research-only use in human patients who are being evaluated for epilepsy surgery. The MEA is implanted subdurally along with the grid of ECoG electrodes and retained for the duration of the clinical monitoring period (approximately 1 week). *In vivo* MEA recordings allow for direct transmission of brain activity from small (<4 mm) patches of tissue. These arrays have a much higher spatial and temporal resolution than standard intracranial surface electrodes due to the use of 96 penetrating microelectrodes across the array (1 μm long electrodes, tapering to 3–5 μm in width at the tip, with 400 μm in between electrodes) and a higher sampling rate (typically 25–30 kHz). In contrast, a grid of ECoG electrodes might cover a large portion of the surface of one or more lobes, with distances between electrodes on the order of 5–10 mm apart [10, 19]. Depth electrodes provide a similar spatial resolution, spanning the length of the hippocampus with 10–12 contacts, each several mm^2 in surface area [8]. Therefore, MEAs provide us with information about seizure dynamics that was not previously available.

Here, we demonstrate the use of signal processing and graph theory techniques to examine data from a multi-electrode array (MEA) recording during a human seizure. Analyzing ictal signals from a MEA allows us to determine the functional connectivity of a mesoscale network. Patients with focal epilepsy often exhibit a clear spatial localization of the source of their seizures, with propagation to distant areas followed by the development of widespread synchrony. Therefore, we can quantify characteristics of networks that are typically assessed qualitatively in clinical settings, and directly compare networks derived from different signal processing techniques. This approach may eventually make it feasible to partially automate the process of evaluation, and will likely yield insights into the activity of underlying local networks that contribute to macroscale seizure dynamics.

Techniques for Defining Networks

The signal analysis techniques we applied here are focused on finding temporal relationships across the signals recorded from spatially separated electrodes in a MEA. The goal is to generate a quantified relationship that depicts the origin and propagation of the ictal phenomena. In particular, the detection of the origin of the ictal events may be a helpful addition to presurgical evaluation of patients with intractable epilepsies. The techniques we apply here are both in the time domain (i.e. cross correlation and Granger causality) as well as the frequency domain (coherence).

Cross Correlation

To determine the cross correlation between two signals X and Y, signal Y is progressively shifted in time relative to signal X by a small lag (typically 1/sampling rate). At each lag, a correlation value between signal X and the shifted signal Y is calculated. Assuming that the signals are stationary and ergodic, we may use the following time average for cross correlation R:

$$R_{xy}(\tau) = E\{X(t)Y(t+\tau)\} \tag{1}$$

where E is the expected values operator and τ is the offset, or time lag.

To define a network, we define each microelectrode in the MEA as a node, while any connection between them is an edge. We determine the cross correlation function between each pair of nodes, then take the maximum correlation value and measure the lag at which this maximum occurs. The maximum value determines the strength of the connection between the two nodes, while the time lag informs the direction. For example, a lag of 0 would indicate complete synchrony, while a negative lag of the cross correlation peak of signal X compared to signal Y would indicate that Y precedes X.

Coherence

Coherence characterizes correlations within the frequency domain rather than the time domain. The power spectra of the two signals are obtained via Fourier transform. The cross spectrum S_{xy}, the Fourier transform of cross correlation, is then calculated and normalized by the products of the autospectra of each signal, S_{xx} and S_{yy}, in order to obtain a coherence value C between 0 and 1 for each frequency ω:

$$C(\omega) = \frac{|S_{xy}(\omega)|^2}{S_{xx}(\omega)S_{yy}(\omega)} \tag{2}$$

The phase coherence ϕ can also be determined by taking the inverse tangent of the real and imaginary components of the cross spectrum at each frequency ω:

$$\phi(\omega) = \tan^{-1}\frac{I}{R} \tag{3}$$

The coherence in the frequency band of interest is then averaged to determine the strength of the edge between the pair of nodes. The average phase coherence in the frequency band of interest is then used to define the direction.

Granger Causality

Granger causality is a concept originally formulated by Norbert Wiener, but first developed in economics [20]. The logic behind it posits that if one signal causes another, knowledge of the causal signal should improve our ability to predict the second signal. Formally, this is calculated by building two vector autoregressive models for each pair of signals [21]. The first "reduced" model is created to estimate signal X using the past of signal X:

$$X(t) = \sum_{j=1}^{p} A_{XX,j} \cdot X_{t-j} + \xi'_t \qquad (4)$$

The coefficients $A_{XX,j}$ are calculated along with the residuals, or errors, ξ'_t up to a set number of lags, p, known as the model order.

A second autoregression, the "full" model, is then calculated in which the past of signal Y is additionally included in the model:

$$X(t) = \sum_{j=1}^{p} A_{XY,j} \cdot X_{t-j} + \sum_{j=1}^{p} A_{XY,j} \cdot Y_{t-j} + \xi_t \qquad (5)$$

If the error in the full model is significantly decreased relative to the error in the reduced model, then signal Y is said to "Granger-cause" signal X. Formally, the Granger causality value F is calculated by taking the log ratio of the residuals variances from each model:

$$F_{Y \to X} = \ln \frac{var(\xi'_t)}{var(\xi_t)} \qquad (6)$$

Additional signals can also be included in both autoregressive models. This is an important distinction from both cross correlation and coherence, which are calculated in an independent fashion for each pair of nodes. This feature of Granger causality may enable us to rule out, for example, whether a third signal Z is affecting some aspect of the behavior of both X and Y in a linked manner, rather than X directly affecting Y or vice versa.

The magnitude of the Granger causality value is meaningful because it increases in proportion to the degree by which error is reduced by in the full model. Thus, when building a network, this magnitude gives us the strength of the edge. Because Granger causality analysis, especially applied to oscillatory signals, tends to result in a large number of bidirectional connections, we can define unidirectional connections by calculating the difference in Granger causality values for each pair of nodes in which the Granger causality in both directions is significant. The larger this difference, the more certain we can be of the directionality.

A few caveats are important to keep in mind. In particular, periodicity, a hallmark of seizure activity, exerts effects on all three measures. Using either the lag (in cross correlation) or the phase coherence is imperfect for determining the directionality of oscillatory signals, especially at higher frequencies. For example, if a sinusoidal signal X is caused by another sinusoid Y, but has a large enough lag or phase relative to Y, then signal X may erroneously appear to precede rather than follow Y as the waveform repeats. As the period of the sinusoids is decreased, the lag necessary for this error to occur becomes smaller and thus more likely. However, the frequencies examined most here (5–8 Hz) are slow enough relative to direct neuronal connections (0–150 ms for monosynaptic AMPA and NMDA connections [22]) that this concern is somewhat mitigated.

Another limitation is that if a signal that independently drives multiple signals on the array is not measured, for example if it lies outside the array, the current analyses may erroneously find a direct connection between the resulting signals. Additionally, if a common reference signal is used, it must be chosen carefully. When a nonzero reference signal is subtracted from each channel, the incorporation of the common elements of the reference signal may induce artificial correlations between channels [23]. In this study, this possibility was minimized by the use of an extracranial reference electrode; because the amplitude of the extracranial EEG is 1–2 orders of magnitude smaller than the signals recorded by the MEA, the influence of the reference on coherence measurements is negligible [24].

A final caveat is that with these techniques, we examine only linear relationships; any nonlinear relationships between nodes would escape our detection methods.

Defining Networks in a Human Seizure

Recordings were obtained from a 96-microelectrode MEA implanted along with subdural ECoG electrodes in a human epilepsy patient. The study was approved by the Institutional Review Board of the Columbia University Medical Center, and informed consent was obtained from the patient prior to participation. (See [25] for additional details on surgical procedures and patient enrollment.) Cross correlation, coherence and Granger causality were applied to a 30-second segment containing a seizure. These analyses were first demonstrated on a subset of six channels, in order to allow for visual comparison of the EEG trace with the networks created (Fig. 11.1a and b). Spectral analysis revealed that the peak frequency of the seizure was 6.6 Hz, in the theta band (5–8 Hz) (Fig. 11.1c).

As expected, cross correlation of the time domain signals and coherence in the theta band, containing the dominant frequency of the seizure, yielded identically structured networks that exhibited reasonable concordance with visual assessment of the EEG (Fig. 11.2a and c). In this example, cross correlation and coherence pinpoint channels 3 and 4 as the main sources of output, with channels 1 and 2

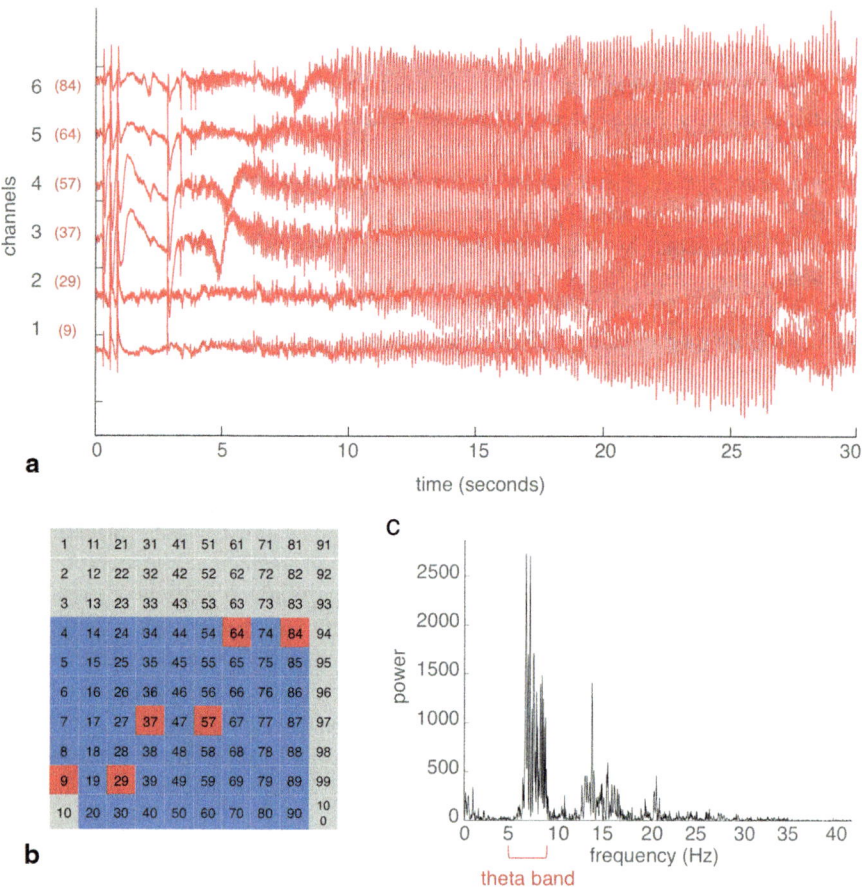

Fig. 11.1 From the MEA recording of a human seizure, a subset of six channels were selected to demonstrate techniques for building networks. **a** Raw EEG traces of channels 1 through 6. The actual channel numbers, referring to the location on the MEA, are shown in *red*. **b** Location of the six channels on the electrode array. *Red* represents the selected subset of six channels, *blue* represents other channels with good quality recordings that were used later, and *gray* refers to channels with artifacts. **c** Power spectrum of the average signal from the six channels. The theta (5–8 Hz) band contained the peak power of the seizure recording.

receiving inputs, and channels 5 and 6 with an intermediate mix of inputs and outputs. Visual inspection of the six channels agrees with this assessment—the seizure begins in channels 3 and 4, spreads to 5 and 6, and finally to 1 and 2.

Plotting the lags derived from cross correlation analysis (lags from fastest (0 ms) to slowest (20 ms and above are displayed as colors from red to blue) shows that many of these connections are fast, operating on a biologically plausible timescale of approximately 10 ms, and with subgroups of channels operating nearly synchronously (Fig. 11.2a).

Coherence analysis was also performed in other frequency bands (Fig. 11.2c), such as high gamma (80–150 Hz) due to its frequently reported relationship with epileptic

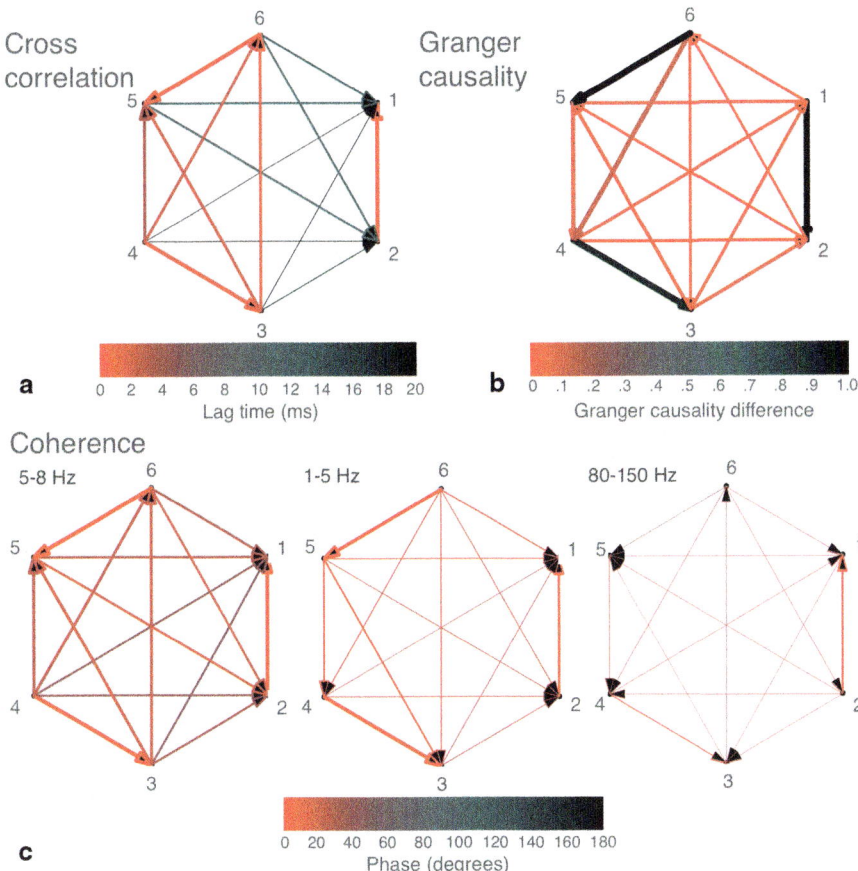

Fig. 11.2 Networks were built from the six-channel recording of a human seizure. **a** Cross correlation-derived network. Width of *arrows* (from narrower to wider) represents value of peak cross correlation, while color represents the lag (in ms) of the peak cross correlation, with faster lags in *red* and slower lags in *black*. **b** Granger causality-derived network. If statistically significant Granger causality was found between two nodes in both directions, the larger of the two was plotted. Width of arrows represents the value of the Granger causality. Color represents the difference between the Granger causality values of each direction (larger—smaller), if both values were statistically significant. **c** Coherence-derived networks in theta (5–8 Hz), delta (1–5 Hz), and high gamma (80–150 Hz) frequency bands. Width represents the mean value of coherence within the indicated frequency band, while color represents the mean phase coherence (in degrees) within the frequency band.

activity [26, 27]. As compared to our previous findings depicted in Figs. 11.2a and c, these yielded less similar networks. Given that the majority of the signal amplitude in the time domain consists of theta frequency signals, this discrepancy of the networks across different frequency bands is not necessarily inconsistent. Most of the connections found in the non-theta frequency bands are of low strength, but the most significant of these may reflect alternative processes either independent of the seizure or underlying a different component of the seizure dynamics.

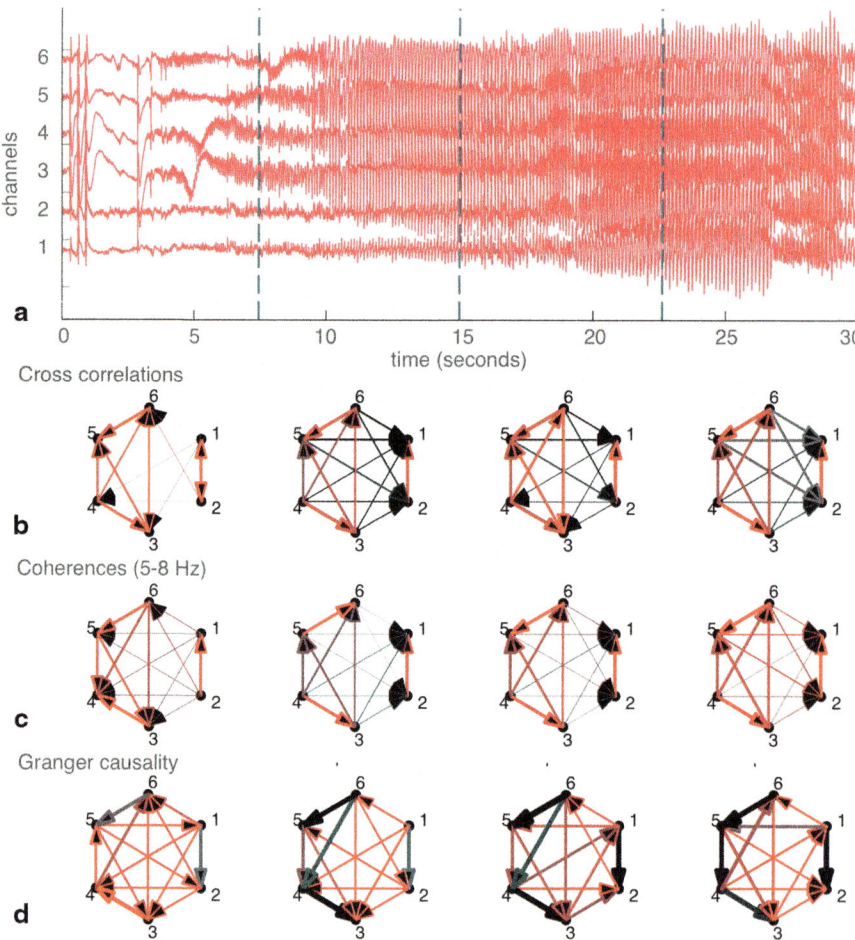

Fig. 11.3 Splitting the data into epochs yields information about how networks change over time. **a** The six-channel recording split into four epochs of equal (7.5s) length. **b** Cross-correlation derived networks for each of the four epochs. **c** Coherence-derived networks, taken in the theta frequency band. **d** Granger causality-derived networks. Widths and colors of *arrows* show strength and "speed" of each connection, as defined in Fig. 11.2.

As compared to the results in Figs. 11.2a and c, Granger causality analysis produced a differently structured network (Fig. 11.2b); however, as mentioned above, this analysis may work best with oscillatory signals when the data is divided into sufficiently small, possibly overlapping epochs.

A still more nuanced approach to constructing networks may be taken by splitting the data into epochs to examine changes in seizure dynamics over time (Fig. 11.3). Here, the data has been segmented into four epochs of 7.5s each and networks constructed for each of the four epochs. The four epochs are of equal length and were arbitrarily selected; however, each contains several general hallmarks of different stages of seizure activity (Fig. 11.3a). The first epoch is representative of seizure onset and includes several heralding spikes. In the second epoch, the seizure

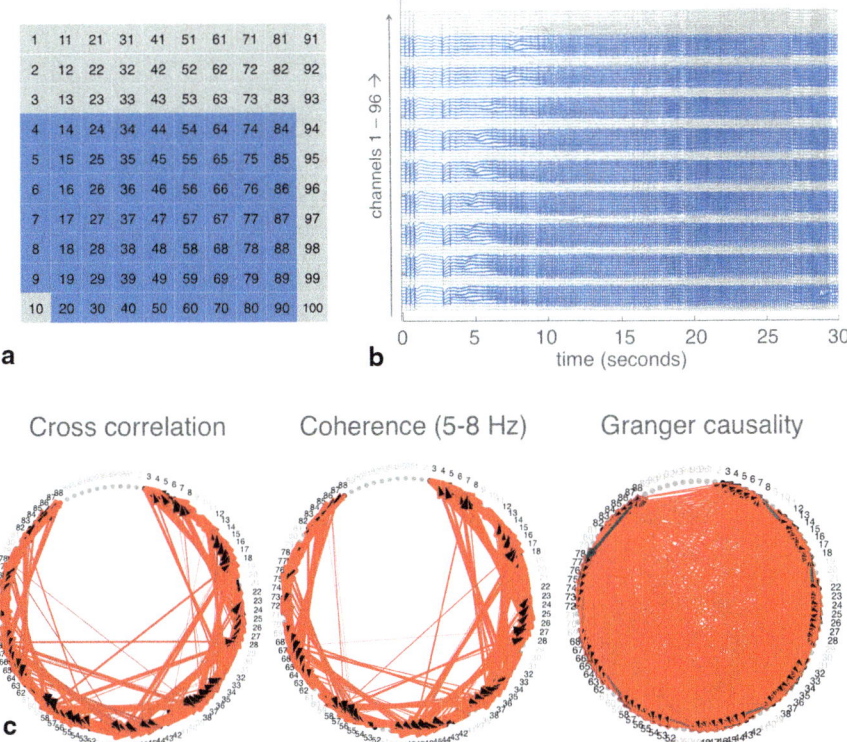

Fig. 11.4 Techniques for defining networks were applied to the full set of MEA recordings. **a** The spatial layout of the MEA. *Blue* indicates the 62 channels with good quality recordings, which were used for analyses, while *gray* indicates the 34 channels with artifacts. **b** The raw EEG trace from the 96 channels, with good channels in *blue* and channels with artifacts in *gray*. **c** Networks built using each of the three techniques. Widths and colors of *arrows* show strength and "speed" of each connection, as defined in Fig. 11.2. Channels with artifacts, which were not included in the analysis, were plotted in *gray*. Threshold for plotting connections determined from both cross correlation and coherence was 0.95. Criteria for plotting connections from Granger causality was as indicated in Fig. 11.2.

further develops and spreads to all channels. The third epoch contains the fully developed seizure, consisting of high amplitude synchronized oscillatory activity. Finally, the fourth epoch is characteristic of seizure offset. For each of the three methods of defining networks, it can be appreciated that networks from the later epochs (in which signal power is largest) more closely resemble the networks derived from the whole dataset (Fig. 11.3b, 11.3c, 11.3d).

Having demonstrated the validity of these techniques at a smaller scale, we then applied them to the entire dataset in which we removed channels with artifacts (62 channels total; 34 channels were removed) (Fig. 11.4a and 11.4b). Even on a larger scale, networks derived from cross correlation and coherence appear to resemble each other (Fig. 11.4c, 11.4d, 11.4e). The Granger causality-derived network, however, found a much larger number of significant connections. This may point to alternative uses for Granger causality as compared to the other techniques—when

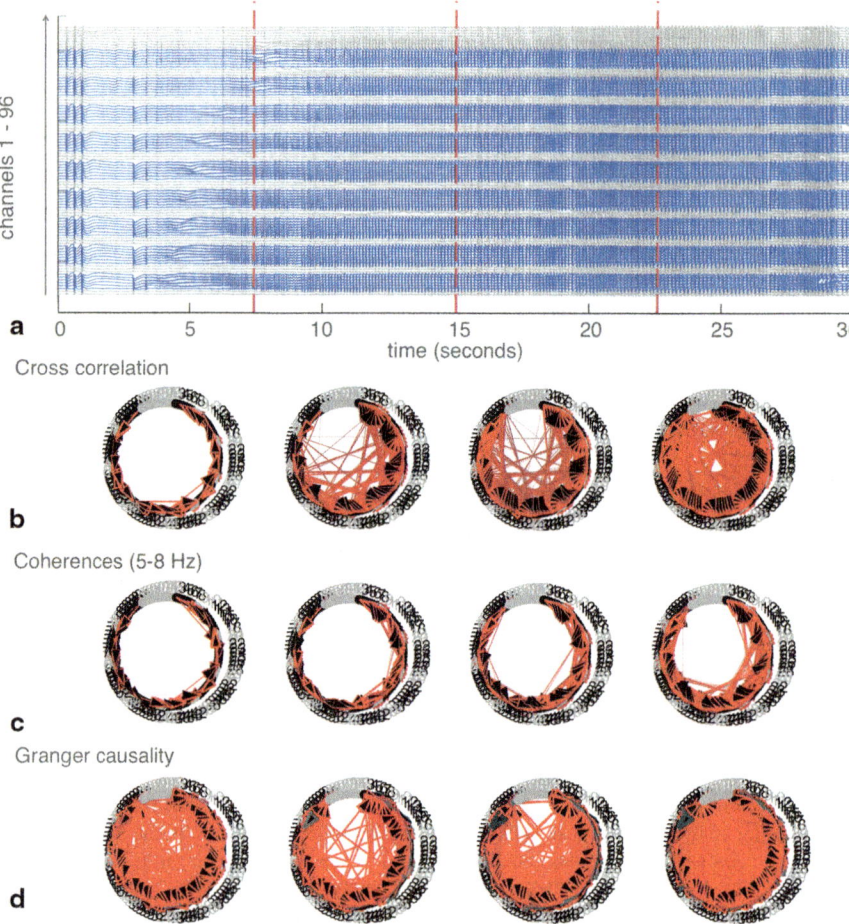

Fig. 11.5 The full recording was split into epochs for further analysis. **a** The *red dashed lines* indicate the division of the recording into four equal-length epochs; channels used for analysis in *blue*. **b** Cross-correlation derived networks for each of the four epochs. **c** Coherence-derived networks, taken in the theta frequency band. **d** Granger causality-derived networks. Widths and colors of *arrows* show strength and "speed" of each connection, as defined in Fig. 11.2.

applied to such large datasets, it may be best for picking up many relationships between nodes. Cross correlation and coherence, on the other hand, can be easily thresholded to pick up only the most dominant connections.

Again, splitting the data into four epochs and applying network analyses to each epoch reveals change in network structures over time (Fig. 11.5). The number of significant connections increases dramatically as the seizure develops. At first, connections are mainly local (adjacent channels are generally numbered consecutively) but over time, more long-range connections appear. Additionally, the lag of these connections appears to decrease until channels are largely synchronous with each other. This also demonstrates that Granger causality results may become more congruent with other techniques when it is applied to smaller epochs of data.

One interesting difference between Granger causality and the other techniques appears in the first epoch, where Granger causality finds a large number of relatively global connections. A possible explanation for this is that while the dominant connections during seizure initiation are sparse and local, many small global connections underlie the start of seizure activity. The three heralding spikes visible in the first few seconds of this epoch may have also driven the Granger causality results while yielding less influence on the other techniques.

Characterizing and Comparing Networks

It quickly becomes apparent that while visual inspection of the different networks is possible with 6 channels, once these techniques are applied to larger datasets and greater numbers of epochs, this task becomes difficult. Fortunately, there exists a large repertoire of graph theoretical measures that can be applied to networks ("graphs") in order to quantify their characteristics.

Degree, Causal Density and Causal Flow

Among the simplest and most intuitive of such measures are degree, causal density and causal flow [28] (Fig. 11.6). The degree of each node is the total number of connections that node makes; the degrees of all the nodes can be plotted as a histogram

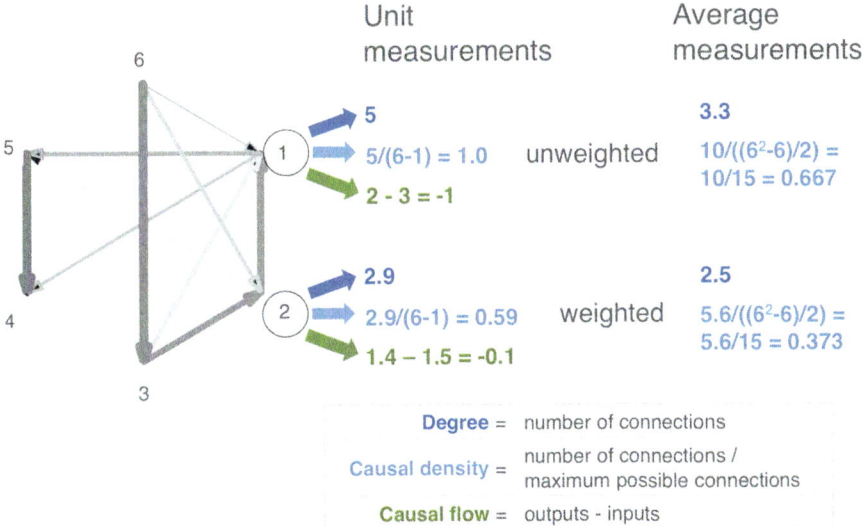

Fig. 11.6 Definition and demonstration of the three graph theory measures used to characterize the networks displayed here: degree (*dark blue*), causality density (*light blue*), and causal flow (*green*). Weighted and unweighted variants, as well as unit vs average variants, are shown.

to show the degree distribution, and can be averaged to obtain the mean degree of the network. Causal density is similar to degree, but the number of connections is divided by the maximum possible number of connections in order to obtain a number between 0 and 1, allowing for comparison between differently sized networks [29]. This can also be calculated as an average value for the network as a whole, or as a unit value for each individual node. Finally, unit causal flow for each node is defined as the number of inputs entering that node, subtracted from the number of outputs leaving the node. All three of these measures can be calculated as either unweighted values, wherein a significant connection is assigned a value of 1 and no connection is assigned 0, or as weighted values, wherein significant connections are weighted by their individual strengths.

Measures to characterize the graphs resulting from neural interactions can be used to characterize any one of the networks created above. Because the location and propagation of signals in space is paramount to analyzing epilepsy data (and indeed neural data in general), the information is displayed in a spatial layout, with the degree, unit causal density, and unit causal flow of each node plotted at the position of each channel on the array.

Several features of the networks become more apparent with these measures. For example, causal flow of the cross correlation-derived network reveals that one side of the MEA sends more outputs than inputs (a "causal source") while the other side receives more inputs than outputs (a "causal sink") (Fig. 11.7).

Applied to multiple epochs and techniques, we can quantify changes over time and differences between networks derived from different techniques. For example, the increase in connections over time (Fig. 11.5) is clearly captured by degree and causal density (Fig. 11.8). Changes in the spatial layout of causal flow over time are also apparent as the seizure develops (Fig. 11.8). Taking the average degree and causal density of each network gives us several overall measures that can be used to compare global network characteristics (Table 11.1).

Interestingly, this pattern of increase in causal density seems, at first glance, in contrast with findings of cross-correlation-derived network analysis of ECoG and depth electrode activity. This study found decreased causal density during most of the seizure, with causal density increasing above preictal levels only immediately after seizure onset and just before seizure termination [17]. However, the authors of this study highlight that their analyses characterize the functional connectivity between macroscopic brain areas, whereas the high amplitude elicited during a seizure suggests an increase in synchrony in the local network. This prediction is supported by the MEA analysis (Fig. 11.7a), again suggesting that the MEA is able to yield a unique set of findings about network dynamics at the mesoscale level, complementary to the macroscale results of ECoG analysis.

Network Similarity

Direct comparison of the exact structure of the network may also be desired. Because a network can be represented as a channel-by-channel matrix of 0's and 1's,

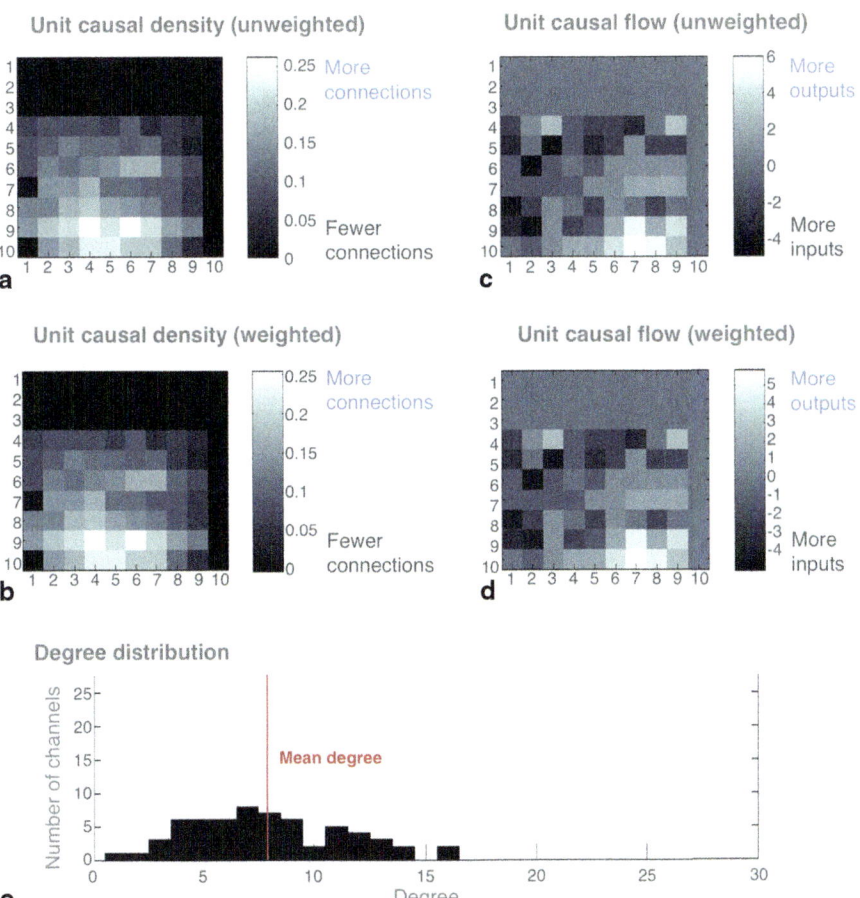

Fig. 11.7 Unit measures were calculated for each node in the cross correlation-derived network. Unit causal density **a, b** and unit causal flow **c, d** both weighted and unweighted variants, are plotted for each node. Nodes are plotted spatially; note the uniformly shaded regions on the *top* and *right side* of each figure, representing electrodes with artifacts which were not included in the analysis. **e** Histogram of degree of each nodes was plotted; *red line* is the mean degree of the network.

with 1 for a significant connection and 0 for a nonsignificant connection, we can quantify the similarity between two networks using a form of the Jaccard coefficient, a general method of comparing binary variables (Fig. 11.9). This measure takes the common overlap between two network matrices and divides it by their size. From two network representations, we can then calculate a new matrix wherein 1 indicates the identical existence of a connection or lack thereof, while 0 indicates a difference between the networks at a given pair of channels. We then take the sum of this matrix, and subtract from it the number of channels (because the diagonal of any two network matrices generated by the techniques described above will always consist of uncalculated autocorrelations, and thus will be the same). This sum is normalized by dividing it by the maximum possible number of similarities between

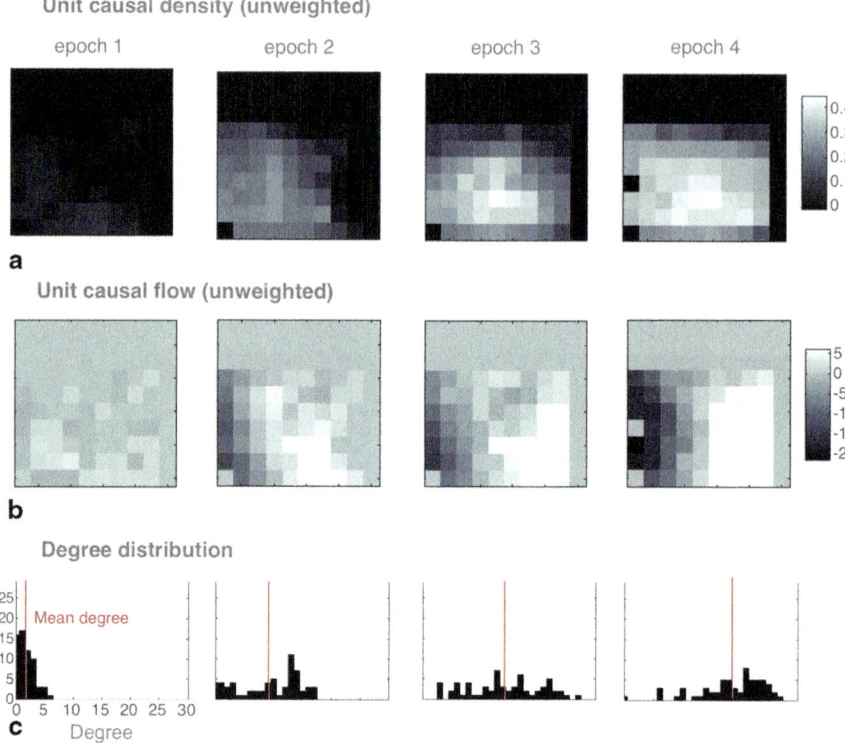

Fig. 11.8 The recording was split into four epochs. Several network measures were calculated for the cross correlation-derived network of each epoch: the unweighted unit causal density (**a**), the unweighted unit causal flow (**b**), and degree distribution (**c**).

Table 11.1 Graph theoretical measures of networks created by each technique

–	Cross correlation	Coherence (5–8 Hz)	Granger causality
Average degree	7.90	7.19	42.06
Causal density (unweighted)	0.032	0.030	0.690
Causal density (weighted)	0.025	0.021	0.004

two networks: (n channels)$^{2-}$ n channels. This will yield a value of network similarity between 0 and 1, where 1 indicates that each network has an identical pattern of connections.

In the example of the 62-channel MEA seizure recording, network similarity was used to compare the outputs of different techniques (Table 11.2) as well as the networks found by one technique across epochs (Table 11.3). As expected, an especially high similarity was found between networks derived from cross correlation and coherence, which agrees with such comparisons made in ECoG-derived networks [17].

11 Epileptogenic Networks

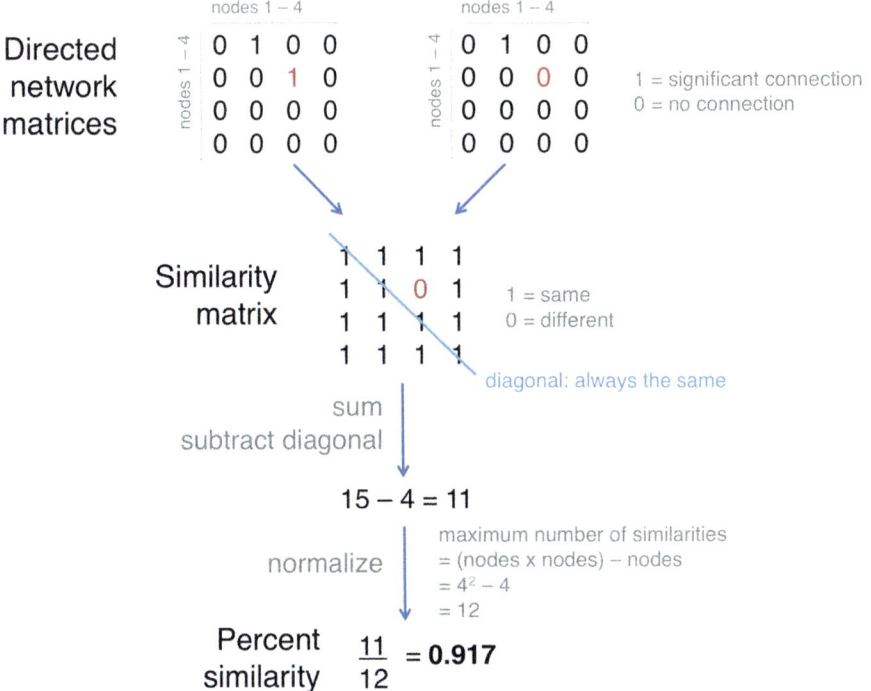

Fig. 11.9 Outline of network similarity calculation.

Concluding Comments

We show that measures such as cross correlation, coherence, and Granger causality can be applied to define networks from intracranial neural activity data obtained from human epilepsy patients. In the context of seizure recordings, these networks can be compared against visual evaluation, the current clinical standard, therefore allowing validation of the relationships found. Graph theoretical measures provide a means of characterizing the networks. Furthermore, our results demonstrate that we can quantify the similarity between networks constructed via different techniques (Table 11.2), between networks derived from consecutive epochs of a dataset (Table 11.3), or between multiple datasets (for example, several seizures from a single patient [17]).

In addition to the several techniques demonstrated here, many other methods for defining networks are available. For instance, the directed transfer function, a frequency-domain analogue of Granger causality, could be used to further clarify the results found by Granger causality analysis [19, 30, 31]. Additionally, several nuances in the execution of each technique allow for wide flexibility in how they are applied. One could divide the signal into overlapping epochs to obtain more detailed time-domain information, or separate the signal into epochs based on physiological events such as seizure initiation, propagation, and termination in order to determine the state of the network during each specific phase.

Table 11.2 Network similarity across techniques

–	Cross correlation	Coherence (5–8 Hz)	Granger causality
Cross correlation	–	–	–
Coherence (5–8 Hz)	0.955	–	–
Granger causality	0.679	0.671	–

Table 11.3 Network similarity across epochs. (cross correlation-derived network)

–	Epoch 1	Epoch 2	Epoch 3	Epoch 4
Epoch 1	–	–	–	–
Epoch 2	0.934	–	–	–
Epoch 3	0.896	0.949	–	–
Epoch 4	0.861	0.913	0.945	–

The graph theory analyses employed here are relatively simplistic, but more sophisticated tools for describing nodes and networks can also be calculated, depending on the network characteristic of interest. Some additional commonly used measures include the average path length and the clustering coefficient, which quantify the interconnectedness of the network [32, 33].

While each of these techniques has its limitations, using multiple techniques to build networks in a set of clinical data allows us to both directly compare the outputs of different approaches and to validate them against the standard of visual evaluation. This may help to justify their use in both clinical and research situations in which visual evaluation is not sufficient. Eventually, one could imagine further validation against clinical data: for instance, patient outcomes following surgical excision of the tissue implicated in seizure generation could be compared to the location of the network "hub" as determined by each technique [11, 18]. Were this effort successful, such computational approaches to evaluate brain electrical activity could become valuable adjuvants to expert visual assessment in the clinic.

Acknowledgements This work was supported by the Dr. Ralph and Marian Falk Medical Research Trust and R01 NS084142-01.

References

1. Ngugi AK, Bottomley C, Kleinschmidt I, Sander JW, Newton CR. Estimation of the burden of active and life-time epilepsy: a meta-analytic approach. Epilepsia. 2010;51(5):883–90.
2. Ngugi AK, Kariuki SM, Bottomley C, Kleinschmidt I, Sander JW, Newton CR. Incidence of epilepsy. Neurology. 2011;77(10):1005–12.
3. Kwan P, Brodie MJ. Early identification of refractory epilepsy. N Engl J Med. 2000;342(5):314–9.
4. Engel J, McDermott MP, Wiebe S, et al. Early surgical therapy for drug-resistant temporal lobe epilepsy: a randomized trial. JAMA. 2012;307(9):922–30.
5. Fiest KM, Sajobi TT, Wiebe S. Epilepsy surgery and meaningful improvements in quality of life: results from a randomized controlled trial. Epilepsia. 2014;55(6):886–92.

6. Wiebe S, Blume WT, Girvin JP, Eliasziw M. A randomized, controlled trial of surgery for temporal-lobe epilepsy. N Engl J Med. 2001;345(5):311–8.
7. Téllez-Zenteno JF, Dhar R, Wiebe S. Long-term seizure outcomes following epilepsy surgery: a systematic review and meta-analysis. Brain. 2005;128(5):1188–98.
8. Duckrow RB, Spencer SS. Regional coherence and the transfer of ictal activity during seizure onset in the medial temporal lobe. Electroencephalogr Clin Neurophysiol. 1992;82(6):415–22.
9. Gotman J, Levtova V. Amygdala-hippocampus relationships in temporal lobe seizures: a phase-coherence study. Epilepsy Res. 1996;25(1):51–7.
10. Towle VL, Carder RK, Khorasani L, Lindberg D. Electrocorticographic coherence patterns. J Clin Neurophysiol Novemb 1999 [Internet]. 1999;16(6):528–47. http://ovid-sp.ovid.com/ovidweb.cgi?T=JS&CSC=Y&NEWS=N&PAGE=fulltext&D=ovftd&AN=00004691-199911000-00005. Accessed 12 Oct 2014.
11. Wilke C, van Drongelen W, Kohrman M, He B. Identification of epileptogenic foci from causal analysis of ECoG interictal spike activity. Clin Neurophysiol Off J Int Fed Clin Neurophysiol. 2009;120(8):1449–56.
12. Adhikari BM, Epstein CM, Dhamala M. Localizing epileptic seizure onsets with granger causality. Phys Rev E [Internet]. 2013;88(3):030701. http://link.aps.org/doi/10.1103/PhysRevE.88.030701. Accessed 12 Oct 2014.
13. Mormann F, Lehnertz K, David P, E Elger C. Mean phase coherence as a measure for phase synchronization and its application to the EEG of epilepsy patients. Phys Nonlinear Phenom. 2000;144(3):358–69.
14. Zaveri HP, Williams WJ, Sackellares JC, Beydoun A, Duckrow RB, Spencer SS. Measuring the coherence of intracranial electroencephalograms. Clin Neurophysiol. 1999;110(10):1717–25.
15. Korzeniewska A, Crainiceanu CM, Kuś R, Franaszczuk PJ, Crone NE. Dynamics of event-related causality in brain electrical activity. Hum Brain Mapp. 2008;29(10):1170–92.
16. Wendling F, Bartolomei F, Bellanger JJ, Chauvel P. Interpretation of interdependencies in epileptic signals using a macroscopic physiological model of the EEG. Clin Neurophysiol. 2001;112(7):1201–18.
17. Kramer MA, Eden UT, Kolaczyk ED, Zepeda R, Eskandar EN, Cash SS. Coalescence and fragmentation of cortical networks during focal seizures. J Neurosci Off J Soc Neurosci. 2010;30(30):10076–85.
18. Wilke C, Worrell G, He B. Graph analysis of epileptogenic networks in human partial epilepsy. Epilepsia. 2011;52(1):84–93.
19. Wilke C, van Drongelen W, Kohrman M, He B. Neocortical seizure foci localization by means of a directed transfer function method. Epilepsia. 2010;51(4):564–72.
20. Granger CWJ. Investigating causal relations by econometric models and cross-spectral methods. Econometrica. 1969;37(3):424–38.
21. Barnett L, Seth AK. The MVGC multivariate granger causality toolbox: a new approach to granger-causal inference. J Neurosci Methods. 2014;223:50–68.
22. Myme CIO, Sugino K, Turrigiano GG, Nelson SB. The NMDA-to-AMPA ratio at synapses onto layer 2/3 pyramidal neurons is conserved across prefrontal and visual cortices. J Neurophysiol. 2003;90(2):771–9.
23. Schiff SJ. Dangerous phase. Neuroinformatics. 2005;3(4):315–8.
24. Zaveri HP, Duckrow RB, Spencer SS. The effect of a scalp reference signal on coherence measurements of intracranial electroencephalograms. Clin Neurophysiol. 2000;111(7):1293–9.
25. Schevon CA, Weiss SA, McKhann G, Goodman RR, Yuste R, Emerson RG, et al. Evidence of an inhibitory restraint of seizure activity in humans. Nat Commun. 2012;3:1060.
26. Jirsch JD, Urrestarazu E, LeVan P, Olivier A, Dubeau F, Gotman J. High-frequency oscillations during human focal seizures. Brain J Neurol. 2006;129(Pt 6):1593–608.
27. Ochi A, Otsubo H, Donner EJ, Elliott I, Iwata R, Funaki T, et al. Dynamic changes of ictal high-frequency oscillations in neocortical epilepsy: using multiple band frequency analysis. Epilepsia. 2007;48(2):286–96.
28. Seth AK. Causal networks in simulated neural systems. Cogn Neurodyn. 2008;2(1):49–64.

29. Seth AK. Causal connectivity of evolved neural networks during behavior. Netw Comput Neural Syst. 2005;16(1):35–54.
30. Kaminski MJ, Blinowska KJ. A new method of the description of the information flow in the brain structures. Biol Cybern. 1991;65(3):203–10.
31. Kamiński M, Ding M, Truccolo WA, Bressler SL. Evaluating causal relations in neural systems: granger causality, directed transfer function and statistical assessment of significance. Biol Cybern. 2001;85(2):145–57.
32. Rubinov M, Sporns O. Complex network measures of brain connectivity: uses and interpretations. NeuroImage. 2010;52(3):1059–69.
33. Watts DJ, Strogatz SH. Collective dynamics of "small-world" networks. Nature. 1998;393(6684):440–2.

Index

A
Albus, J., 190, 215
Alzheimer's disease (AD), 2, 3, 6, 8, 9, 34, 35, 222
 diagnosis of, 21
 hypothesis of, 225
Amari, S.I., 143
Anderson, W.S., 179
Anxiety, 22, 23, 31, 32
Arousal state, 11, 109, 113, 115, 119

B
Badawy, R.A.B., 174
Basal ganglia, 4, 9, 10, 21, 61, 72, 74–76
 computational, model of, 77, 96
Behrens, T., 1
Benjamin, O., 177
Bhattacharya, B.S., 265
Binning, 50, 185, 288–290
Blenkinsop, A., 179
Blumenfeld, H., 176
Borisyuk, R.M., 33
Brain dynamics, 2, 108–110, 120, 179, 181, 266
Bressloff, P.C., 143, 172
Buszaki, G., 2, 177, 178

C
Canavan, S.V., 3, 13
Canolty, R.T., 1
Cerebellum, 13, 190–192, 207, 215
Coherence, 3, 22, 34, 109, 179, 223, 295, 297–299, 309
Computational modelling, 173, 178, 223, 226, 241, 260, 262

Conductance based model, 4, 45, 47, 50–54, 168, 169
Coombes, S., 173
Cowan, J.D., 33, 143, 171, 177
Coyle, D., 3, 6, 9
Crick, F.H., 17
Cross correlation, 295, 297–300, 304, 306, 309
Crunelli, V., 230

D
Davod, O., 180, 144
De Jong, L.W., 264
De Wall, H., 223
Deco, G., 170
Denham, M.J., 33
Destexhe, A., 165, 169, 171, 176
Diekman, C., 279
Disease, 2, 5, 20, 34
Disorders, 2, 5, 12, 13, 17, 20, 22, 194, 266, 278
Doiron, B., 145, 148
Dovzhenok, A., 75
Dupuy, J-P., 17
Dynamical causal modeling (DCM), 44
Dynamical diseases, 21

E
Electroencephalography (EEG), 3, 6, 9, 10, 45, 55, 150, 163, 180, 222, 294, 299
 dynamics of, 224, 226
Electrophysiology, 3, 9, 47
Engel, J., 179
Epilepsy, 3, 12, 13, 21, 161, 163, 173, 174, 179, 180,

EPSP-spike complex, 197, 202
Érdi, P., 2, 3, 6, 8, 9

F
Faugeras, O., 179
Feedforward inhibition, 169
Feng, J., 92
Finkel, L.H., 20
Freeman, W.J., 180
Friston, K.J., 180
Fröhlich, F., 226, 235

G
Gorzelic, 76
Granger causality, 179, 279, 295, 298, 299, 304, 305, 309
Granular layer, 3, 13, 48, 190, 191, 206, 207, 214, 215
Graph theory, 179, 295, 296, 310
Grimbert, F., 179
Guo, Y., 3, 10, 74, 75, 80, 81, 85, 96

H
Hauptmann, C., 74
Hellwig, B, 152
Heteroclinic orbits, 153–156
Horn, D., 226
Hughes, S.W., 230
Hutt, A., 2–4, 6, 11

I
Input, 10, 16, 25, 47, 48, 53, 55, 58, 61, 77–79, 190, 197, 211, 214, 237, 300
Izhikevich neuron model, 229, 231, 279

J
Jansen, B.H., 55, 171, 173, 177, 179
Jeewajee, A., 34
Jelic, V., 223
Jirsa, V.K, 177, 180
Jiruska, P., 178
Jong, L.W., 264

K
Kalitzin, S.N., 177
Kernels, 48, 54, 144, 146, 153, 155, 156
Kirk, I.J., 33
Kitano, K., 279
Knockout mutations, 212
Kobayashi, R., 279
Koenig, T., 223, 246
Kronauer, R.E., 116

L
Large-scale network model, 98, 195
Liley, D.T.J., 144, 172
Little, S., 75, 88, 97
Local field potential (LFP), 3, 9, 13, 26, 33, 45, 55, 72, 76, 195, 209, 226
Lopes Da Silva, F.H., 21, 163, 175, 240
Lourens, M., 76

M
Magnetoencephalography (MEG), 9, 54, 56, 163, 171, 180, 294
Malenka, R.C., 225
Marr, D., 190, 215
Marten, F., 176
Mattson, R.H., 164
McCormick, D.A., 176
McNaughton, N., 33
Message passing interface (MPI), 227
Microscale, 166–169, 176
Model validation, 7, 10, 96, 266
Modeling, 2, 3, 5, 11, 27, 32, 110, 115, 193, 209, 214, 278
Models, 3, 4, 165, 194, 195
Mokeichev, A., 279

N
Networks, 169, 296, 305
    definition of, 296, 297, 299
Neural field model (NFM), 4–6, 9, 10, 45, 50, 110, 150, 172
    conductance-based, 54–59
    Amari model, 143
    Robinson model, 144, 145
Neural field theory, 54, 109, 112
Neural mass model (NMM), 47, 48, 50, 55, 109, 171, 177, 180, 238
Neural oscillation, 21, 22, 171
Neuromodulation, 111
Neurons, 77, 78
Neurophysiology, 2, 4, 115
Neuroscience, 2, 19

P
Pair, 2, 15, 85, 96, 280–282, 290, 297
Parkinson's disease (PD), 7, 9, 10, 96
Parkinsonian network, 60, 81, 84, 85, 89, 94
    model design of, 80
Patterns, 2, 10, 13, 21, 22, 72, 76, 154, 163, 177, 204, 209, 261, 279, 290
Phillips, A.J.K., 115
Potential, 2, 9, 10, 20, 22, 26, 38, 48, 51, 87, 174, 195, 204, 211, 232, 234
Puckeridge, M., 131

Q
Quantitative systems pharmacology, 20, 21, 23, 38

R
Richardson, M.P., 178, 179
Rit, V.G., 55, 171, 173, 177, 179
Robinson, P.A., 5, 6, 10, 116, 144, 149, 150, 172, 176
Rosen, R., 18
Rubin, J.E., 74, 75, 81, 85

S
Seizures, 13, 21, 126, 127, 162, 163, 173, 178, 181, 295
Sejnowski, T.J., 165
Simulations, 3, 8, 9, 20, 45, 49, 96, 134, 169, 205, 207, 214, 240, 278
Soltesz, I., 173
Spikes, 167, 197, 280
St Hilaire, M.A., 116
Staley, K., 173
Stellwagen, D., 225
Stimulation, 26, 27
Suresh, J., 279
Synaptic compensation mechanisms, 225, 242, 262, 266
Synaptic loss, 224, 228, 235–238, 249, 261, 263, 267
Systems, 2, 12, 16, 18, 19, 31, 44, 46, 174, 253

T
Tass, P.A., 97
Tejeda, J., 166
Terman, D.H., 74–76
Theta rhythm, 22, 23, 26, 33–34, 36
Toppin, K., 3, 10
Traub, R.D., 21, 178

V
Validation, 5, 9, 11, 32, 33, 37, 38, 96, 97, 309
van Drongelen, W., 179
Villette, V., 34
Visser, S., 173, 179
Volman, V., 178

W
Wang, X.J., 33, 180, 264
Wang, Z., 264
Watson, J.D., 17
Wendling, F., 178–180
Wiener, Norbert, 298
Wilson, H.R., 22, 33, 143, 171, 177
Wolkenhauer, O., 18
Wright, J.J., 144

Y
Yeung, M., 32

Z
Zlokovic, B.V., 225

CPSIA information can be obtained at www.ICGtesting.com
Printed in the USA
BVOW11*0449031115

425405BV00001B/9/P